# CYANOBACTERIA FROM PADDY FIELDS

## *The Author*

**Dr. Chittaranjan Sen** is currently working as a senior most Associate professor in the department of Botany, Charuchandra College, University of Kalcutta in West Bengal. He obtained his master degree and doctoral degree from VISVA BHARATI university. Earlier he served as HOD of Botany Department in his College from 2002 to 2012,He is a member of many learned bodies in India. In 2015, He visited Spain to attend and deliver a talk on Role of Cyanobacteria as Biofertilizer in Paddy Fields of West Bengal, in Barcelona University. He started his research carrier as an Algal Taxonomist, later on he worked on Taxonomy and Ecology of paddy fields cyanobacteria of Gangetic West Bengal. At present he is working on physiology and ecology of Blue-green algae from paddy fields of Lower Gangetic Plain of West Bengal. His area of special interest on research is role of cyanobacteria as biofertilizer in paddy fields. Dr. Sen published large number of research papers in reputed international and national journals on different aspects of blue- green algae. He attended and delivered talks in different symposiums and presented papers in seminars. He was the Organising Secretary of a National Seminar held in 2002,in the Botany Department of Charuchandra College. Dr. Sen has taught students at all levels, including teaching for many years different branches of Botany specially Phycology, Microbiology, Plant Physiology and Biochemistry for undergraduate students at Charu Chandra College. His dedication to teaching and his enthusiasm for cyanobacterial research shine through his work.

# CYANOBACTERIA FROM PADDY FIELDS

*by*

**Dr. Chittaranjan Sen**

**2025**

**Daya Publishing House®**
*A Division of*
**Astral International Pvt. Ltd.**
**New Delhi – 110 002**

ISBN: 978-93-5919-942-9

*Publisher's Note:*

*Every possible effort has been made to ensure that the information contained in this book is accurate at the time of going to press, and the publisher and author cannot accept responsibility for any errors or omissions, however caused. No responsibility for loss or damage occasioned to any person acting, or refraining from action, as a result of the material in this publication can be accepted by the editor, the publisher or the author. The Publisher is not associated with any product or vendor mentioned in the book. The contents of this work are intended to further general scientific research, understanding and discussion only. Readers should consult with a specialist where appropriate.*

*Every effort has been made to trace the owners of copyright material used in this book, if any. The author and the publisher will be grateful for any omission brought to their notice for acknowledgement in the future editions of the book.*



*Published by* : **Daya Publishing House®**
*A Division of*
**Astral International Pvt. Ltd.**
– ISO 9001:2015 Certified Company
4736/23, Ansari Road, Darya Ganj
New Delhi-110 002
Ph. 011-43549197
E-mail: info@astralint.com
Website: www.astralint.com

*Printed at* : **Replika Press Pvt Ltd**

*Dedicated in*
*the memory of my Teacher*

**Dr. Dilip Gupta**

***-: Fomer Reader :-***

*Department of Botany Kalyani University,*
*West Bengal*

# Preface

Phycology remained for many decades a discipline almost completely isolated from the rest of biological science. The scope of using microbes in agricultural practice has enlarged tremendously during the last 30 years or so. Among all micro-organisms Cyanobacteria have unique features that render them as most ideal organisms for use in agricultural fields.

I have made an attempt to arrange diverse areas of Cyanbacterial studies in a coherent manner. The book has four parts, in part 1, I have concentrated on basic concepts of Cyanobacteria. This part of the book is primarily designed for under-graduate and post-graduate students of Botany. In part 2 and part 3 some portions of my PhD Thesis on Paddy Fields Cyanobacteria of Lower Gangetic Plain of West Bengal were incrporated. In part 2 Taxonomical studies and in part 3 Ecological observations were included. These two parts of the book may be helpful for the researchers of Agriculture and Botany working on Taxonomy and Ecology of rice-fields Cyanobacteria. The part 4 of the present book is designed primarily for researchers , though it may also serve as introductory text to those preparing for their Master's degree in Plant Biotechnology and Botany.

I am indebted to my teacher and esteemed guide, late Dr. Dilip Gupta of Kalyani University, West Bengal, who had been a source of continuous encouragement and inspiration to me. Without the understanding and encouragement of my wife and daughter, I would never have been able to undertake the task of writing the manuscript of this book.

My thanks are due to Sri. Tapash Chakrobarty of BSI, Howrah, West Bengal who undertook with pleasure the task of typing the manuscript of the present book. My thanks are due to all persons of Astral International Pvt. Ltd. of New Delhi who are concerned with the publication of this book.

**Chittaranjan Sen**

# Content

## PART III : ECOLOGICAL STUDIES OF BLUE-GREEN ALGAL FLORA OF PADDY FIELDS

## PART – IV : ANALYSIS & ASSESSMENT

# PART – I
# GENERAL ACCOUNT OF CYANOBACTERIA

# INTRODUCTION

Cyanobacteria (Blue-green algae, Cyanophytes, Cyanoprokaryotes) are the most fascinating organisms in the Earth's biosphere. Their origin in the Early Precambrian was one of the most important steps in evolution (Schopf 1974a, 1074b, 1993, 1996). They are prokaryotic in nature, distinct from other prokaryotic organisms in their ability to carry out photosynthesis by using two photosystems reaction like that of eukaryotic organisms. However, unlike eukaryotes they do not contain organized nucleus and DNA is dispersed in the cytoplasm in the form of a filamentous system; contains chlorophyll a, carotenoids and biliproteins. Other interesting and important features of cyanobacteria are complementary chromatic adaptation, the presence of gas vacuoles, morphological variations, the formation of specialized cells heterocysts in some filamentous forms, formation of mucilaginous sheath surrounding trichomes, pigmentation of sheath, formation of specialized cells akinetes (resting cells) which germinate in favourable conditions to form new trichomes. They are distributed throughout the oceans, fresh waters and soil environments, almost found everywhere in this planet. The reserve material – cyanophycin, a copolymer of aspartate and arginine would appear to be unique to this group. The pigments are chlorophyll a in the thylakoid membrane, and the phycobiliproteins, which are water soluble chromoproteins found in phycobilisomes attached to the outer surface of the thylakoid membranes. Carotenoids present in Cyanophyceae differ from those of eukaryotic algae in having echineone (4-Keto-β-carotene) and myxoxanthophyll without lutein is also present in cell. Four phycobiliproteins are observed in cyanophyceae – C- phycocyanin, allophycocyanin, C-phcoerythrin and phycoerythrocyanin. All cyanobacteria contain e-phycocyanin and allophycocyanin, whereas e-pycoerythrin and phycoerythrocyanin occur only in few species.

It is an enigma in nature that the percentage of nitrogen in the atmosphere is near about 79% and its relative non-availability to plants and animals. So, crop production has to depend largely on synthetic nitrogen fertilizers, the production of which is based on non-renewable fossil fuels. The modern day intensive crop cultivation requires the use of chemical fertilizers but fertilizers are not only in short supply but they are expensive in developing countries. Therefore, the current trend is to explore the possibility of supplementing chemical fertilizers with organic one more particularly biofertilizers of microbial origin. The present day global interest in biological nitrogen fixation is a direct consequence of this necessity to provide some economic assistance to the small and marginal farmers and to introduce a judicious combination of linear and cyclic fertilization of soils. Cyanobacteria represent a self-supporting system capable of carrying out both photosynthesis and nitrogen fixation, the energy requirement for the latter process being supplied by the sun. Rice is perhaps the only major cereal crop to whose nitrogen economy the $N_2$ fixing cyanobacteria make a significant contribution. If we use these organisms in rice fields which form an ideal environment for them, our dependence on costly nitrogen fertilizers may be significantly reduced. Recent research has shown the feasibility of using blue-green algae as a biological input in rice cultivation and of obtaining additional yields for which the incremental input cost will be low.

Cyanobacteria are known to liberate a wide array of extracellular substance like plant growth regulators, vitamins, amino-acids, sugars and other metabolites which have direct or indirect impact on plant growth and yields (Mandal *et al.*, 1999; Prasanna *et al.*, 2008). Besides their well-established role as nitrogen supplements and tolerance to desiccation, cyanobacteria can be key players in carbon segregation and improving efficiency of nitrogen use and crop yields (Watanabe and Yamamato, 1971; Rao and Burns, 1990). Diverse habitats of micro-organisms are noted in paddy field eco-system. These habitats could exhibits biologically distinct features which are microenvironments physioc-chemically different to each other. Such dissimilarities of the habitat should influence the diversity and structure of microbial communities in the structure of microbial communities in the paddy field eco-system as a whole and may support various microbiological processes occurring in paddy fields which are agronomically and bio-geochemically important (Kimura, 2000; Kirk, 2004). BGA inoculation in paddy fields, popularly called as "Algalization" helps to provide an environmentally safe agro-ecosystem contributing to economic viability in cultivation of paddy reducing cost inputs of energy (Pabbi, 2008).

## DISTINGUISHING FEATURES OF CYANOPHYCEAE

- The individual cells are prokaryotic in nature. The nucleus is incipient type and they lack membrane bound organelles.
- Complete absence of flagella.
- Locomotion is generally absent, but when occurs, it is of gliding type.
- Presence of microfibrils in the cell wall and is differentiated into four layers. Composed of mucopeptide along with amino acids, carbohydrates and fatty acids.
- The principal pigments are chlorophyll a, c-phycocyanin and C-phycoerythrin. In addition, other pigments like β-carotene and different xanthophylls like myxoxanthin and myxoxanthophyll are also present.
- Membrane bound chromotophores are absent, Pigments are found embedded in thylakoids.
- The storage foods are cyanophycean starch and cyanophycin granules (protein in nature).
- Many filamentous forms possess specialized cells known as heterocysts supposed to be the site of $N_2$ fixation.
- The protoplast of the cell is divisible into peripheral, pigmented region called chromoplasm and central colourless region called centroplasm.
- In most planktonic forms gas vacuoles have been reported.
- Mucilage is secreted by all members of cyanobacteria. Some members of chroococcales are surrounded by their individual mucilaginous sheath, whereas others are surrounded by a common mucilaginous envelope.

- Reproduction takes place by vegetative and asexual methods. Vegetative reproduction takes place bycell division, fragmentation etc. Asexual reproduction takes place by endospores, exospores, akinetes, nanospores etc.
- Sexual reproduction is completely absent in cyanophyceae. But genetic recombination has been reported in two genera.
- Branching of the filament is observed in some members. Branching is of two types – true and false branching. True branching is again of three types – lateral (*Hapalosiphon*), dichotomous (*Colteronema*) and reverse V-shaped (*Mastigocladus*).

## OCCURRENCE

Cyanobacteria can grow in all kinds of habitats from tropics to the polar regions. They inhabit most parts of the world where plants grow. *Nostoc, Oscillatoria, Anabaena, Scytonema, Gloeotrichia, Rivularia*, etc. are common fresh water forms that generally occur in stagnant waters. They are found in freshwater, sea water, lakes, stream, on or inside the soil, on moist rocks and walls and even on snow. Genera like *Gloeocapsa, Nostoc, Scytonema* remain attached in the form of mucilaginous clumps on the moist rocks. Several cyanobacteria like *Chroococcus, Gloeocapsa, Nostoc, Scytonema, Stigonema*, etc. live symbiotically as algal partners (phycobiont) in lichens. Symbiotic relationship is shown by different blue-green algae. Some members grow endophytically (inside the plants) in the leaves and stems of *Gunnera*, thalli of *Azolla, Anthoceros* and roots of *Cycas*. Some of the blue-green algae reported from hot spring are *Chroococcus, Microcystis, Oscillatorial, Phormidium, Plectonema* and *Scytonema* (Prescott, 1969). On the other hand, some cryptophytic species *Phormidium* occur in freeze arctic lakes on snow and certain others collected from Antarctica, have been shown to survive freezing. *Gloeocapsomorpha, Girranella, Nostocities*, etc., have been reported as fossils-species of *Microcystis* and certain other planktonic forms multiply so prolifically that they impart blue-green colour to water bodies, appearing as pea-soup. Such "water blooms" are more or less perennial in tropical areas. Many other cyanobacteria like *Anabaena, Oscillatoria, Anabaenopsis, Arthrospira*, etc., can form water blooms in fresh-water bodies *Trichodesmium erythraeum* causing characteristic red colour of Red Sea.

## RANGE OF THALLUS IN CYANOBACTERIA

Cyanobacteria exhibit a wide range of thallus organization. The thallus may be of unicellular or multicellular, colonial and filamentous types.

- **Uncellular Forms:** Unicellular genera are *Anacystis, Chroococcus, Gloeocapsa* (Fig. 1A), *Synechoccus*, etc. The cells may be oval or spherical.
- **Multicellular Colonial Forms:** Colonial members are *Aphanocapsa, Aphanothece, Gloeothece, Microcystis* (Fig 1B), etc. The actual shape of a coccoid colony largely depends on the numbers of planes (one, two, three or more) in which the cells of the colony when divide either alternately or in three planes and thereby they form spherical, cubical (*Eucapsis alpine* Fig 1C), squarish (*Merismopedia*) or irregular (*Microcystis* Fig. 1B) colony.

- **Multicellulor Filamentous Forms:** Subsequent evolution in thallus organization resulted in the formation of a row of cells called a trichome (eg., *Anabaena, Oscillatoria, and Scytonema* (Fig. 1. F,D and Fig.2B). When a trichome is surrounded by a sheath, it is called a filament, eg., *Lyngbya, Nostoc* and *Phormidium* Fig. 2E, Fig 1F and Fig. 2F). The filamentous forms may be uniseriate and unbranched (as in *Lyngbya, Nostoc and Oscillatoria*), truly branched in Stigonema Fig. 2G, *Mastigocladus*), or falsely branched (as in *Scytonema, Calothrix* and *Tolypothrix* Fig. 2B and C). The filament may contain single trichome (*Lyngbya, Phormidium* etc.) or several trichomes (as in Microcoleus (Fig. 1E).

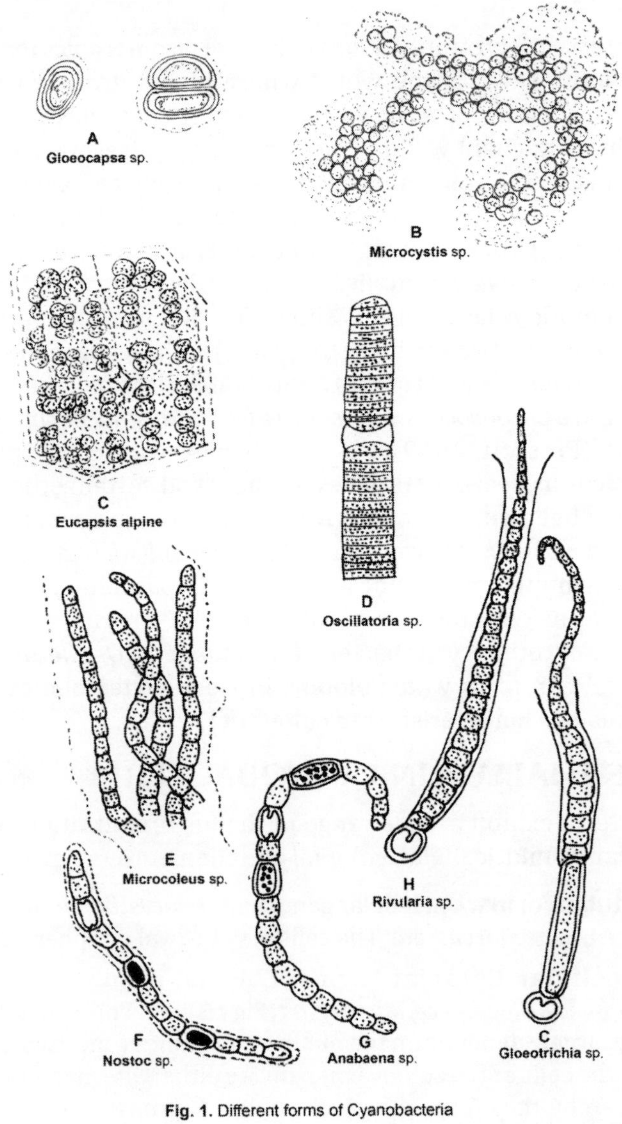


Fig. 1. Different forms of Cyanobacteria

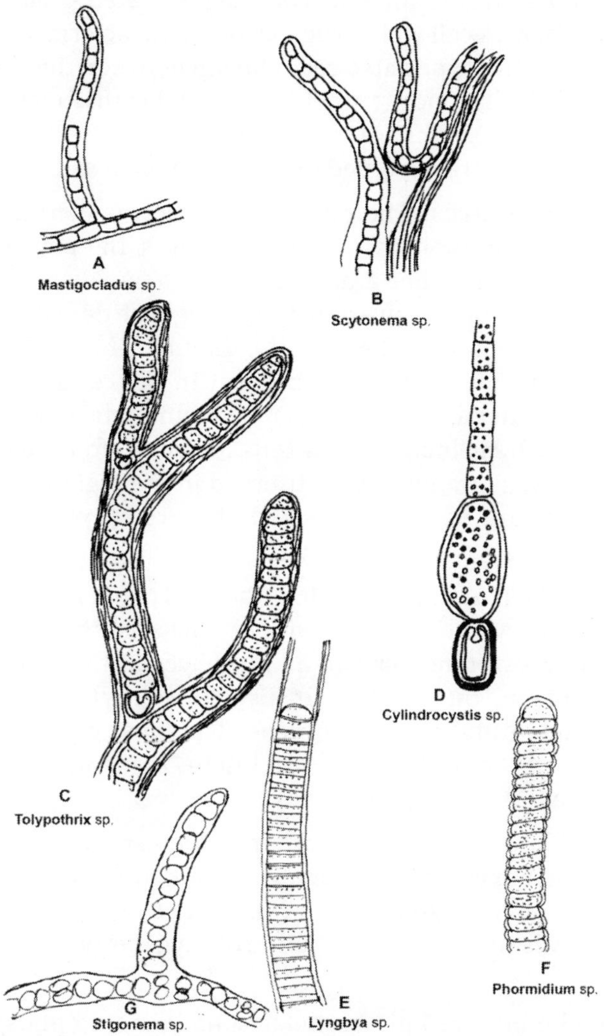


**Fig. 2.** Different forms of Cyanobacteria

## CELL STRUCTURE

A Cyanophycean cell appears surrounded by a distinct cell wall, which remains enveloped by a mucilaginous sheath. Both these represent the cell envelope. The protoplast of the cell is differentiated into an outer or peripheral coloured region called chromoplasm and inner central colourless region is called centroplasm.

**Cell wall and Sheath:** The cell wall is a rigid and complex structure, usually made up of four layers. Carr and Whitton (1973) stated that four layers are L I, L II, L III and L IV. The Layer LI is an electron transparent zone and the second layer is a mucopolymer, made up of alanine, glucose amine, muramic acid, glutamic acid and $\alpha$-diaminopimelic acid. The structure of L III and L IV layers depend on environmental condition and the amount of mucilage secreted (Smarda *et al.*, 1979). There is a close similarly in cell wall structure of cyanobacteria and that of Gram negative bacteria indicating a possible relationship between the two. The sheath outside the cell wall is fibrillar and often coloured. Red sheaths found in algae grown on highly acidic soils, Blue sheaths are found in algae of basic soils (Drouett, 1978). Yellow and Brown sheaths are observed in habitats of high salt content.

**Protoplasmic structures:** The peripheral region is distinguished by the presence of thylakoids and their associated structures, the phycobilisomes and glycogen granules. Ribosomes - like granules (70s type) are dispersed throughout the cytoplasmic matrix but are present in the highest density in the central region around the nucleoplasm (Allen, 1984). Circular fibrils of DNA which are not associated with histone protein are present in the centroplasm area. The centroplasm has been named "Open nucleus" by Baumgertel (1920) and suitably named by West (1916) as "incipient nucleus". Reserve food materials, in the form of large cyanophycin granules, are observed in the region of chromoplasm. Cyanophycin granules are bodies composed of stored protein in the form of polypeptides.

**Pigments:** According to Chapman (1962) the chief pigments are chlorophyll-a, β-carotene and C-phycocyanin. Some quantity of myxoxanthin, myxoxanthoplyll and little amount of E carotene, flavicin and C- phycoerythrin are also present. Chlorophyll-b is absent, but present in Prochlorophyta group which was previously not included under Myxophyceae. The most common xanthophyll is myxoxanthophyll, other xanthoplylls are lutein, zeaxanthin in oscillaxanth, myxorhodin, phycoxanthin and aphanizophyll.

Carboxysomes (Polyhedral bodies) occur in the central part of cells which are similar to the carboxysomes of bacteria and contains enzyme Ribulose-1,5-bisphasphote carboxylase/oxygenase (Skleryk *et al.*, 1997). Cyanobacteria photosynthesis involves two photosystems and of oxygenic type like that of higher plants.

Polyphosphate bodies are lipid granular contain stored phosphate present in older cells but found in young growing cells. Electron microscope study on blue-green algal cell observed no membrane bound organelles of the categories chloroplast, mitochondria, nucleus, golgi bodies and endoplasmic reticulum. α-granules or Polyglucon granules are common in blue-green algae located between thylakoids in actively photosynthesizing cells. B-granules are small dense osmophilic bodies scattered in thylakoids. Ribosomes are of 70s types like bacteria.

A gas vacuole is composed of gas vesicles, which are nothing but hollow cylindrical tubes with conical ends present in the cytoplasm of all cyanobacteria except the order chamaesiphonales. Gas vesicles do not have typical protein - lipid membranes being composed of protein ribs or spirals arranged similarly the hoops on a barrel. By applying pressure to of the gas vacuole (gas vesicles) is to reduce the density of the blue-green and therefore allow the cells to stratify at the best level for photosynthesis in bodies of water (Walsby, 1978). Thus in many blue-greens they collapse and form gas vesicles to control their density and thus their position in the water column.

Some irregularly appeared complex invaginations of plasma membrane are called as mesosomes. They are typically roughly spherical and are about 0.5 μm in diameter. Mesosomes are observed in *Anacystis nidulans, Phormidium luridum, Mastigocladus laminosus* and in species of *Calothrix* and *Gloeotrichia*.

Intrathylakoidal Granules are spherical bodies formed by the end of the two oppressed membranes of a thylakoid dilating out at the end. They are covered by a unit membranes and are about 0.5 μm in diameter. Found in two cyanobacteria *Microcystis* sp. and *Nostoc commune*.

Pankratz and Bowen (1963) observed cylindrical bodies in cells of *Symploca muscorum*. They have been reported in *Oscillatoria, Lyngbya, Spirulina platensis, Trichodesmium erythraeum* etc. They may be located anywhere in the cell having many of the bodies. They are most commonly located near the plasma membrane adjacent to cross walls. It has been suggested that they may provide a micro-anaerobic niche within the cell for the function of certain enzymes such as the nitrogenase which are oxygen sensitive.

In some blue-greens tubular structures are observed called as Striated Microtubules. They have been called striated microtubules because in longitudinal sections, they show osmophilic striations at right angles to the longitudinal wall, which has a wavy appearance. Found in *Anabaena minutissima, Calothrix marchica* and *Nostoc muscorum*.

In some members of *Anabaena* and *Nostoc* Microtubule like inclusions are present which are variable in diameter. They are found only in the small cuboidal cells of hormogonia.

Polyhedral Body associated tubules are tubular inclusions occur in groups and are always associated with one face of the polyhedral bodies. They have been reported in some species of *Anabaena*. They are responsible in orderly transport of materials into and out of the polyhedral bodies.

Jensen and Ayala (1976b) observed Sleeve Bodies in two *Anabaena* isolates consists of a short cylinder 32 nm in diameter, the wall of the sleeve has thick osmophilic striations perpendicular to the longitudinal axis of the sleeve.

Microplate-microtubule array are inclusion bodies of irregular occurrence consist of an arched plate which is generally parallel and close to the plasma membrane. Jensen and Ayala (1980) reported presence of these inclusions in two *Nostoc* isolates and in one *Anabaena* isolate.

Microfilaments are inclusions which are wavy and up to several micrometers in length. They occur in bundles in *Cylindrospermum* isolates.

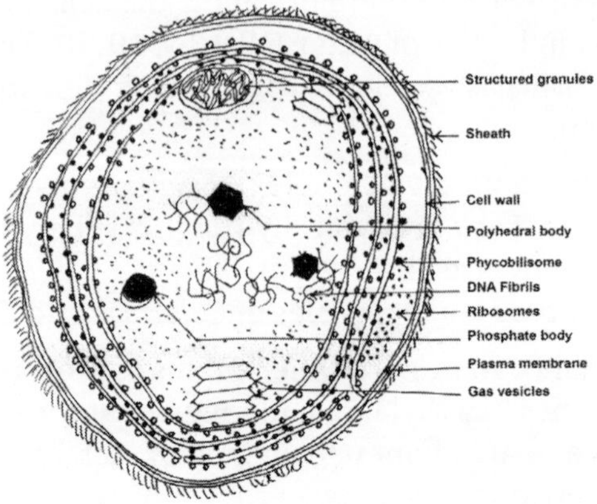


**Fig. 3A.** Diagram of the fine structural features of a cyanobacterial cell

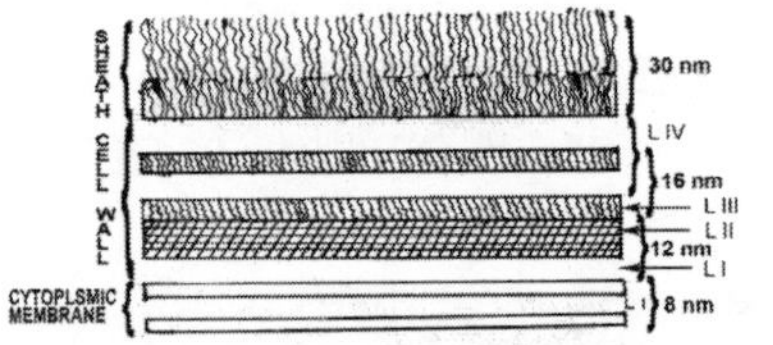


**Fig. 3b.** Cyanobacterial Cell Wall (After Jost)

Paracrystalline bodies range from cuboidal to complex aggregates reported in *Oscillatoria chalybea*, *Gleotrichia echinulata* and in isolates of *Anabaena* and *Nostoc*. Because of their morphology, they have been compared to some storage proteins of higher plants and the crystalloid core of microbodies of plant cells.

Granular bodies are generally found between the thylakoids in few *Nostoc* and *Anabaena* isolates. It is suggested as storage material because the granules resemble glycogen granules present in cyanobacteria cells.

Spherical bodies of medium electron density are observed in the areas which are electron transparent. They res*emble known storage bodies cyanophycin granules and noted in some isolates of Nostoc, Calothrix* and *Anabaena*.

Spherical spoke like inclusions have been reported in *Gloeotrichia* sp. and *Calothrix scopulorum*. They are probably a lipid material similar to the usual lipid inclusions (β-granules).

In *Nostoc muscorum* spheroids inclusions occur in clusters of 10 or more which are found within the expanded thylakoids.

## HETEROCYST

These are specialized distinct cells of some filamentous members of Nostocales (except Oscillatoriaceae) and Stigonematales which can be distinguished from the remaining cells by their thick walls and pale-yellow homogeneous matrix. They arise from somewhat enlarged vegetative cells, and are characterized by the presence of pore at one or both poles. They may be terminal in *Gloeotrichia*, intercalary in *Nostoc*, *Scytonema* etc., lateral in *Nostochopsis*. A terminal heterocyst, when formed at the base of the filament, is called basal (as in *Tolypothrix distorta*), but when it terminates some lateral branches it is called pedicellate heterocyst (as in *Mastigocladus testarum*) (Fig. 4B-4D). In terminal heterocyst only one pore is present on the side it is attached with the other cell of the filament, but two pores are observed one in each pole. Heterocysts with three pores are present in *Brachytrichia balani* (Fig. 4A)

Heterocysts are thickwalled, pale yellow and barrel shaped structures. The wall is differentiated into an outer fibrous, a middle homogeneous and an inner thin lamellar layer. Pores are found at one of both the poles, in polar region the wall layers become thicker and the pores are plugged with refractive material, called polar granules or nodules (Fig. 4). Through the pores of heterocysts pass the protoplasmic connections, which help the heterocysts remain connected with the neighbouring cells of the filament. Heterocysts contain dense and homogeneous cytoplasm high concentration of carotenoids, hence yellowish in colour. Principal light absorbing pigments phycobilins and chlorophyll-a are absent and there is no $CO_2$ fixation or $O_2$ production within heterocysts. They lack $O_2$ - evolving photosystem II. Thylakoids are tightly packed, with reduced interlamellar spaces. They are observed along the periphery in a complex reticulate manner. Structural granules, crystalline granules and small granules are associated with peripheral lamellae, these granules are gradually decrease in number and ultimately disappear as the heterocyst matures.

The clearest difference between heterocysts and vegetative cells is the possession by the former of a thickened cell envelope. This can be resolved by electron microscopy into three layers - an inner most laminated layer, an outer fibrous layer and sandwiched between, a homogeneous layer (Lang and Fay, 1971; Fay and Lang, 1971). In addition, there is considerable re-arrangement of the thylakoids and deposition of an amorphous material at the two pores adjoining neighbouring vegetative cells (Wilcox *et al.*, 1973b). This amorphous material is rich in arginine and has the ultra-structural appearance and solubility properties of cyanophycin which is a nitrogenous reserve food. The innermost of the heterocyst envelope layers consists of glycolipids and surrounds the heterocyst protoplast, except at the junction with the neighbouring vegetative cell. The homogeneous layer is largely polysaccharide, the extra heterocyst layers, the glycolipid, have been considered to form a barrier to limit the ingress of oxygen, which would otherwise impair the functioning of nitrogenase. It has been demonstrated that diffusion of both $N_2$ and $O_2$ is considerably reduced in heterocysts. Nevertheless, sufficient $N_2$ can still enter the cell to supply the requirements of fixed nitrogen, while the reduction of $O_2$ ensures the activity of nitrogenase (Walsby, 1985).

During the transition of vegetative cell to heterocyst a series of stages can be defined, based on ultra-structural changes, such as the formation of homogenous, laminated and fibrous layers, observed by electron microscopy. Wilcox et al. (1973b) defined seven stages of heterocyst development, the first five of which were related to pro-heterocyst formation and the final two, stages of maturation. The mature heterocyst has paler and less granular cytoplasm than vegetative cells, with thickened walls and a refractile polar granule at the point of attachment to neighbouring vegetative cells, there being two in intercalary and only one in terminal and basal heterocysts. Pro-heterocysts lack the thick wall and polar granules but tend to be paler and less granular than vegetative cells and often larger. It has been found that pro-heterocysts can dedifferentiate and resume vegetative growth, heterocysts can not.

Fogg (1949) noted that fixed nitrogen can inhibit formation of heterocyst, so there being an inverse relationship between heterocyst frequency and the concentration of combined nitrogen in the medium. In *Anabaena cylindrica* ammonia is the best suppressor of heterocyst development, while nitrate is very poor. The continuous presence of pro-heterocysts in a culture does not imply, however, that growth is limited by the availability of fixed nitrogen, but shows that the trigger for pro-heterocyst formation is more sensitive to nitrogen availability than is growth. The development of heterocysts is initiated before growth is seriously affected by nitrogen supply. Fogg (1949) concluded that the formation of a heterocyst from a vegetative cell occurred when "the concentration within it of a specific nitrogenous inhibitory substance, probably ammonia or some simple derivative of ammonia, falls below a critical level".

The study of cyanobacterial nitrogen fixation and differentiation by classical methods of bacterial genetics has been considerably hampered due to lack of a means of gene transfer in filamentous cyanobacteria (Delaney et al., 1976; Herdman, 1982). Introduction of molecular cloning, however, provided new techniques for the genetic analysis of cyanobacteria. To establish a cloning system two things are necessary – i) a suitable host into which DNA can be transferred and ii) a vector to carry the DNA, containing at least one marker to render it traceable. In contrast to the filamentous cyanobacteria, a number of unicellular cyanobacteria have been shown to undergo genetic information. A large number of cyanobacteria have been shown to contain plasmids, this characters encouraged many workers to investigate the possibility of establishing a gene cloning system in unicellular cyanobacteria.

**Functions of heterocysts:** Earlier it was considered that they are dead cells, some considered them as spore like structures. It has been observed that after removal of heterocysts from the trichome sporulation is inhibited. The vegetative cells away from them do not sporulate but adjacent cells sporulate earlier than those away from them.

Fay *et al.* (1968) considered heterocyst as N2-fixing organ, now it has been suggested that heterocysts are the main sites of N2-fixation in blue-green algae, particularly under aerobic conditions. The normal vegetative cells have the capacity to fix nitrogen, under anaerobic conditions, but in aerobic conditions the enzyme nitrogenase becomes inactive. Thickwalled heterocysts provide anaerobic environment, which is necessary to activate nitrogenase. Within heterocysts photosystem II is absent, as a result due to lack of photolysis $O_2$ does not evolved, only photosystem I is observed which generates assimilatory power ATP required in nitrogen fixation. According to Stewart (1970) the main physiological roles of heterocysts are: i) control of sporulation, ii) involvement in growth and N2-metabolism and iii) act as secondary reproductive units.

Large number of heterocystous and some non-heterocystous cyanobacteria have the capacity to fix atmospheric nitrogen into ammonium in cells using ATP as a source of energy and either molybdenum or vanadium as a cofactor (Attridge and Rowell, 1997). The ammonium is then incorporated into amino-acids, such as glutamic acid and alanine. Cyanobacteria produce oxygen during photosynthesis, it is very important because the key enzyme for $N_2$-fixation nitrogenase is very much sensitive to inactivation by oxygen. Three different mechanisms are found in cyanobacteria for separating nitrogenase from oxygen.

- N2-fixation by heterocystous cyanobacteria in presence of light.
- N2-fixation by non-heterocystous cyanobacteria in the dark or in absence of light.
- N2-fixation by *Trichodesmium* a filamentous, non-heterocystous cyanobacteria in presence of light.

A. Heterocystous forms lack photosystem II and therefore, unable to evolve oxygen during light reaction. So, required ATP for N2-fixation are formed by cyclic photophosphorylation. Potts et al., 1992 stated that heterocysts contain a protein named cyanoglobin a form of myoglobin that scavenges oxygen present, preventing nitrogenase enzyme.

B. Non-filamentous non-heterocystous cyanobacteria can fix nitrogen in the dark when photosynthesis is not producing oxygen which is inhibiting nitrogenase. Under alternating light and darkness, most non-heterocystous cyanobacteria effect an almost complete temporal separation of N2-fixation and photosynthesis. Under these conditions, cultures of Gloeothece continue to reduce acetylene for at least 28 days. This pattern seems not to be caused by synchronous cell division since the doubling time of Gloeothece under alternating light and darkness was 55-90 hours. It must provide protection for nitrogenase from photosynthetically evolved $O_2$.

C. The cycle of nitrogen fixation in *Trichodesmium* is entrained by an endogenous circadian clock (Capone et al., 1997). *Trichodesmium* a planktonic flora of tropical seas are estimated to be responsible for fixing one quarter of the total nitrogen in the world's ocean (Bergman and Carpenter, 1991). Within the filament, some cells are able to fix nitrogen, while others do not. Nitrogen fixing cells are adjacent to one another and the cells have a denser thylakoid network with minimum number of gas vacuoles and cyanophycin granules (Fredrikson and Bergman, 1997).

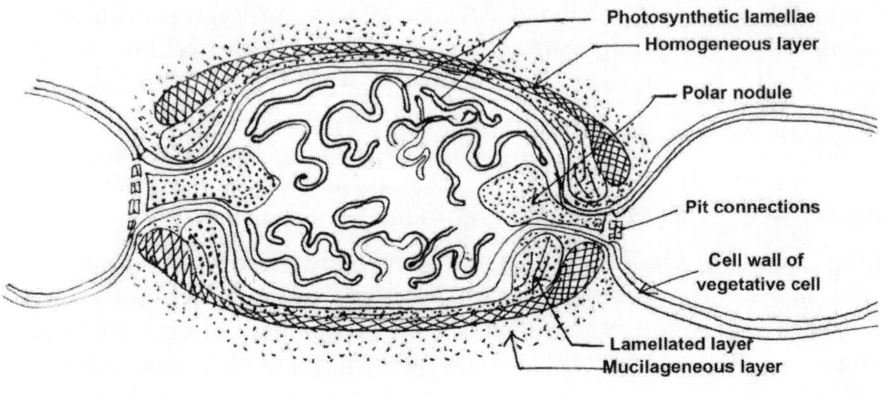


**Fig. 4.** A Mature Heterocyst

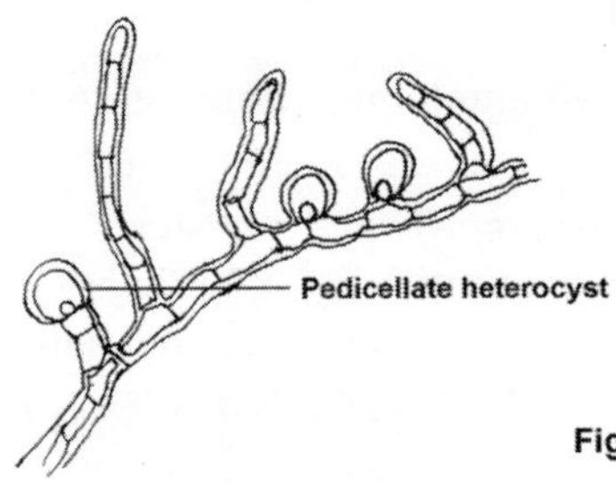


**Fig. 4D.** Pedicellate heterocysts in
**Mastigcladus testarum**
(After Iyengar and Desikachary)

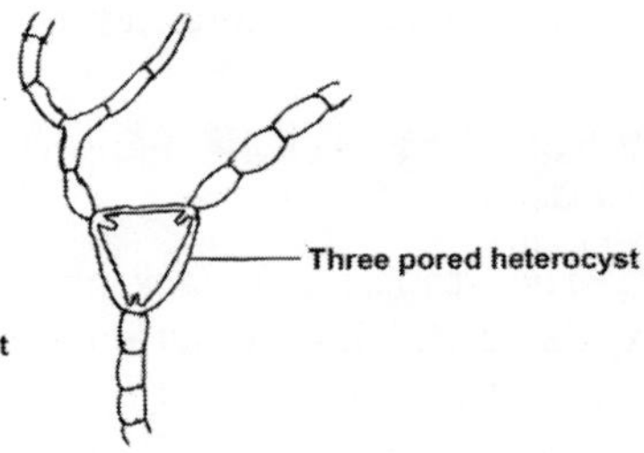


**Fig. 4A.** Three-pored heterocysts in
**Brachytrichia balani**
(After Iyengar and Desikachary)

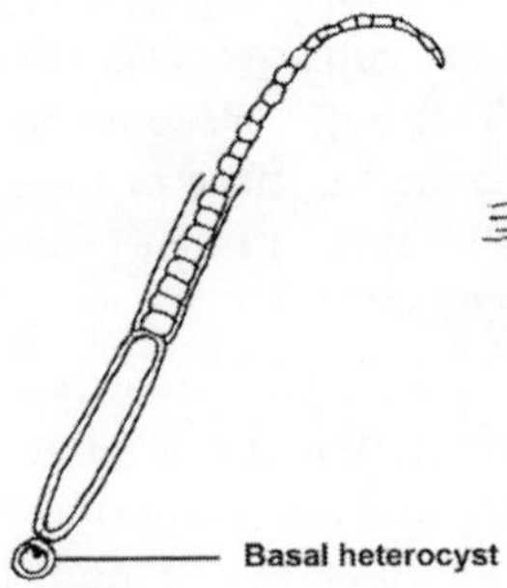


**Fig. 4C.** Basal heterocysts in
**Gloeotrichia** sp.
(After Iyengar and Desikachary)

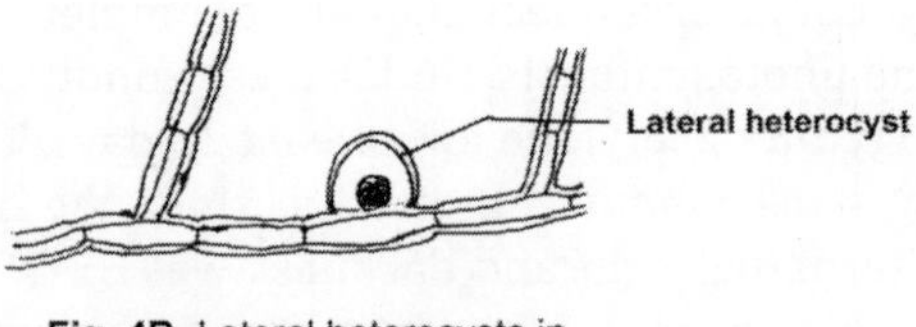


**Fig. 4B.** Lateral heterocysts in
**Nostochopsis lobatus**
(After Desikachary)

The blue-green algae, fixing nitrogen from the air, must possess an efficient mechanism to check oxygen inhibition of the enzyme nitrogenase. Heterocysts provide a good site for an efficient anaerobic functioning of nitrogenase by giving thick wall. This enzyme nitrogenase present in heterocyst reduces nitrogen gas into ammonia, which in turn is metabolized to glutamine by the enzyme GOGAT (glutamine oxalo-gluterate aminotransferase). Other amino acids are formed from glutamine and glutamate by various transamination reactions (Fig. 5).

According to R.N. Singh, the fertility of rice fields in India is increased tremendously by species of *Anabaena*, *Tolypothrix* and *Aulosira*. Wolk 1996, Golden and Yoone, 1998 have studied the genes involved in early differentiation and pattern of heterocyst formation. They have identified several genes, such as ntc A, hanA, hetR, hetC, hatA, hatB and hatS. It has been found that the frequency of heterocyst production is affected by the following external factors:

- Development of heterocyst is inhibited in the presence of blue and green light, whereas the white and red light supports it.
- C:N ratio of cells play an important role in the differentiation of vegetative cells into heterocysts.
- Concentration of phosphate salts in the medium stimulates heterocyst development.
- Absence of Mg and Fe ions in the medium inhibits heterocyst development.
- Presence of combined nitrogen in the medium inhibits development of heterocyst and the formation of heterocyst is inversely related to the amount of nitrogen in the medium.

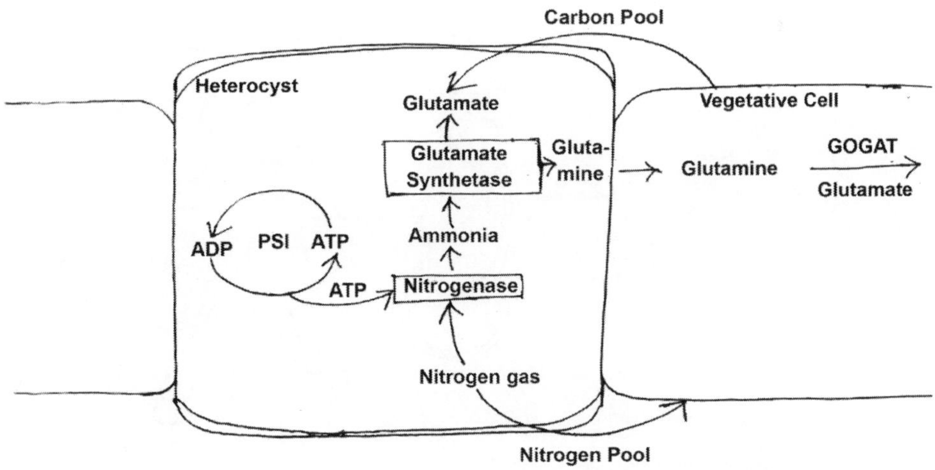


**Fig. 5.** Schematic representation of nitrogen transfer from heterocyst to vegetative cell

# REPRODUCTION

Cyanobacteria reproduce both by vegetative and asexual means. Sexual reproduction does not take place.

Vegetative reproduction performs through fragmentation (*Oscillatoria*), fission (*Synechococcus*), hormogonia formation (*Nostoc, Oscillatoria*), hormospores (*Westiella*) and by planococci and by palmella stage.

Asexual reproduction occurs by the formation of hormogonia, endospores, exospores, akinet and nannocytes.

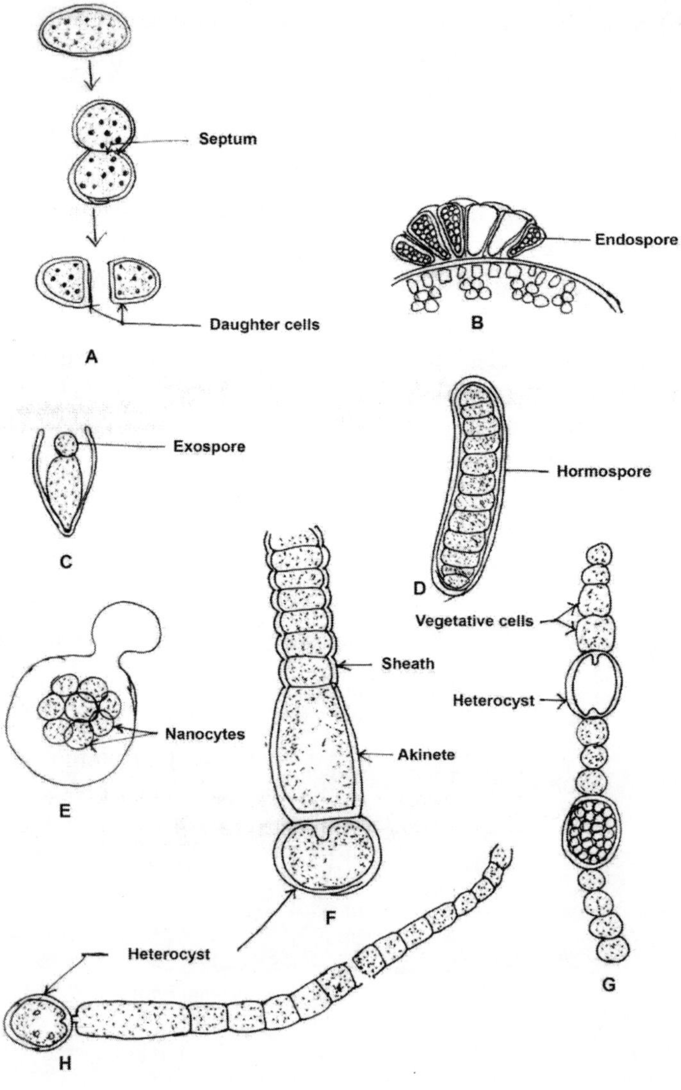


**Fig. 6.** Different types of vegetative and asexual reproduction of Cyanobacteria

**Hormogonia:** Observed in filamentous cyanobacteria, which are short pieces of trichome that become detached from the parent filament and move away by gliding and ultimately develop a separate filament. In some forms like *Oscillatoria* and *Cylindrospermum* hormogonia are produced at the tips of special branches. (Fig. 6D)

**Akinetes:** Resting spores, resistant to unfavorable environmental conditions, occur in the members of Nostocales and Stigonematales. Mature, Akinetes are larger than vegetative cells, contain protoplast full of reserve foods and specialized thick cell wall. Akinetes of *Nostoc* lose 90% of their photosynthetic respiratory capabilities, as compared to vegetative cells. On germination Akinetes produce new filaments and they are considered to have similar function to endospores of bacteria. (Fig. 6F)

**Endospores:** Endospores are formed by internal division of the protoplast, resulting in an irregular mass of spores. Noted in *Dermocarpa* sp. (Fig. 6B)

**Exospores:** Exospores are also formed by internal division of protoplast. These spores are abstricted serially from the open ends of sporangia (*Chamaesiphon*). (Fig. 6C)

**Nannocytes:** These are naked protoplast formed by repeated and successive divisions. No well-marked cell enlargement is observed in such cases. They are known as nannocytes because of their naked nature, noted in *Gloeocapsa*, *Aphanothece*, etc, (Fig. 6E).

## SIMILARITIES OF CYANOBACTERIA WITH RED ALGAE

- The blue (c-phycocyanin) and red (c-phycoerythrin) pigments present in cyanobacteria are chemically similar to those found in Rhodophyceae (r-phycocyanin and r-phycoerythrin).
- In both the groups flagellated motile cells are absent.
- Fatty acid synthesis mechanisms are similar.
- Presence of pit connections in Stigonemataceae is a character also noted in members of Rhodophyceae.
- In both case, photosynthetic thylakoids occur singly and widely separated.
- In both case, the principle constituents of mucilage are sulfated galactoses, uronic acid, glucose and xylose.

## SIMILARITIES OF CYANOBACTERIA WITH BACTERIA

- Both are prokaryotes, i.e., absence of a definite and organized nucleus, cell organelles like golgi body, mitochondria, plastid etc.
- Peptidoglycan and diaminopimelic acid are present in cell wall of both groups.
- Close similarity between capsule of bacteria and mucilage sheath in cyanobacteria.

- Both are sensitive to antibiotics.
- Both lack true sexual reproduction; Bacteria fulfill their requirements of sexuality by means of conjugation, have been reported in certain cyanobacteria.
- Many metabolic processes (e.g., sulfur and nitrogen metabolism) are similar in both groups.
- Some bacteria can fix nitrogen which is an important function of many cyanobacteria.
- Some bacteria (*Thiothrix*) are structurally similar to hormogonia, found in the blue-green algae.
- *Beggiatoa*, a sulphur bacterium, resembles with *Oscillatoria* in shape and movement.

## ECOLOGY OF CYANOBACTERIA

They are found in freshwater lakes, ponds, water reservoirs, rivers, streams marsh lands etc.; as terrestrial in soil, walls, stones, rocks, barks of trees and plant pots; found in brackish water and marine habitats. They may form blooms e.g. *Microcystis, Trichodesmium erythraeum* (responsible for the red colour of the "Red-Sea"), *Anabaena, Spirulina, Anabaenopsis* etc. *Anabaena* and *Nostoc* form symbiotic association with bryophyte such as *Anthoceros*, pteridophyte like *Azolla*, Gymnosperm such as *Cycas*, Angiosperm such as *Trifolium alexandrium* and Fungi to form Lichen.

All species of BGA are photosynthetic, some are facultative phototrophic anaerobes. Some cyanophyceae like *Oscillatoria limnetica* are capable of photosynthesizing under both aerobic and anaerobic conditions. Some forms, under anaerobic conditions, in the presence of sulphur, derive electrons by reducing sulphur. Dissolved $CO_2$ reacts with $H_2S$ in the presence of light and chlorophyll to produce sugar, water and sulphur.

Bacteriocin is a proteinaceous antibiotic secreted by *Nostoc* sp. That can kill related strains of alga. *Scytonema hofmanni* is known to secrete cyanobacterin, a chlorinated gammalactone. These antibiotics play a role in inhibiting the growth of competing organisms. *Microcystis* has been observed to inhibit growth of *Staphylococcus* (a bacterium) and *Closterium* (a green alga).

Blue-green algae are considered as the first colonizers of marshlands. They are important in ecological succession as they hold soil and dust particles as they dry up. Blue-greens are capable of reclamation of saline-alkali soils by preferential absorption or adsorption of sodium. Investigations on soil indicators of land-use have shown that certain cyanobacteria like *Cylindrospermum* sp., *Calothrix* sp., *Pseudoanabaena* sp., *Scytonema* sp. and *Thricomus* sp. are highly sensitive to disturbance and as such, their absence indicate highly disturbed soils.

Kaushik, 1997 observed that the growth of some BGA in soil seems to influence the physical and chemical properties of the soil. The ability of Cyanobacteria such as *Oscillatoria, Scytonema* etc. can bind sand and soil particles and helps in prevention of oil erosion. They help to maintain the moisture content of the soil by reducing direct evaporation of water from open lands.

A number of blue-green algae especially belonging to the Chroococcales, Oscillatoriaceae, Nostocaceae occur ordinarily as planktonic forms, while a few others may occur in plankton in one or the other stages of their life-history, though ordinarily can not be called planktonic forms. Several species of the cyanobacteria grow in abundant quantities so as to colonizes entire body of water and in some cases these are caused by a single species. Such growths are called Water blooms. A large number of species are reported which can form water blooms like- *Microcystis aeruginosa, M. scripta, M. viridis, Nostoc carneum, N. linekia, Nodularia spumigena, Anabaena catenula, A. catenula* var. *affinis, A. cirinalis, A. circinalis* var. *cyrtospora, A. flos-aquae, A. macrospora, A. macrospora* var. *robusta*; *Gloeotrichia echinulata, G. natans*; *Spirulina gomontiana, S. thompsoniana*; *Arthrospira platensis*; *Oscillatoria agardhii, O. prolifica, O. rubescens, O. planctoniea; Lyngbya limnetica, L. hieronymusii.* In India especially blooms caused by *Microcystis aeruginosa* and *Trichodesmium* sp. are very common.

Cyanobacteria occur in many saline conditions, the chief of which is the sea. More than 80 species are known in India from saline situations out of which 50 species are marine. The members of the Chamaesiphonales and the Pleurocapsales are dominant marine forms while a number of species belonging to other orders are also marine. Blue-green algae have been reported from salt pans growing in brine, in many countries, where they form as it were a mat at the bottom. This mat formation is very important in manufacture of salt in solar salt works. After manufacture when the salt is removed from the beds care is taken to see that the algal mat is not disturbed or removed. Two cyanobacteria have been identified, namely, *Microcoleus chthonoplastes* and *Oscillotoria* sp., and some other members like *Microcoleus tenerrimus, Aphanocapsa marina, Lyngbya aestuarii, Phormidium* sp., *Anabaena torulosa, Aphanocapsa* sp., *Spirulina* sp. etc. as minor partners.

Cyanobacteria can grow on calcareous substrata both as epilithic and endolithic forms. *Calothrix* sp. and *Pleurocapsa* sp. are the commonest epilithic forms, where as a number of genera are endolithic like *Plectonema, Mastigocoleus, Hyella* etc.

In shallow water few blue-greens can form aegagrophilous balls in the same way as the *Cladophora* balls. Species of *Tolypothrix, Scytonema* and *Chroococcus* as well as *Schizothrix lacustris* are found in these balls.

Cyanobacteria of the littoral zone occur as a black encrusting film on rocks at the upper limit of the high tide zone. The brackish zone of cyanobacteria is composed of species of *Calothocix, Phormidium, Nodularia, Rivularia* and *Gloeothece* (Little, 1973). Most of the cyanobacteria present in littoral zone are nitrogen fixing and they have a significant role in the productivity of rocky shores and coral reefs (Mague and Holm-Hansen, 1975). Cyanobacterial cells with less than 2 μm in diameter are considered as photosynthetic picoplankton. The coccoid forms *Synechococcus* and *Synechocystis* are the major picoplankton and they are more evident in nutrient poor offshore water where large phytoplanktons are unable to grow properly due to low concentrations of nutrients. Picoplanktons are concentrated at the bottom of the zone, because they grow best under these conditions. They contain large amounts of phycoerythrin that allows to absorb blue-green light penetrating into deep water (Fogg, 1986). *Trichodesmium* is a major component of the Caribbean sea plankton, comprising 60% of the total chlorophylla in the upper 50 meter and about 20% of the primary production. It is an important source of nitrogen, fixing 1.3 mg of nitrogen per sq. m per day (Carpenter and Price, 1977). The cells produce gas vacuoles which cause the cells to accumulate at the surface of water, a phenomenon called as "sea sawdust" by sailors. They move up in the water column by means of gas vescicles, and move down in the water column by carbohydrate loading (Romans *et al.*, 1994).

Fresh water blooms of cyanobacteria are formed different species of *Anabaena, Microcystis, Gloeotrichia, Aphanizomenon, Lyngbya, Oscillatoria* etc. Fresh water blooms are found all over the year, but they grow vigorously in late summer and in early autumn.

*Mastigocladus laminosus* occurs in thermal springs all over the world whereas other cyanobacteria such as *Synchococcus lividus* and *Oscillatoria terebriformis* are also found in restricted areas. Genera like *Porphyrosiphon, Microcoleus, Schizothrix, Scytonema* and *Plectonema* grow as primary colonizers in the soil. They help to maintain moisture in the soil, bind sand and soil particles to prevent soil erosion, fixes atmospheric nitrogen into the soil, supply growth substances to higher plants. The cosmopolitan terrestrial cyanobacterium *Nostoc commune* is able to tolerate acute water stress and can survive in the air-dry state for many years. This cyanobocterium is known as anhydrobiotic organism.

*Microcolues* chthonoplastes grows on salt marshes and flat muddy area and can migrate to the surface of the mud at a rate of about 7 mm per day.

Cyanobacteria are used as human food and animal food supplements. In countries like China, Japan *Spirulina* industry is well developed. Besides *Spirulina* pills and capsules, there are also pastries, blocks and *Spirulina*-filled chocolate blocks. Some blue-greens like *Aphanothece sacrum, Nostoc commune, N. verrucosum, Brachytrichia* etc. have been used as side dishes in hotels of Japan.

## ROLE OF CYANOBACTERIA IN ORIGIN OF LIFE

The planet earth was originated 4.5 billion years ago in our solar system. Between 3.5–4.0 billion years ago, after several events of chemical evolution and modification in the earth's atmosphere the first life, amino acid and nucleic acid, called ecobiont or protobiont, was originated in abiogenesis phase and protobiology phase was started. Life probably originated when the earth's atmosphere was oxygen free and consist of methane, nitrogen, ammonia and water vapour with high percentage of hydrogen. Probably between 3500–4000 million years ago simpler organic compounds like amino acids were formed under the influence of UV radiations and electric discharge as sources of energy. Ultimately, the earth cooled down and provide pools of water. This phase when many organic compounds were produced but there was no life in the earth. This phase is known as abiogenesis.

The droplet called coacervates, may absorb water, swell and undergo the processes of selectivity, recognition, interaction etc. This phase was the origin of amino acid which on heating formed nucleic acid, this is called as eobiology or protobiology phase. The geological and biological evidences suggest that the first life in this planet were probably some heterotrophic anaerobic bacteria. The life has evolved into three domains: age of Eobacteria, Cyanobacteria and Eukaryotes which ultimately evolve into land plants.

Age of Eobacteria began before 2.9–2.8 Gyrs ago when the origin of cells and anoxygenic photosynthesis generated the most primitive prokaryote chlorobacteria, they have anoxygenic photosynthesis, chemosynthesis using hydrogen or $H_2S$ as hydrogen donor to fix carbon. This is also called chlorobacterial age.

After the Eobacterial age, the cellular innovation and microbial radiation dramatically transformed the earth's surface and the glycobacterial revolution initiated on oxygenic age called age of cyanobacteria. During this age, between 2.8–0.9 billion years ago, cyanobacterial photosynthesis appeared and prodominate in the ecosystem and the biosphere was differentiated into aerobic and anaerobic zones. That is why, cyanobacteria are also called first oxygen producing organism of this planet. The Eobacteria and Cyanobacteria both have always two distinct lipid surface membranes whereas posibacteria consists only one surface membrane. For this reason, the Procambrian period could be properly termed as Age of Prokaryotes and the later half of this Eon, was the Age of Cyanobacteria (Schopf, 2006). The cyanobacteria have a greater diversity of morphological phenotype, many members have extremely long generation times as compared to the heterotrophic bacteria. The genome size is apparently longer in most complex cyanobacteria than *E. coli*, *Bacilli* and smaller unicellular cyanobacteria, have efficient DNA-repair systems. The prebiontic atmosphere was highly reducing and hence hydrogenase, nitrogenase, acetylene, cyanide, Cyanogens, nitrites etc. have high probability in cyanobacteria. They have specialized thick walled cells heterocysts, a novel device, to protect the oxygen sensitive nitrogenase. The cyanobacterial oxygen liberation grew the ozone layer and later 0.9 Gyr ago after rise of planktons another age started known as age of Eukaryotes and Archaebacteria. The evolution of green algae plastid is of cyanobacterial origin as a result of endosymbiosis which is supported by the discovery of the Prochlorophyta (Lewin, 1976). Cyanobacteria are the source of Glaucophyle to simple plastids. Members of the Glaucophyta have endosymbiotic cyanobacteria in the cytoplasm instead of chloroplasts. This symbiotic association are thought to represent intermediates in the evolution of the chloroplast. The evolutionary event that led to the chloroplast in green algae and red algae occurred as follows - A phagocytic protozoan took up a cyanobacterium into a food vesicle. The cyanobacterium was not digested by the protozoan but lived as an endosymbiont. Protozoan was benefited by getting photosynthate from the endosymbiont alga and the alga received a protected stable environment. A mutation in the endosymbiont resulted in a loss of the wall and the membrane of the food vesicle of the host became the outer membrane of the chloroplast envelope. The cell membrane of the endosymbiont cyanobacterium became the inner membrane of the chloroplast envelope. Ultimately by the rearrangement of the thylakoid membrane and transformation of polyhedral bodies resulted a pyrenoid to complete the transition to a true chloroplast as in chlorophyceae and Rhodophyceae (Fig. 7).

The Euglenophyta and Dinophyta are the only algal groups which have one membrane of chloroplast endoplasmic reticulum (CER). CER resulted when a chloroplast from a eukaryotic alga was taken up into a food vesicle by phagocytotic euglenoid or dinoflagellate. Instead of digestion, in this case, the chloroplast was retained in the cytoplasm of the host as an endosymbiont. The euglenoids probably evolved by the capture of a chloroplast of a green alga, as the chloroplasts of the green algae and euglenoids are very similar. It is thought that chloroplast of the dinoflagellates is derived from chloroplast of an endosymbiotic red algae (Fig. 8)

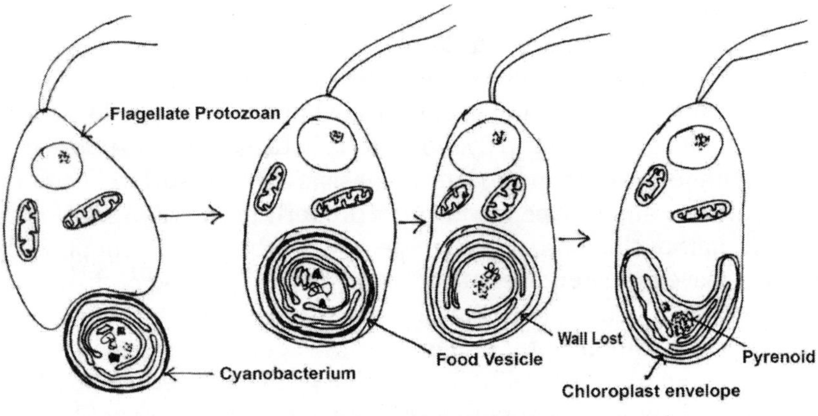


**Fig. 7.** A cyanobacterium was taken up by a protozoan into a food vesicle. An endosymbiotic association was formed and through evolution the endosymbiotic cyanobacterium evolved into a chloroplast surrounded by two membranes of the chloroplast envelope (Courtesy: Robert Edward Lee)

By the process of secondary endosymbiosis two membranes of CER is evolved in algae. A phagocytic protozoan took up a eukaryotic photosynthetic alga into a food vesicle. This alga was not digested by the host, instead established as an endosymbiote with the food vesicle. The endosymbiotic alga benefited from the acidic environment in the food vesicle where they got inorganic carbon in the form of $CO_2$ which is needed by RUBP carboxylase enzyme for fixation of carbon. The host benefited from photosynthate of endosymbiotic alga. The membrane of the food vesicle eventually fused with endophosmic reticulum of the host, as a result ribosomes appeared on the outer surface of this membrane, which became the outer membrance of the chloroplast ER (Fig. 9). The type of CER that exists in the Heterokontophyta and the Prymnesiophyta resulted from further reduction. The nucleomorph was lost totally and the host taken over the formation of storage product. The cells resulted had two membranes of chloroplast envelope surrounding the chloroplast.

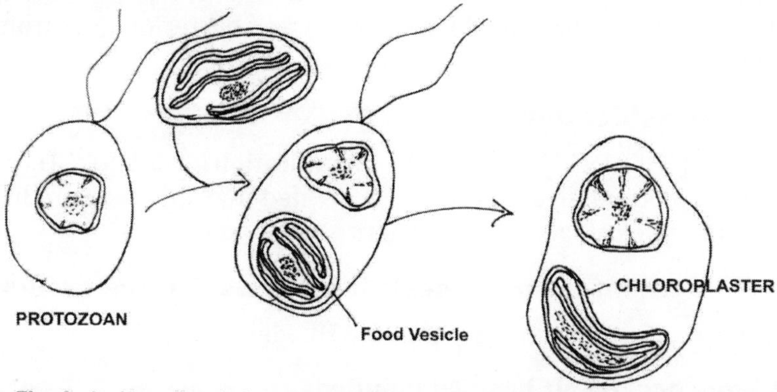


**Fig. 8.** A chloroplast being surrounded by a single membrane of CER. Initially a chloroplast was taken up by a phagocytic protozoan - A food vesicle is formed - Endosymbiosis resulted - Food vesicle membrane converted to a single membrane of CER (Courtesy: Robert Edward Lee)

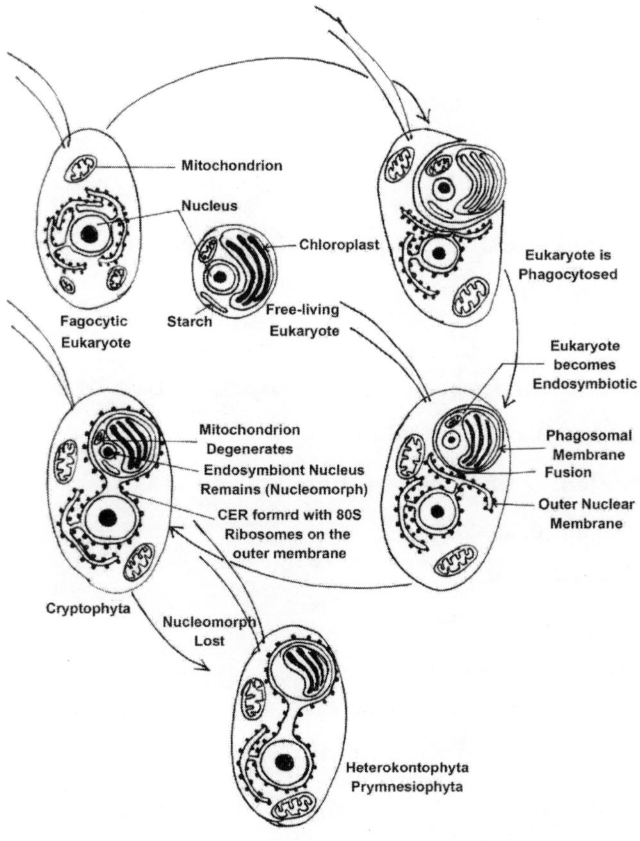


**Fig. 9.** Evolution of Algae with two membranes of Chloroplast Endoplasmic Reticulum (adopted from Brec Clay) [Courtesy: Robert Edward Lee]

## CLASSIFICATION OF CYANOBACTERIA

Blue-green algae or Myxophyceae or Cyanophyceae popularly known as Cyanobacteria has been classified by algologists and bacteriologists from time to time. The popular ones are as follows:

### 1. Fritsch's Classification

F.E. Fritsch (1935–1945) proposed his classification in his book "The structure and Reproduction of the Algae" (Vol. 2). He divided myxophyceae on the basis of hormogones into two categories, five orders and 17 families.

### Category 1. (without hormogones): It comprises three orders:

I. Chrroococcales; II. Chamaesiphonales; III. Pleurocapsales

**Order-I. Chroococcales: It has two families**

Family (a) Chroococcaceae

Family (b) Entophysalidaceae

**Order-II. Chamaesiphonales: It is divided into four families.**

Family (a) Dermocarpaceae

Family (b) Chamaesiphonaceae

Family (c) Endonemataceae

Family (d) Siphonemataceae

**Order-III. Pleurocapsales: It comprises of two families**

Family (a) Pleurocapsaceae

Family (b) Hyellaceae

### Category 2. (with hormogones): It comprises two orders

I. Nostocales and II. Stigonematales

**Order-I. Nostocales: It has five families:**

Family (a) Oscillatoriaceae

Family (b) Nostocaceae

Family (c) Microchaetaceae

Family (d) Scytonemataceae

Family (e) Rivulariaceae

**Order-II. Stigonematales: It is divided into five families.**

Family (a) Capsosiraceae

Family (b) Nostocopsidaceae

Family (c) Pulvinulariaceae

Family (d) Loefgreniaceae

Family (e) Stigonemataceae.

**2. Fritsch (1945) and Desikachary (1959)** divided the blue-green algae into the same five orders although Fritsch used the word "Myxophyceae (Cyanophyceae)" for the class and did not believe its inclusion in any phylum whereas Desikachary used the word "Cyanophyceae" for the class and included it under phylum Cyanophyta. An outline of the classification of blue-green algae, as proposed by **T.V. Desikachary (1959),** is given below.

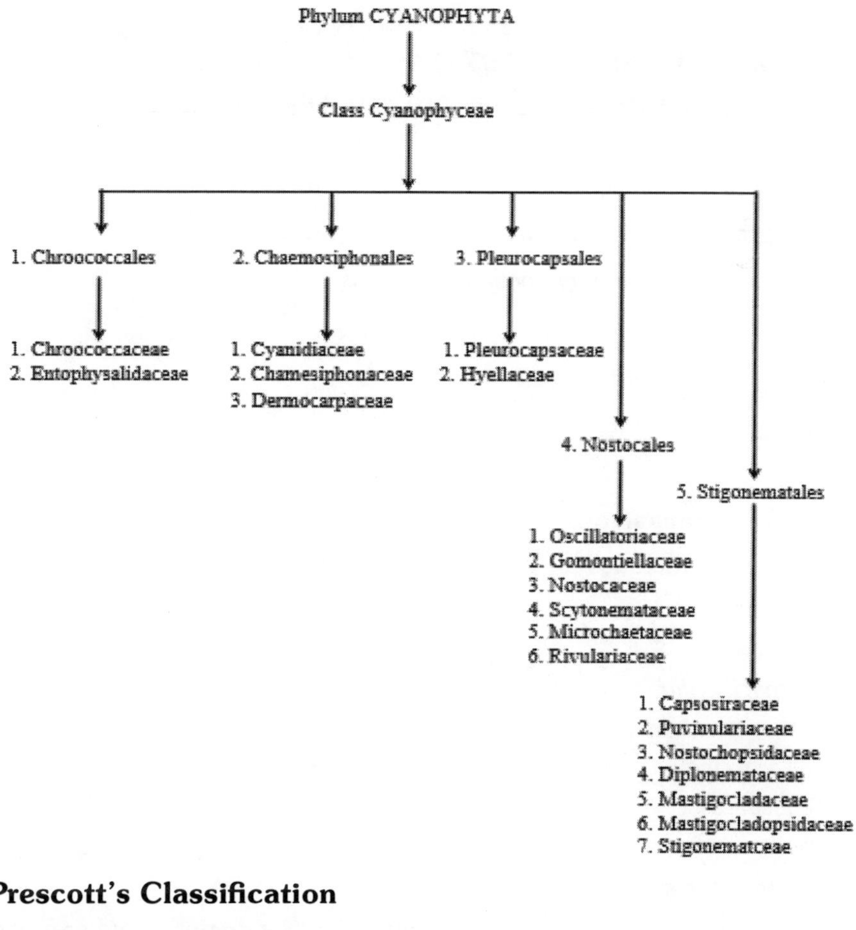


## 3. Prescott's Classification

**Prescott (1969) divided Cyanophyta into two tribes:**

**Coccogoneae and Hormogoneae.**

**Phylum: Cyanophyta**

**Class: Myxophyceae**

**Tribe I: Coccogoneae**

Order-I: Chroococcales

Order-II: Chamaesiphonales

**Tribe II: Hormogoneae**

Order I: Oscillatoriales

Order II: Nostocales

Order III: Scytonematales

Order IV: Stigonematales

Order V: Rivulariales

## 4. Classification by Stanier (1977)

**R.Y. Stainer and Cohen-Bazire (1977)** divided cyanobacteria into six subgroup.

Subgroup Chroococcacean.

This subgroup includes the unicellular rods and cocci.

- *Synechococcus including Anacystis nidulans*
- *Synechocystis*
- *Chamaesiphon*
- *Gloeobacter*
- *Microcysytis*
- *Aphanothece*

### Subgroup Pleurocapsalean

This subgroup includes unicellular cyanobacteria bound into aggregates by shared outer wall layers.

- *Dermocarpa*
- *Dermocarpella*
- *Chroococcidiopsis*
- *Myxosarcina*
- *Xenococcus*
- *Pleurocapsa*

### Subgroup Oscillatorian

This subgroup includes the trichomatous cells that do not form heterocysts.

- *Oscillatoria*
- *Spirulina*
- *Pseudoanabaena*
- *Lyngbya*
- *Phormidium*

- *Plectonema*
- *Trichodesmium*

**Subgroup Nostocacean**

This subgroup is heterocystous and includes organisms forming trichomes of even width.

- *Anabaena*
- *Nostoc*

**Subgroup Rivularean**

This subgroup is heterocystous and includes organisms forming trichomes that taper, with heterocysts that form at the thicker of the trichome.

- *Calothrix*

**Subgroup Stigonematalean**

This subgroup is heterocystous and includes organisms forming trichomes that branch.

- *Fischerella*
- *Chlorogloea or Chlorogloeopsis*

*Genera of uncertain affiliation*

- *Aphanothece*
- *Aphanocapsa*
- *Cylindrospermum*

## 5. Classification by Rippka *et al.*

**Rippka, R., J. Deruelles., John B. Waterbury, Michael Herdman and R.Y. Stanier (1979)** proposed a new scheme for classifying cyanobacteria. They recognized five sections out of which two sections, I and II, as "Unicellular"; cells single or forming colonial aggregates held together by additional outer cell wall layers. Other three sections, III to V, they describe as "Filamentous; a trichome (Chain of cells) which grows by intercalary cell division".

## SECTION I (Unicellular)

Unicellular cyanobacteria with cylindrical to ovoid cells that reproduce by binary fission or by budding.

*Synechococcus* Nageli 1849

- Thylakoids present
- Sheath absent

*Gloeothece* Nageli 1849

- Thylakoids present
- Sheath present

*Gloeobacter* Rippka, Waterbury & Cohen Bazire 1974

- Thylakoids absent
- Sheath absent

Unicellular cyanobacteria with spherical cells that divide in two or three successive planes at right angles to one another

*Synechocystis* Sauvageau 1892

- Thylakoids present
- Sheath absent

*Gloeocapsa* Kutzing 1843

- Thylakoids present
- Sheath present

Unicellular cyanobacteria that reproduce by forming successive spherical buds from one pole of the ovoid cell.

*Chamaesiphon* Braun & Grunow 1895 emend. Geitler 1925

## SECTION II (Unicellular)

Unicellular cyanobacteria that reproduce by multiple fission

Reproduction only by multiple fission

*Dermocarpa* Crouan & Crouan 1858

- Motile baeocytes

*Xenococcus* Thuret 1880

- Immotile baeocytes

Reproduction by both binary fission and multiple fission

*Dermocarpella* Lemmermann 1907

- Binary fission yields pear-shaped structure of one or two basal cells and one apical cell
- Subsequent multiple fission of apical cell yields motile baeocytes

*Myxosarcina* Printz 1921

- Binary fission yields cubical cellular aggregates
- Subsequent multiple fission yields motile baeocytes

Chroococcidiopsis Geitler 1933

- Binary fission yields cubical cellular aggregates
- Subsequent multiple fission yields immotile baeocytes

Pleurocapsa group

- Binary fission yields irregular cellular aggregates (pseudofilamentous)

- Subsequent multiple fission yield motile baeocytes

## SECTION III (Filamentous)

Filamentous non-heterocystous cyanobacteria that divide in only one plane

### Trichome helical

*Spirulina* Turpin 1827

- Cells in trichome isodiametric, cylindrical or disc-shaped
- Little or no constriction between adjacent cells
- Reproduction by transcellular trichome breakage
- Trichome motile
- Trichome not ensheathed or thinly ensheathed

### Trichome straight

*Oscillatoria* Vaucher 1803

- Cells in trichome disc-shaped and not separated by deep constrictions
- Reproduction by transcellular trichome breakage
- Trichome motile, not ensheathed or thinly sheathed

LPP group A

- Cells in trichome disc-shaped and not separated by deep constrictions
- Reproduction by transcellular or intercellular trichome breakage
- Motility restricted to sheathless or thinly sheathed hormogonia

*Pseudoanabaena* Lauterborn 1915

- Cells in trichome isodiametric or cylindrical
- Deep constrictions between adjacent cells
- Reproduction by transcellular or intercellular trichome breakage
- Trichome motile, not ensheathed
- Cells contain polar gas vesicles

LPP group B

- Cells in trichome isodiametric or cylindrical
- Variable degree of constriction between adjacent cells
- Reproduction by transcellular or intercellular trichome breakage
- Do not fit the entire description:
    - Deep constrictions between adjacent cells
    - Trichome motile, not ensheathed
    - Cells contain polar gas vesicles
- May be with or without sheath, motile or immotile

## SECTION IV (Filamentous)

Filamentous heterocystous cyanobacteria that divide in only one plane

Reproduction by random trichome breakage; some germinate from akinetes; trichomes indistinguishable from mature vegetative trichomes

*Anabaena* Bory de St Vincet 1822

- Heterocysts intercalary or terminal
- Position of akinetes variable
- Vegetative cells spherical, ovoid or cylindrical

*Nodularia* Mertens 1822

- Heterocysts intercalary or terminal
- Position of akinetes variable
- Vegetative cells disc-shaped

*Cylindrospermum* Kutzing 1843

- Heterocysts exclusively terminal and form at both ends of the trichome
- Akinetes are always adjacent to heterocysts
- Vegetative cells isodiametric or cylindrical

Reproduction by random trichome breakage; some germinate from akinetes; form hormogonia distinguishable from mature trichomes by absence of heterocysts and one or more of the following characters: rapid gliding motility, small cell size, cell shape and gas vacuolation

*Nostoc* Vaucher 1803

- Hormogonia give rise to young filaments with terminal heterocysts at both ends of the cellular chain.
- Vegetative cells spherical, ovoid or cylindrical
- Akinetes (if produced) are not initiated adjacent to heterocyst and are often formed in chains

*Scytonema* Agardh 1824

- Hormogonia give rise to young filaments, with terminal heterocyst at only one end of the cellular chain
- Mature trichome tapers from base, which bears a terminal heterocyst, to apex.
- Vegetative cells disc-shaped, isodiametric or cylindrical.

## SECTION V (Filamentous)

Filamentous heterocystous cyanobacteria that divide in more than one plane

Reproduction by random trichome breakage, by formation of hormogonia and (if produced) by germination of akinetes

*Chlorogloeopsis* Mitra & Pandey 1966

- Hormogonia composed of small cylindrical cells which enlarge become spherical
- Heterocysts develop in terminal and intercalary positions
- Cell in mature trichome divide in more than one plane
- Associated detachment of groups of cells leads to irregular

*Gloeocapsa* - like aggregates containing terminal heterocysts

- Hormogonia are produced within such aggregates

*Fischerella* Gomont 1895

- Hornogonia composed of small cylindrical cells which enlarge and become rounded
- Heterocysts develop almost exclusively in an intercalary position.
- Cells in mature trichome divide in more than one plane to produce a partly multiseriate trichome with lateral uniseriate branches.
- Heterocysts in the primary trichome are predominantly terminal or lateral.
- Hormogonia are produced from the ends of trichomes or from lateral branches.

## 6. Classification Proposed by R.E. Lee

R.E. Lee (2008) classified algae in two groups Prokaryota and Eukaryota which were further divided into divisions. Prokaryota has just one division Cyanophyta, whereas Eukaryota were further divided on the basis of nature of chloroplast membrane. Lee divided division Cyanophyta into the flowing three orders:

- **Chroococcales:** Single cell or cells loosely bound into gelatinous irregular colonies, e.g., *Gloeothece, Microcystis, Aphanocapsa* etc.
- **Oscillatoriales:** Filamentous cyanobacteria, e.g., *Oscillatoria, Spirulina, Lyngbya, Arthrospira* etc.
- **Nostocales:** Filamentous cyanobacteria with heterocysts, e.g., *Nostoc, Anabaena, Calothrix, Aulosira* etc.

Following five orders roughly agrees with that proposed by Rippka *et al.* (1974).

**Order 1. Chroococcales:** unicellular, held together in palmelloid colonies by mucilage e.g., *Gloeothece, Microcystis, Synechococcus, Synechocystis* etc., are some of the genera included in this order,

**Order 2. Chamaesiphonales:** Epiphytic cyanobacteria with a polar-oriented thallus are included in this order. Some genera are unicellular, whereas others are multicellular; Reproduction by endospores or exospores e.g., *Chamaesiphon, Dermocarpa* etc.

**Order 3. Pleurocapsales:** Thallus is composed of an erect and a prostrate system (heterotrichous), filaments without heterocysts e.g., *Pleurocapsa*.

**Order 4. Nostocales:** Filamentous cyanobacteria without true branching, heterocysts are formed, thalli not heterotrichous e.g., *Nostoc, Lyngbya, Spirulina, Anabaena, Aulosira, Rivularia, Calothrix, Tolypothrix, Scytonema* etc.

**Order 5. Stigonematales:** Exhibit true branching with a tendency toward multiseriate thalli, differentiated into upright and prostrate filaments (heterotrichous) e.g., *Stigonema*.

## GAIDUKOV PHENOMENON

Certain cyanobacteria could change their pigmentation in response to the wavelength of light. The pigmentation in cyanobacteria is modulated in a way that is complementary to the wavelengths of the light in the environment. That is why it is also called as complementary chromatic adaptation. This phenomenon was first observed by Gaidukov (1902) in *Oscillatoria sancta*. Therefore, it is known as Gaidukov phenomenon. He observed that *O. sancta* assumed a red coloration after growth under green light and a blue-green tint after growth under orange light. Their exposure to red light increases the synthesis of phycocyanin whereas exposure to green light increases the synthesis of phycoerythrin. An inverse correlation is observe between light intensity and pigment content: the less light energy available, the more photosynthetic pigments are synthesized by the cells. This effect of light wavelength on the pigment content of the cells is termed as complementary chromatic adaptation, which is restricted to some cyanobacteria. In this type of phenomenon, changes in cell pigmentation in response to specific spectral illuminations result from modifications of the relative amounts of the red coloured phycoerythrin and the blue-coloured phycocyanin, with a predominance of phycoerythrin in green light grown cells and of phycocyanin in red-light grown cells. Only cyanobacteria are able to synthesize phycoerythrin can undergo complementary chromatic adaptation. Among the phycoerythrin producers three physiological groups can be distinguished. (N. Tandeau de Marsae, 1977).

Group I, no complementary chromatic adaptation, phycoerythrin and phycocyanin synthesis is independent of light wave length; Group II, only the synthesis of phycoerythrin is regulated by light wavelength. Group III, type cyanobacteria, after transfer of green-adapted cells to red light, phycoerythrin synthesis is reduced or completely arrested, whereas the expression of a specific set of phycocyanin genes (PC-2) is "induced". The expression of this set of genes is turned off after transfer of red-adapted cells to green light, while phycoerythrin synthesis occurs. The regulatory processes involved in complementary chromatic adaptation are controlled by a photoreceptor pigment system which probably acts at the transcriptional level.

## CIRCADIAN RHYTHM IN CYANOBACTERIA

Circadian rhythms are physical, mental, and behavioral changes which follow a 24-hour cycle. These natural processes respond primarily to light and dark and affect most living things in the world, including plants, animals and microbes. Circadian rhythms are a series of endogenous autonomous oscillators generated by the molecular circadian clock which acting on coordinating internal time with the external environment in a period of about one day, which are also reported in cyanobacteria.

Cyanobacteria operate most simplest type of circadian clock, controlled by three proteins, known as Kai A, Kai B and Kai C. These tiny organisms regulate oxygenic photosynthesis, which requires sunlight during day time, and fix nitrogen at night. A detail study of the cyanobacterial circadian rhythm has been done by William *et al.* (2002) in *Syanechococcus elongatus*. They have identified a group of genes responsible for controlling the rhythm. Interactions among these three proteins maintain the circadian rhythm at about 24 hours. The circadian rhythm is created by Kai C adding and removing phosphate chemical groups to itself over the course of a 24-hour cycle. Kai A helps Kai C to add phosphate groups while Kai B counteracts the activity of Kai A. Other proteins track signals from sunlight and translate them to Kai-proteins in order to tune the clock properly. At the start of a cycle, Kai A binds to Kai C and stimulates Kai C to add phosphate groups to itself. As the cycle started, Kai B also binds to Kai C, and Kai C begins to remove it phosphate groups. After that, Kai C and Kai B are separated and the cycle started again. Kai C uses energy stored in ATP molecule. By breaking ATP molecule, the energy is released to drive the other chemical reactions. In cyanobacterial circadian cycle consumption of energy by the hydrolysis of ATP, the addition and removal of phosphate called as phosphorylation and dephosphorylation, are the key chemical reactions.

In nature, the clock is reset daily by cycles of different environmental factors like light and darkness, temperature and humidity. The major portal into the clock is the protein kinose CiK A. Due to change in environment CiK A changes the configuration of the protein Kai A which causes phosphorylation of the protein Kai C, resetting the clock. $N_2$ fixation, photosynthesis etc. set their cycles when Kai C interacts with the protein Sas A (*Synechococcus* adaptive sensor).

## CYANOTOXINS

Cyanobacteria are ecologically most significant prokaryotic group. They are present in marine and freshwater habitats. In any water ecosystem, they served as important producers, they are the source of several secondary metabolites and cluster of toxic compounds called as cyanotoxins. In comparison to other algae, many members of blue-green algae may accumulate to surface scums, often called "blooms" with high concentration of cells in water. The toxins produced by cyanobacteria are responsible for causing livestock poisonings and detrimental to human health. Cyanobacterial blooms and the accumulation of several cyanotoxins in water bodies pose severe ecological repercussions with high risk to aquatic organisms and global community health. Cyanobloom formed by cyanobacteria is a serious global problem which affects the water quality due to the formation and accumulation of different cyanotoxins. Cyanobacterial bloom formation due to excessive growth of certain cyanobacteria subsequently the production of toxic compounds have been reported by Rastogi *et al.* (2014) in many eutrophic to hypertrophic lakes, ponds and rivers throughout the world.

Up to 1980s detection of cyanotoxin was primarily done with mouse bio-assay method, mainly to judge the safety of drinking water supplies. But this bio-assay method is very costly and dependent on few approved laboratories. That is why, it is not suitable for large screening or monitoring programs. Now effective methods of chemical analysis are available for the known toxins. Sensitive immune-assays and enzyme assays are commercially available for micro-cystins and saxitoxins. A list of cyanotoxins, cyanobacteria known to produce the toxins and mechanism to toxicity is given in the following table.

**Table 1. Cyanotoxins and their toxicity**

| ***Cyanotoxins*** | ***LD50 (i.p. mouse) of pure toxin*** | ***Taxa known to produce the toxin(s)*** | ***Mechanism of toxicity*** |
|---|---|---|---|
| *Protein-phosphatase-blockers* (cyclic peptides with ADDA) | | | Block protein Phosphates by covalent binding and cause hemorrhaging of the liver, cumulative damage may occur |
| Microcystins in general (ca. 60 known congeners) | 45->1000 μg/kg | *Microcystis, Planktothrix, Oscillatoria, Nostoc* | |
| Microcystin-LR | 60 (25-125) μg/kg | *Anabaena, Anabaenopsis* | |
| Microcystin-YR | 70 μg/kg | *Hapalosiphon* | |
| Microcystin-RR | 300-600 μg/kg | | |
| Nodularin | 30-50 μg/kg | *Nodularia spumigena* | |
| ***Neurotoxins*** | | | |
| Anatoxin-a (alkaloid) | 250 μg/kg | *Anabaena, Oscillatoria, Aphanizomenon, Cylindrospermum* | Block post-synaptic depolarization |
| Anatoxin-a (s) (unique organophosphate) | 40 μg/kg | Known only from 2 species of *Anabaena* | Block acetyl-cholinesterase |
| Saxitoxins (carbamate alkaloids) | 10-30 μg/kg | *Aphanizomenon, Anabaena, Lyngbya, Cylindrospermopsis raciborskii* | Block sodium channels |
| ***Cytotoxin*** Cylindrospermopsin (alkaloid) | 2100 μg/kg/d 200 μg/kg/d | *Cylindrospermopsis raciborskii* | Blocks protein synthesis; substantial cumulative toxicity |

Three neurotoxins (anatoxin-a, anatoxin-a (s) and saxitoxins) with different modes of blocking neuronal signal transmission have been identified and isolated. Cylindrospermopsin is a general cytotoxin which inhibits protein synthesis and a group of toxins called microcystins causing inhibition on enzyme protein phosphatases were also isolated. These toxins were named after the organism from which they were isolated first, but it has been observed that these toxins are found in many species and some species contain more than one toxin. The above table shows an overview of the most important cyanotoxins known and their mode of acute action (Turner *et al.*, 1990; Sivonen and Jones, 1999; Kuiper-Goodman *et al.*, 1999).

**Neurotoxins:** All three neurotoxins (Table 1) have the potential to cause death in human. Anatoxin-a and Anatoxin-a (s) can cause cramps and saxitoxins are able to create paralysis. Anatoxin-a (s) is the only known naturally occurring organophosphate cholinesterase- inhibitor which causes heavy salivation (s stands for salivation), cramps, tremor, diarrhea, vomiting and may cause death also. Microcystins are common in lakes and rivers but saxitoxins and Anatoxin-a (s) are not so common, though it has only been found in a small number of *Anabaena* blooms in North America.

**Microcystins:** Most widespread cyanotoxins, chemically they are cyclic heptapeptides (consisting of seven amino acids linked by peptide bonds) whereas nodularin (a cyclic pentapeptide toxin) present in brackish water cyanobacteria. Rinehart *et al.*, 1994; Sivonen and Jones (1999) reported 60 structural analogues of microcystin. Nodularin, produced by *Nodularia spumigena* as brackish water species, has very close and similar structure to microcystins. Microcystins are found in some forms of cyanobacteria like *Microcystis* sp., *Anabaena* sp., *Oscillatoria agardhii, O. rubescens* etc. Microcystins are responsible for causing liver damage and they can promote tumour growth also.

**Cylindrospermopsin:** It is a general cytotoxin which blocks protein synthesis, alkaloid in nature and first isolated from *Cylindrospermopsis raciborskii*. Crude extracts of the cyanobacterium also causes lungs injury, damage of adrenal glands and intestine. *Cylindrospermopsis raciborskii* is found mainly in tropical and sub-tropical zones,

**The most common toxic cyanobacteria are**

*Microcystis sp.*

*Planktothrix (Oscillatoria) rubescens*

*Planktothrix (Oscillatoria) agardhii*

*Anabaena spp.*

*Some species of Oscillatoria*

*Cylindrospermopsis raciborskii*

*Synechococcus spp.*

*Gloeotrichia spp.*

*Lyngbya spp.*

*Nostoc spp.*

*Schizothrix spp.*

*Synechocystis spp.*

*Nodularia spumigena (found in brackish or marine environments)*

**Control of Cyanoblooms:** A number of approaches like chemical, physical and biological came into consideration to reduce the harmful cyanobactrial blooms.

Some chemicals such as algicides, inhibitors can be used to control cyanoblooms, however, use of these chemicals can re-contaminate water bodies. Bloom control by physical methods generally involves mechanical removal techniques. The use of new and advanced technologies can eliminate cyanoblooms and minimize the water pollution. Stimulated waterfalls or fountains may be effective to control the blooms in small water bodies. Hydraulic jet cavitation may be a good approach to manage cyanobacterial water bloom (Janĉula *et al.*, 2014).

Biological control of cyanoblooms is very much effective because this approach is not expensive do not cause pollution of water bodies also. Zhang *et al.* (2012) reported that the gastropod *Radix swinhoei* can ingest cyanobacteria and survive well in the water reservoirs with cyanobacterial blooms without losing fecundity. The combined use of snails (*Radix swinhoei*) and a submerged plant (Potamogeton lucens) in eutrophic waters can eliminate cyanoblooms by minimizing the eutrophication (Zhang *et al.*, 2014). Different species of bacteria are used now a days for the biodegradation of cyanobacteria which may be considered as very efficient process to control cyanobacterial toxins as well as blooms.

# PART II
# TAXONOMICAL STUDIES OF BLUE-GREEN ALGAE FROM LOWER GANGETIC PLAIN OF WEST BENGAL

# INTRODUCTION

Cyanobacteria are morphologically diverse and exist in different forms including unicellular, filamentous, planktonic or benthic and colonial ones. They can thrive in a wide range of ecological habitats, ranging from marine, freshwater, to terrestrial environments. All cyanobacteria are capable of oxygenic photosynthesis, some filamentous forms have evolved specialized cells called as heterocysts to carry out nitrogen fixation. Taxonomic survey work on blue-green algae of West Bengal has not been made extensively by former workers though work on a few district flora has been made. Only fragmentary work on the taxonomy of blue-green algae of West Bengal has so far been done by Martens (1870, 1871), Prain (1905), Bruhl and Biswas (1922, 1924); *Carter (1923)*, Biswas (1925, 1926, 1927, 1932 and 1942) and Banerji (1936, 1938) before the political partition of Bengal. After independence of India when politically West Bengal was recognized, the available works on blue-green algal flora were done by Gupta (1965, 1975), Sinha and Mukherjee (1975, 1984), Mukhopadhyay and Chatterjee (1981), Pal and Santra (1984, 1985), Maity and Santra (1985), Sen and Gupta (1987, 1993), Gupta and Sen (1987), Santra *et al.* (1987, 1988), Sen and Gupta (1998), Sen (2005), Naskar *et al.* (2008), Chakraborty *et al.* (2010), Ghosh *et al.* (2012), Mukhopadhyay and Naskar (2015), Banerjee *et al.* (2021), Dey and Chakraborty (2020).

It has been proposed that cyanobacteria could be the vital bio-agents in ecological restoration of degraded lands (Singh, 2014). Cyanobacteria can contribute to about 20-30 kg N $ha^{-1}$ as well as the organic matter to the soil, quite significant for the economically weak farmers unable to invest costly chemical nitrogen fertilizer (Issa *et al.*, 2014). Algal metabolites can play an important role in soil decontamination and fertilization; plant protection against biotic and abiotic stress factors; and plant development. In addition, microalgae and cyanobacteria also secrete phytohormones, which can cause plant growth. Taking into account their potential benefits for the development of sustainable agriculture, both biomass and extracts from microalgae and cyanobacteria are commercially available (Goncalves, 2021).

Though the ecological aspects in relation to paddy fields have been recognized in work in Indian states other than West Bengal, rarely such attempt has been made in West Bengal. Such work includes by Singh (1942); Aiyer (1965), Pandey (1965), Grover and Pandhol (1975), Prasad and Mehrotra (1980), Goel (1982), Venkataraman (1981), Singh (1981a, b and 1982, Jain and Sharma (1984), Goel *et al.* (1984), Bisoyi and Singh (1986), Singh and Singh (1987), Misra and Kaushik (1989a), Singh (2014, 2015b).

The importance of blue-green algae in rice and other cultivated fields needs no emphasis. It was to record the algae common in paddy fields and on other areas; presence of $N_2$ fixing forms on paddy fields; the seasonal variations in the frequency of occurrence in rice field, their distributional pattern, association and succession in relation to the growth stages of rice; effects of different fungicides and pesticides on their occurrence, role of cyanobacteria as biofertilizers, the effect of salinity on cyanobacterial diversity in the rice fields and other ecological aspects. As such work covering the entire state of West Bengal is not feasible, the areas beside the course of Ganges were taken into account. It covered the chief rice growing districts with alluvial soil. The district left unattended, are chiefly dry, arid regions often with lateritic soil and low rainfall.

## PRESERVATION AND PROCEDURE

Collections of available blue-green algae were made. They were washed and preserved in 2% formalin. The algae obtained in culture were also preserved similarly. Observations under optical microscope were made. Camera lucida drawings of the materials were prepared carefully and photomicrograph of some representative members were taken. Necessary measurements were taken under high power of microscope using stage and ocular micrometer.

The detailed methods of collection, isolation from soil and preservation are given in the Section A of Part III of the documentation.

## MAIN GEOGRAPHICAL FEATURES

Present study on the blue-green algal flora includes those obtained from the soil of West Bengal under the influence of the river Ganges (Ganga) and relatively close to the river. The Ganga is the main river of West Bengal and has entered this State near Rajmahal. After progressing through Malda district it is divided into two branches near Dhulian in Murshidabad district. One branch under name Padma has entered Bangladesh and the other branch Bhagirathi flowing southwardly, reached the Bay of Bengal. The southern part of Bhagirathi has taken the name Hooghly river in district Hooghly. The lower gangetic plain under consideration is the gangetic delta of West Bengal on the eastern side of the Bhagirathi and the Hooghly river and a part of gangetic plain on the western side of the river. This area is situated approximately between 87°36' to 89° E longitude and between 21°48' to 24°43' N latitude. Between the northern and southern limit of the area, the tropic of cancer has crossed the district Burdwan and Nadia.

Accordingly the collections of blue-green algae are made from lower gangetic plain which includes the following areas in different districts from south to north. On the western side of the Ganga

- In the district of Midnapore only a small area is considered, beside the Hooghly river at the eastern part of Tamluk subdivision.
- In Howrah district the land bound by the Hooghly river in the east and the Damodar in the west has been surveyed.
- In the Hooghly district the surveyed area is bound by the Hooghly river in the east and the Damodar river in the west. It includes the two subdivisions and excludes Arambagh subdivision.
- The two subdivisions - Kalna and Katwa of Burdwan district are considered which are nearer to the river Ganga.

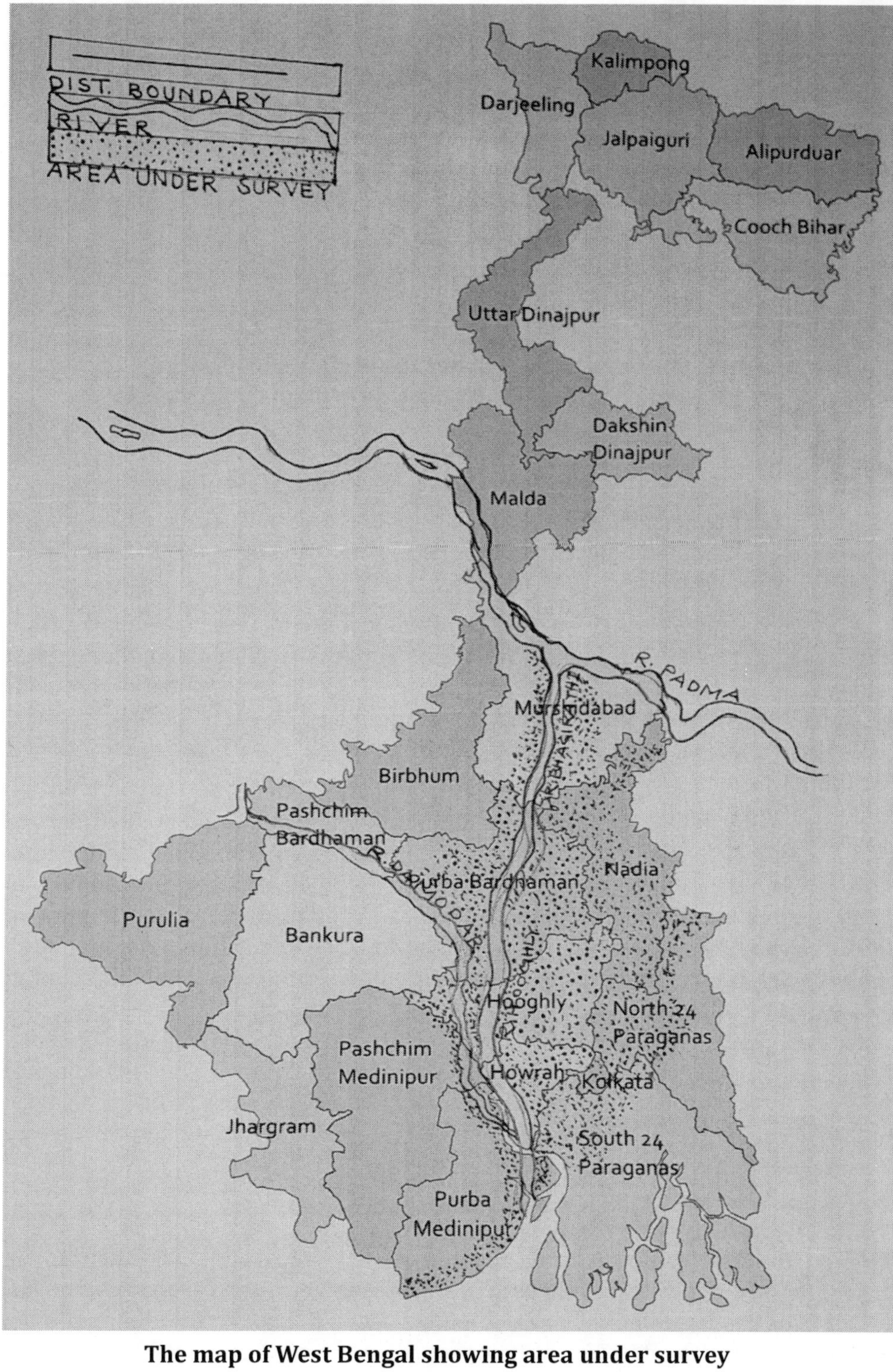


**The map of West Bengal showing area under survey**

- Greater Kalcutta.
- North 24-Parganas district.
- Part of South 24-Parganas district.
- Nadia district.
- Of the district Murshidabad, only the Berhampur subdivision has been taken into consideration.

The southern part of the district South 24-Parganas has not been included under the present study, though it is a part of gangetic delta. Being the seaward side of the delta, there is a vast stretch of tidal forests and swamp land of Sunderban. The influence of the saline sea water is more on this part of delta.

The area under survey chiefly includes the lower Gangetic delta which is the western part of the largest delta of the world, which lies mostly in Bangladesh. The rivers of West Bengal areas, being sluggish, have been described as dead or dying. The water passes down them to the sea, and new deposits of the delta is practically absent. So, the western side of Gangetic delta in West Bengal has, however, remained practically unchanged since the 18th century. The area is composed of repeated alterations of clays, sands and marls, with recurring layers of peat, lignite and beds of what were once forests.

Gangetic delta of West Bengal is a well-developed region from ancient time because of its geographical features which encourage the habitation for the peoples. In this vast area varieties of localities are present including metropolitan city, well-developed urban areas and the rural areas. Cultivated lands, fallow lands, tanks, pools etc. are also present. Consequently, blue-green algae grow in varieties of habitat in this region.

Fresh water blue-green algae are found in ponds, tanks and pools. They are obtained either attached to the substratum near the bank or on some aquatic plants or on decaying plants in a pond. They are also obtained as floating colonies or as water blooms. In a populated area of villages or towns they frequently grow in drains, in shallow puddles, in accumulated rain water, on dripping wall of house, on moist dripping bricks, in any other stagnant water, as mucilaginous scum in shallow water filled pits or as roadside slime. Moist ground, water logged soil, moist ridges of earth around agricultural land, moist or flooded paddy field etc. also help blue-green algae to grow. Some species seldom grow on the moist bark of the trees. Agricultural soil is a good substratum for the growth of blue-green algae and different species remain within the soil out of the sight and grow profusely when favourable environment returns.

## CLIMATE

The climate of the lower Gangetic plain of West Bengal under consideration in the present study is fairly humid with moderate temperature. The tropic of cancer passes over this State and the monsoon is a direct influence on the climate. So the climate of West Bengal is Tropical monsoon. During summer (April to 2nd week of June) humid heat prevails in most part of the area and characteristic strong wind blows very often in a south-east direction mostly in the afternoon. Summer is followed by monsoon (2nd week of June-July) with the heaviest downpore. Monsoon is succeeded by a post-monsoon period with occasional rainfall which lasts up to the middle of October. The beginning and the termination of the rainy seasons are generally marked by oppressive, cloudy and sultry weather. The winter (November-February) is moderately cool and dry. A short spell of autumn and spring is observed during October and March, respectively. May is the hottest month when the mean daily maximum temperature is 39.79°C and January is the coldest month with an average minimum temperature of 10.4°C. The average annual rainfall of West Bengal is about 1750 mm, about 77% of which occurs during June-September. Humidity is high from June-September and the humidity decreases progressively in the following months.

## NOTES ON THE FLORA

### Statistical synopsis of the flora

In general, the number of species of blue-green algae in lower Gangetic West Bengal is significant and more so in the paddy fields. The ratio between the blue-green algae and those other than blue- green in rice field is 1:0.37. Among the blue-green algae, the multicellular species are much more than the unicellular forms and the ratio between the unicellular and multicellular species is 1:5.5 approximately.

**Table 1.1: Statistical synopsis of unicellular and multicellular forms of blue-green algae**

<table>
<tr><th rowspan="2"></th><th colspan="2">Unicellular</th><th colspan="2">Multicellular</th><th rowspan="2">Total</th></tr>
<tr><th>No.</th><th>%</th><th>No.</th><th>%</th></tr>
<tr><td>Families</td><td>3</td><td>33.33</td><td>6</td><td>66.67</td><td>9</td></tr>
<tr><td>Genera</td><td>7</td><td>26.92</td><td>19</td><td>73.08</td><td>26</td></tr>
<tr><td>Species</td><td>19</td><td>15.45</td><td>104</td><td>84.55</td><td>123</td></tr>
</table>

**Table 1.2: Statistical synopsis of the blue-green algal flora**

<table>
<tr><th rowspan="3"></th><th colspan="2">Unicellular</th><th colspan="4">Multicellular</th><th rowspan="3">Total</th></tr>
<tr><th rowspan="2">No.</th><th rowspan="2">%</th><th colspan="2">Non-heterocystous</th><th colspan="2">Heterocystous</th></tr>
<tr><th>No.</th><th>%</th><th>No.</th><th>%</th></tr>
<tr><td>Families</td><td>3</td><td>33.33</td><td>2</td><td>22.22</td><td>4</td><td>44.44</td><td>9</td></tr>
<tr><td>Genera</td><td>7</td><td>26.92</td><td>8</td><td>30.77</td><td>11</td><td>42.31</td><td>26</td></tr>
<tr><td>Species</td><td>19</td><td>15.45</td><td>63</td><td>51.22</td><td>41</td><td>33.33</td><td>123</td></tr>
</table>

**Table 1.3: Statistical synopsis of algal flora obtained from paddy fields**

| | ***Blue-green algae*** | | | | ***Other algae*** | | ***Total*** |
|---|---|---|---|---|---|---|---|
| | ***Non-heterocystous*** | | ***Heterocystous*** | | **No.** | **%** | |
| | **No.** | **%** | **No.** | **%** | | | |
| Genera | 11 | 27.50 | 8 | 20.00 | 21 | 52.50 | 40 |
| Species | 42 | 43.75 | 28 | 29.17 | 26 | 27.08 | 96 |

Again, among the multicellular blue-green algal species the ratio between the non-heterocystous and hetero-cystous members is 2:1. A total of 123 species of blue-green algae belong to 26 genera and the proportion of genera to species is 1:4.7 approximately (Tables 1.1, 1.2 and 1.3). Possibly it is due to varied ecological condition over a large area.

### Citations, Keys, Order of Families etc.

The work of Rabenhorst, Geitler, Desikachary, Gomont, Rao, C.B., Bharadwaja, Ghose, Bruhl and Biswas and Biswas have been cited frequently together with those of others which have been referred in 'Literatures consulted.

(i) a general key to the families, (ii) key to the genera of each family and (iii) key to the species in different genera have been provided. The keys have been prepared chiefly on the basis of microscopic characters. All keys have been kept as simple as possible.

Desikachary's order of families has been followed for the sake of convenience. As regards splitting of the families, the classification by the same author has been adopted in general. For the measurements, decimal and metric systems are used.

## CLASSIFICATION

DIVISION: CYANOPHYTA

Class: Cyanophyceae Sachs

(Lehrb. d. Botanik, 4 Aufl., 1874)

1. Plant unicellular ........................................................................................ 2

1. Plant multi cellular ..................................................................................... 3

2. Free floating as colony; without differentiation into base and apex; endospores or exospores absent, reproduction mainly by cell division ....... 1. ***Order Chroococcales***

2. Attached; having differentiation into base and apex; endospores or exospores present..... ........................................................................ 2. ***Order Chamaesiphonales***

3. Having no differentiation into trichome and filament, often forming parenchymatous thalli ............................................................ 3. ***Order Pleurocapsales***

3. Differentiated into trichomes and filaments..............................................4

4. Unbranched or with false branching ........................................ 4. ***Order Nostocales***

4. With true branching ........................................................ 5. ***Order Stigonematales***

Order **CHROOCOCCALES** Wettstein

(Handb. Syst. Botanik, 3, Aufl., 79, 1923)

1. Unicellular or in colony, not forming any filamentous organization ........................................................................................................................ 1. ***Chroococcaceae***

1. Cells organised in linear series assuming pseudofilamentous growth..........................................................................................................................2. ***Entophysalidaceae***

Family **CHROOCOCCACEAE** Nageli

(Gatt. einzell. Algen, 44, 1848)

1. Definite number of cells in a colony ........................................................ 2

1. Cells innumerable in a colony ........................................................ 3

2. 2 to 8 cells in a colony, with a number of concentric envelops ........................................................................................................................ 1. ***Gloeocapsa***

2. 2 to 4, up to 16 cells in a colony, concentric enveloping sheath absent ........................................................................................................................ 2. ***Chroococcus***

3. Colony microscopic, formed by closely aggregated cells ........................................................................................................................ 3. ***Microcystis***

3. Colony macroscopic, formed by loosely arranged cells ........................................ 4

4. Cells spherical ........................................................ 4. ***Aphanocapsa***

4. Cells ellipsoidal or cylindrical ........................................................ 5. ***Aphanothece***

Family **ENTOPHYSALIDACEAE** Geitler

(Synopt. Darst. Cyano., Beih. bot. Cbl., 41 (Abt. 2): 235, 1925; Kryoptogamenflora, 292, 1932)

Cells in a more or less homogeneous gelatinous thallus in distinct rows ........................................................................................................................***Chlorogloea***

Order **CHAMAESIPHONALES** Wettstein

(Handb. Syst. Botanik, 3, Aufl., 79, 1923)

Family **DERMOCARPACEAE** Geitler

(Synopt. Darst. Cyano., Beih. bot. Cbl., 41(2): 247, 1925)

Endospores formed by division of protoplast in three directions ........................................................................................................................ ***Dermocarpa***

Order **PLEUROCAPSALES** Geitler

(Synopt. Darst. Cyano., Beih. bot. Cbl., 41(2): 238, 1925)

Family **PLEUROCAPSACEAE** Geitler

(Synopt. Darst. Cyano., Beih. bot. Cbl., 41(2): 238, 1925)

Colony three dimensional, made of closely packed cells; endospores present ....... ........................................ ***Myxosarcina***

Order **NOSTOCALES** Geitler

(Synopt. Darst. Cyano., Beih. bot. Cbl., 41(3): 252, 1925)

1. Trichomes without heterocysts; spores absent ........................................ 1. Family ***Oscillatoriaceae***

1. Trichomes with heterocysts; spores present ........................................ 2

2. Trichomes with false branching ........................................ Family ***Scytonemataceae***

2. Trichomes unbranched ........................................ 3

3. Trichomes attenuated into a terminal hair .................... 3. Family ***Rivulariaceae***

3. Trichomes without any terminal hair ........................................ 4. Family ***Nostocaceae***

Family **OSCILLATORIACEAE** Kirchner

(Schizophyceae in Engler and Prantl, Nat. Pflan- zenfam., I Ed., 1a, 61, 1898)

1. Trichomes without a sheath........................................ 2

1. Trichomes with sheath ........................................ 3

2. Trichomes more or less straight, not spiral ........................................ 1. ***Oscillatoria***

2. Trichomes regularly spiral ........................................ 4

3. Trichomes having individual sheath ........................................ 5

3. Several trichomes in a common sheath ........................................ 6

4. Cross-wall distinct ........................................ 2. ***Arthrospira***

4. Cross-wall if present obscure ........................................ 3. ***Spirulina***

5. Sheath firm; trichome free in a sheath ........................................ 4. ***Lyngbya***

5. Sheath mucilaginous; more or less confluent with trichome ......... 5. ***Phormidium***

6. Filaments unbranched or sparsely branched ........................................ 6. ***Microcoleus***

6. Filaments branched ........................................ 7. ***Schizothrix***

Family **NOSTOCACEAE** Kutzing

(Phycologia generalis, 203, 1843)

1. Heterocyst generally terminal with an adjoining spore ........... 1. ***Cylindrospermum***

1. Heterocyst generally intercalary ........................................ 2

2. Filaments free or as fascicles, firm persistent sheath around trichome distinct........................................ 2. ***Aulosira***

2. Filaments in definite colony, sheath inconspicuous or as diffluent mucilage ....... 3

3. Filaments in a formless gelatinous mass, not contorted though may be spiral .. ..........................................................................................................................3. ***Anabaena***

3. Filaments in definite colony, controted, curved or entangled .................4. ***Nostoc***

Family **SCYTONEMATACEAE** Rabenhorst

(Fl. Eur. Alg., 2: 246, 1865)

1. False branching generally geminate ................................................ 1. ***Scytonema***

1. False branching arising next to terminal heterocyst........................ 2. ***Tolypothrix***

Family **RIVULARIACEAE** Rabenhorst

(Fl. Eur. Algarum, 2: 200, 1868)

1. Heterocyst absent, trichomes on a base of irregularly arranged cells .................. ........................................................................................................................ 1. ***Amphithrix***

1. Heterocyst present, trichomes attenuated ...................................................... 2

2. Spores always absent ................................................................ 2. ***Rivularia***

2. Spores present or rarely so ........................................................................ 3

3. Filaments radial or so, in a solid spherical thallus, sheath at the base firm; basal spore commonly formed next to heterocyst .................................... 3. ***Gloeotrichia***

3. Filaments in small bundle, mostly erect, sheath firm generally extended much above the basal part; spores often absent .......................................... 4. ***Calothrix***

Order **STIGONEMATALES** Geitler

(Synopt- Darst. Cyano. Beih. bot. Cbl., 41: 252, 1925)

Family **STIGONEMATACEAE** Kirchner

(Schizophyceae in Engler and Prantle, Die naturlichen Pflanzenfam. I, 1a, 80, 1898)

Filaments prostrate, irregularly laterally branched having as many rows of cells as in main filament ........................................................................... **Stigonema**

PLATE 1

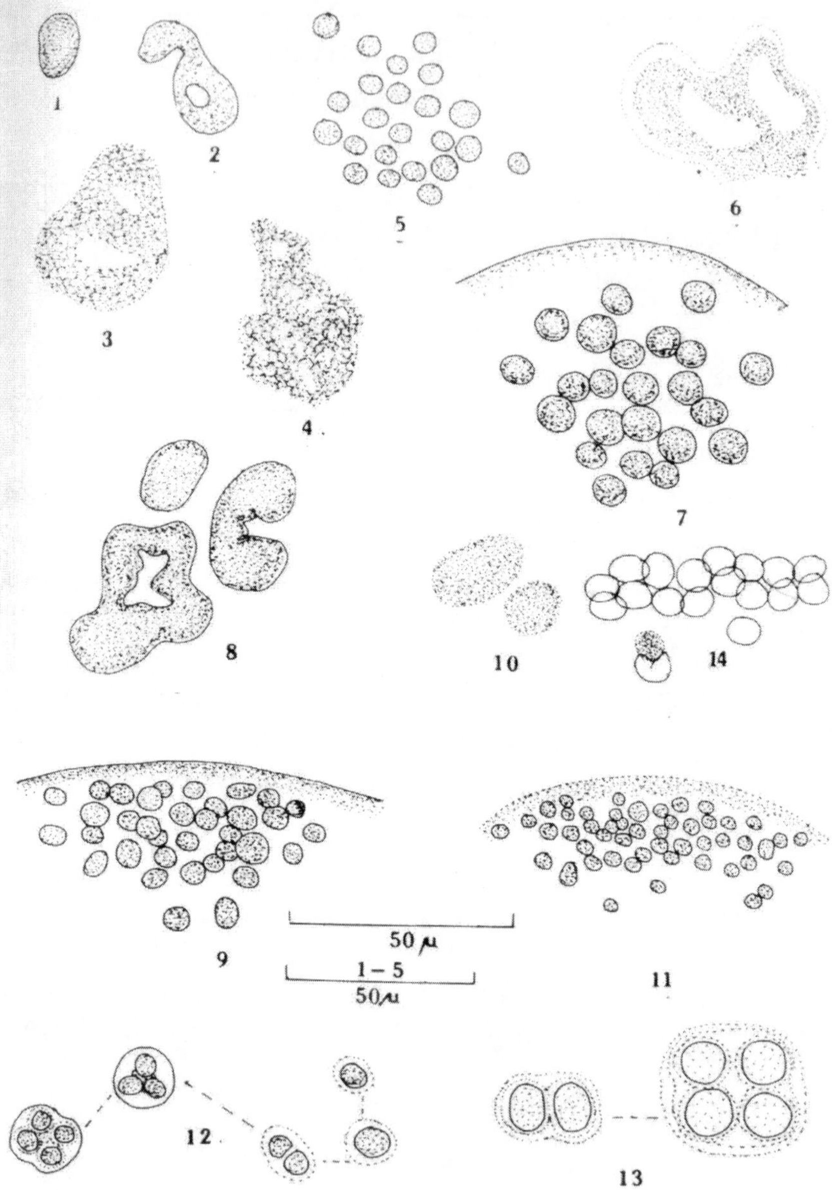


**Plate 1: 1-5. Microcystis aeruginosa Kutz.; 6 & 7. M. robusta (Clark) Nygaard.; 8 & 9. M. viridis (A. Br.) Lemm.; 10 & 11. M. flos-aquae (Wittr.) Kirchner; 12. Chroococcus minutus (Kutz.) Nag.; 13. Chr. turgidus (Kutz.) Nag. ; 14. Chlorogloea fritchii Mitra.**

PLATE 2

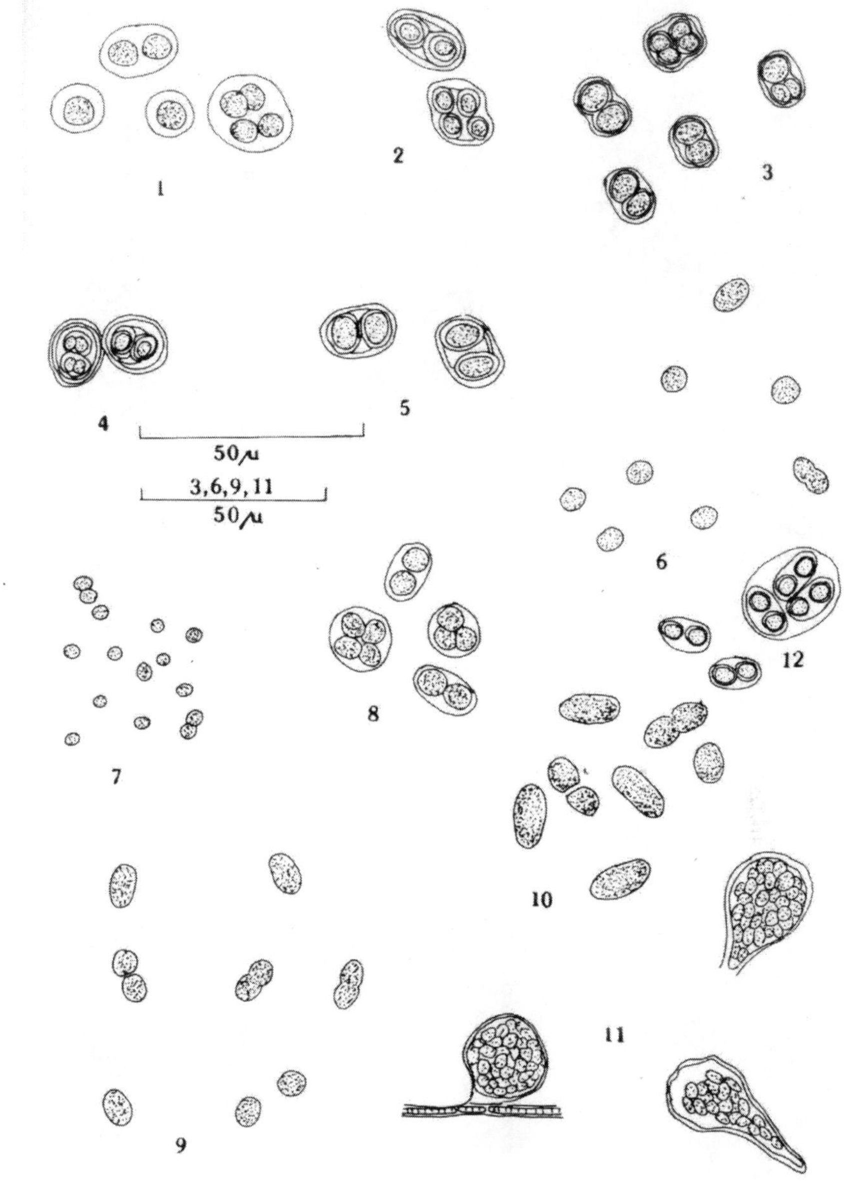


**Plate 2: 1. Chroococcus pallidus Nag.; 2. Gloeocapsa polydermatica Kutz.; 3. Gl. decorticans (A. Br.) Richter; 4. Gl. quaternata (Breb.) Kutz.; 5. Gl. rupestris Kutz.; 6. Aphanocapsa grevillei (Hass.) Rabenh.; 7. A. montana Cramer; 8. A. biformis A. Br.; 9. Aphanothece stagnina (Spreng.) A. Br.; 10. A. pallida (Kutz.) Rabenh.; 11. Dermocarpa leibleiniae (Reinsch) Born et. Thur.; 12. Gloeocapsa gelatinosa Kutz.**

PLATE 3

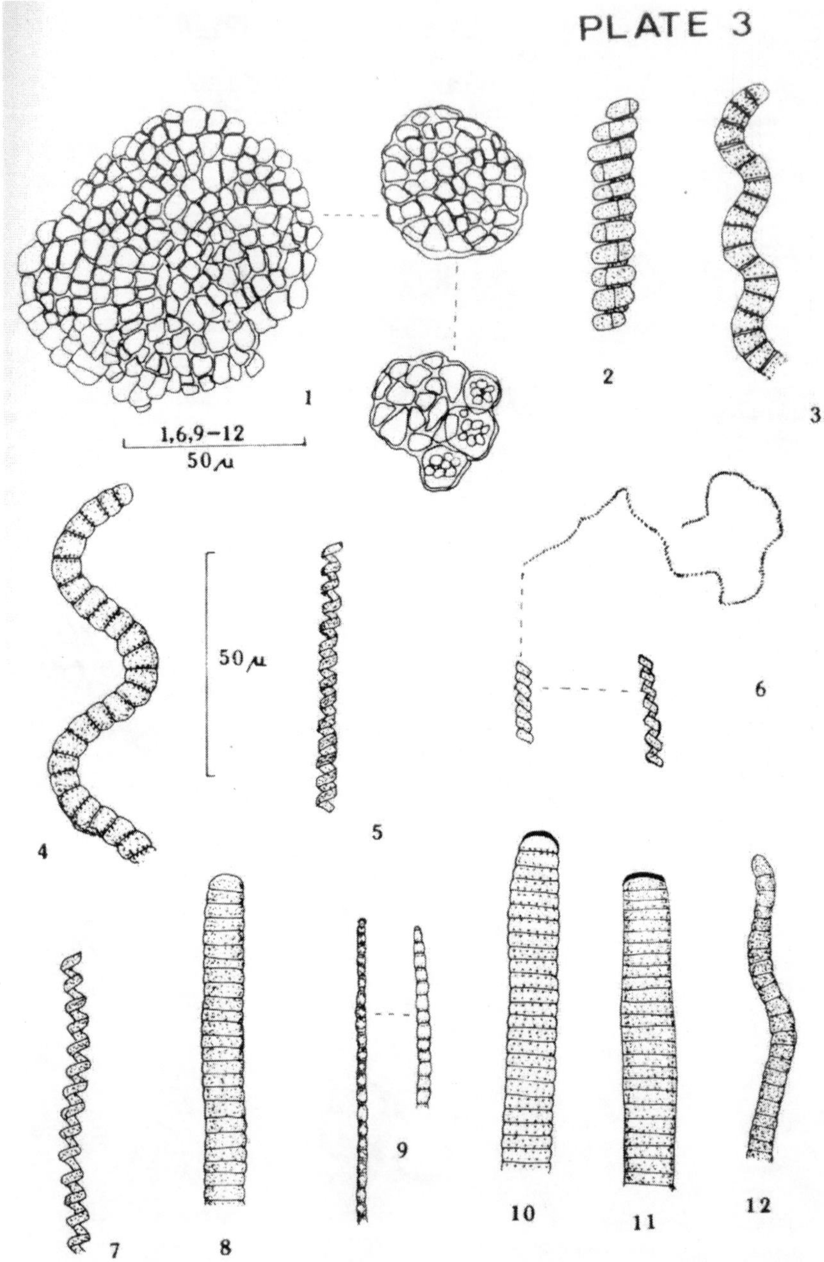


**Plate 3: 1. Myxosarcina spectabilis Geitler; 2. Arthrospira spirulinoides Ghose; 3. A. jenneri Stizenb. ex Gom.; 4. A. platensis (Nordst.) Gom.; 5. Spirulina major Kutz. ex Gom.; 6. Sp. subsalsa Oerst. ex Gom.; 7. Sp. meneghiniana Zanard. ex Gom.; 8. Oscillatoria annae Van Goor; 9. O. schultzii Lemm. ex Forschb.; 10. O. sancta (Kutz.) Gom.; 11. O. limosa Ag. ex Gom.; 12. O. chalybea (Mertens) Gom.**

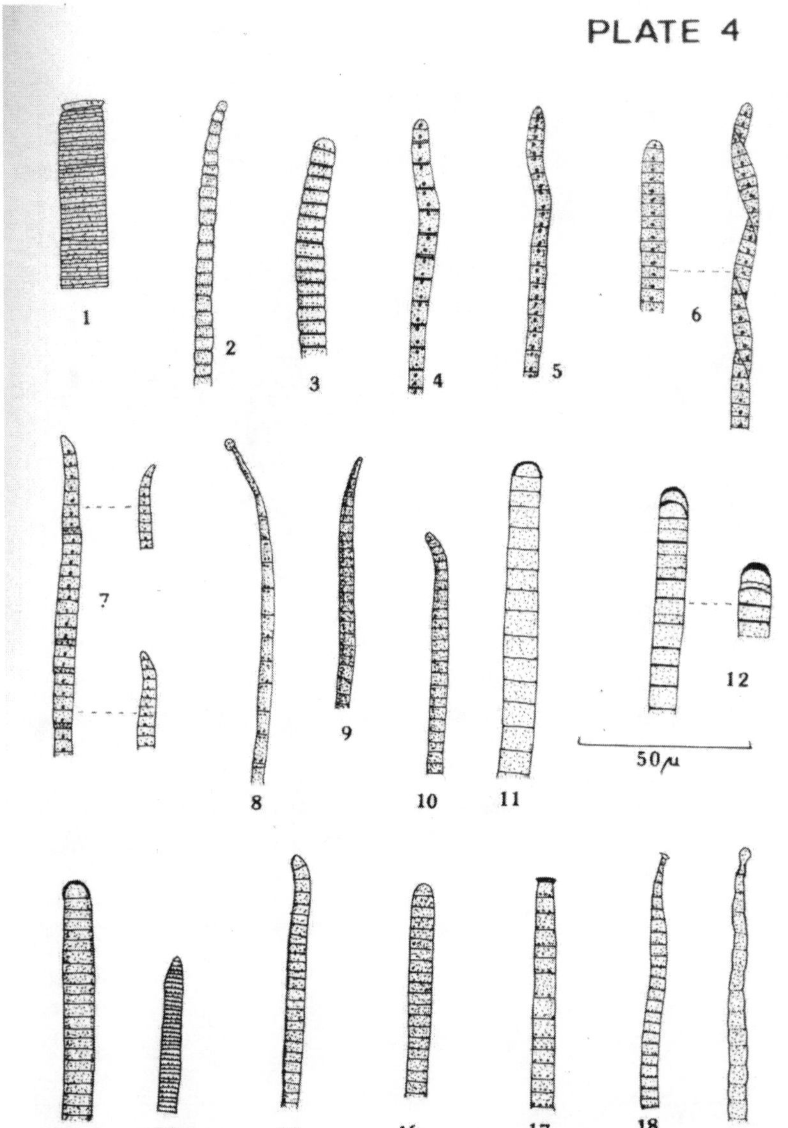


**Plate 4: 1. Oscillatoria princeps Vouch. ex Gom.; 2. O. formosa Bory ex Gom.; 3. O. proteus Skuja; 4. O. pseudogeminata G. Schmid var. unigranulata Biswas; 5. O. amphibia Ag. ex Gom.; 6. O. terebriformis Ag. ex Gom.; 7. O. salina Biswas; 8. O. splendida Grev. ex Gom.; 9. O. animalis Ag. ex Gom.; 10. O. acuta Bruhl et Biswas, orth. mut. Geitler; 11. O. irrigua (Kutz.) Gom.; 12. O. rubescens DC. ex Gom.; 13. O. simplicissima Gom.; 14. O. obscura Bruhl et Biswas; 15. O. brevis (Kutz.) Gom.; 16. O. subbrevis Schimidle; 17. O. prolifica (Grev.) Gom.; 18. O. agardhii Gom.; 19. O. amoena (Kutz.) Gom.**

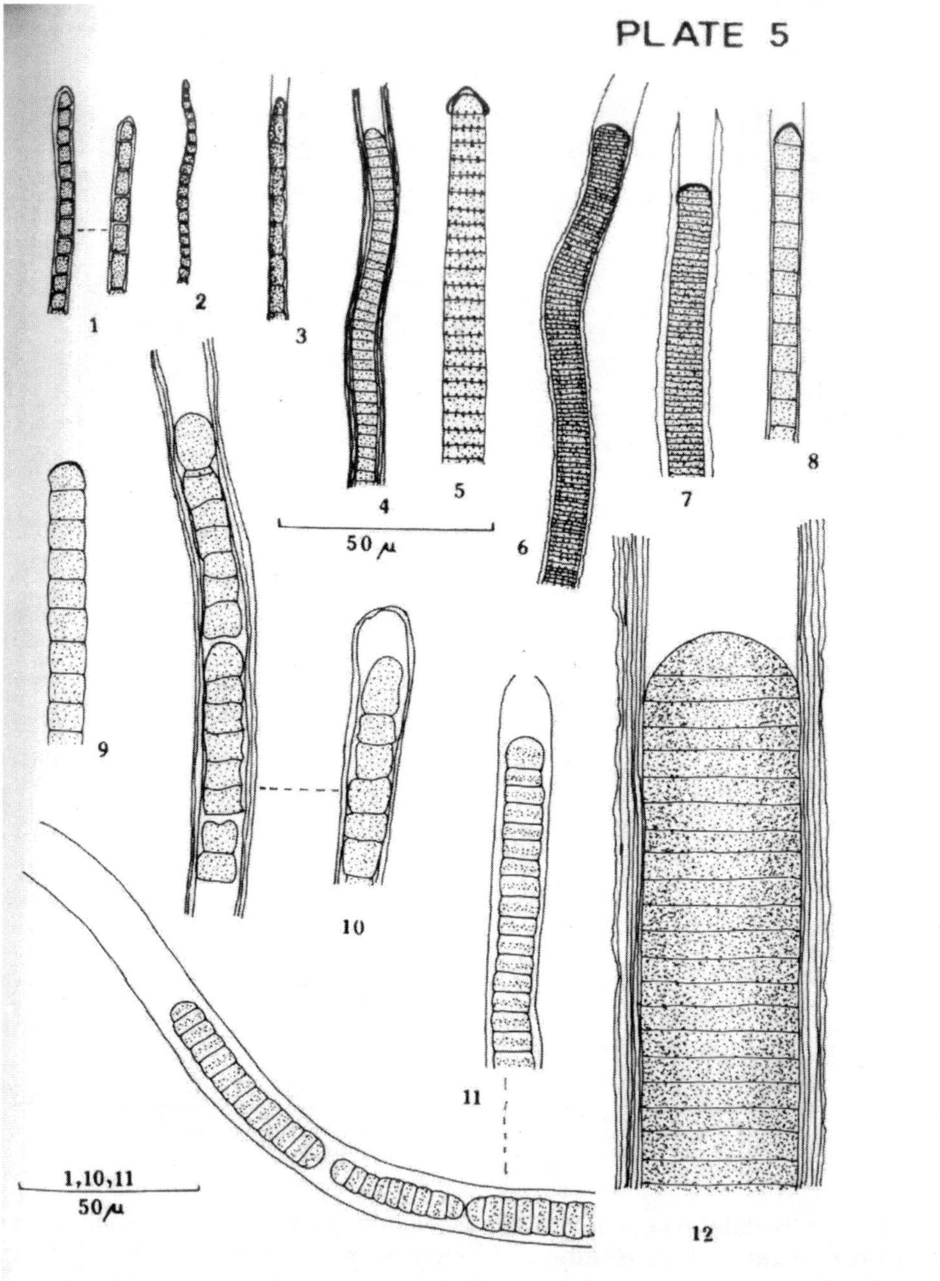


**Plate 5: 1. Phormidium molle (Kutz.) Gom.; 2. P. fragile (Menegh.) Gom.; 3. P. tenue (Menegh.) Gom.; 4. P. ambiguum Gom.; 5. P. subfuscum Kutz. ex Gom.; 6. P. anomala Rao, C.B.; 7. P. stagnina Rao, C.B.; 8. P. corium var. capitatum Gardner; 9. P. retzii (Ag.) Gom.; 10. Lyngbya baculum Gom.; 11. L. putealis Mont. ex Gom.; 12. L. majuscula Harvey ex Gom.**

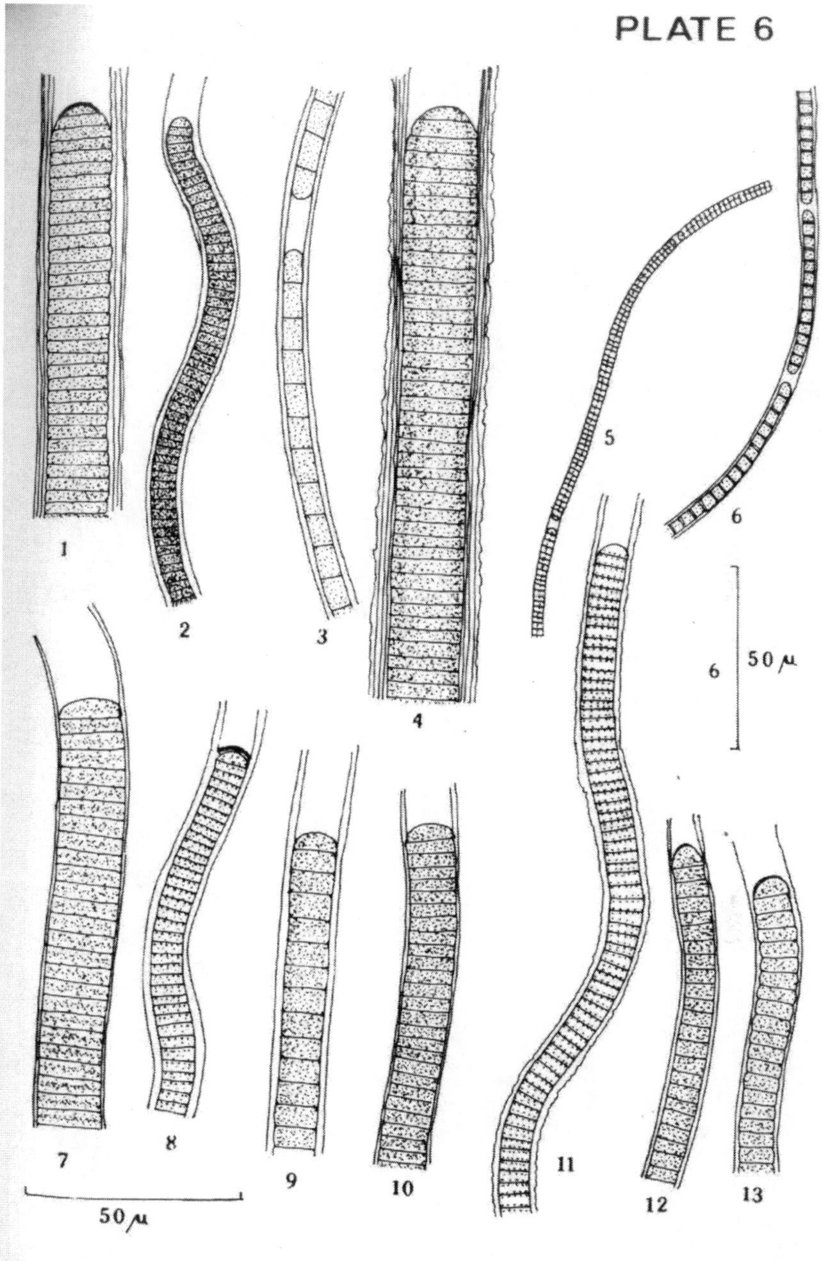


**Plate 6: 1. Lyngbya major Menegh. ex Gom.; 2. L. spiralis Geitler; 3. L. rubida Fremy; 4. L. confervoides C. Ag. ex Gom.; 5. L. lagerheimii (Mob.) Gom.; 6. L. digueti Gom.; 7. L. birgei Smith, G.M.; 8. L. stagnina Kutz.; 9. L. ceylanica Wille; 10. L. hieronymusii Lemm.; 11. L. martensiana Mengh. ex Gom.; 12. L. aerugineo-coerulea (Kutz.) Gom.; 13. L. gracilis (Menegh.) Rabenh.**

# PLATE 7

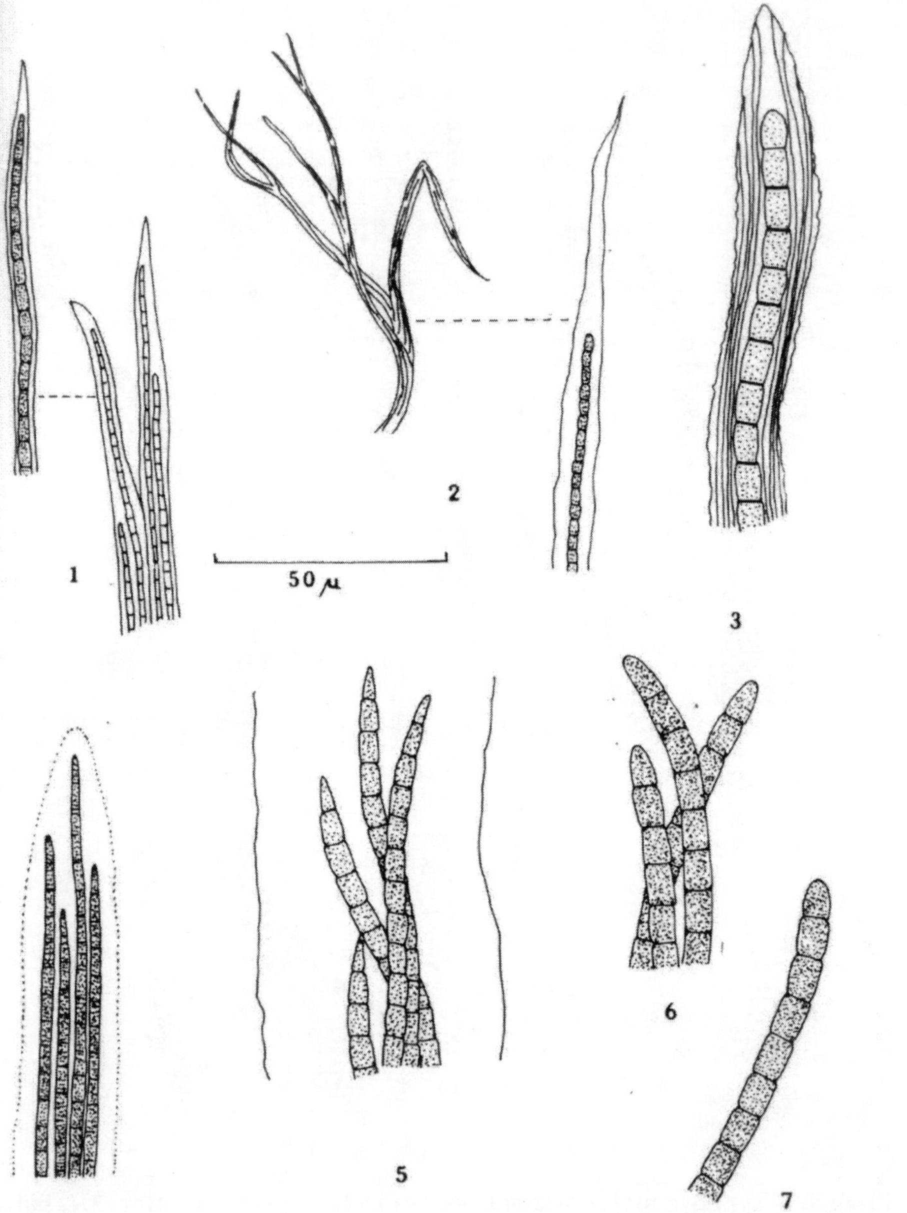


**Plate 7: 1. Schizothrix arenaria (Berk.) Gom.; 2. Sch. lacustris A. Br. ex Gom.; 3. Sch. telephoroides (Mont.) Gom.; 4. Microcoleus acutissimus Gardner; 5. M. chthonoplastes Thur. ex Gom.; 6. M. paludosus (Kutz.) Gom.; 7. M. subtorulosus (Breb.) Gom.**

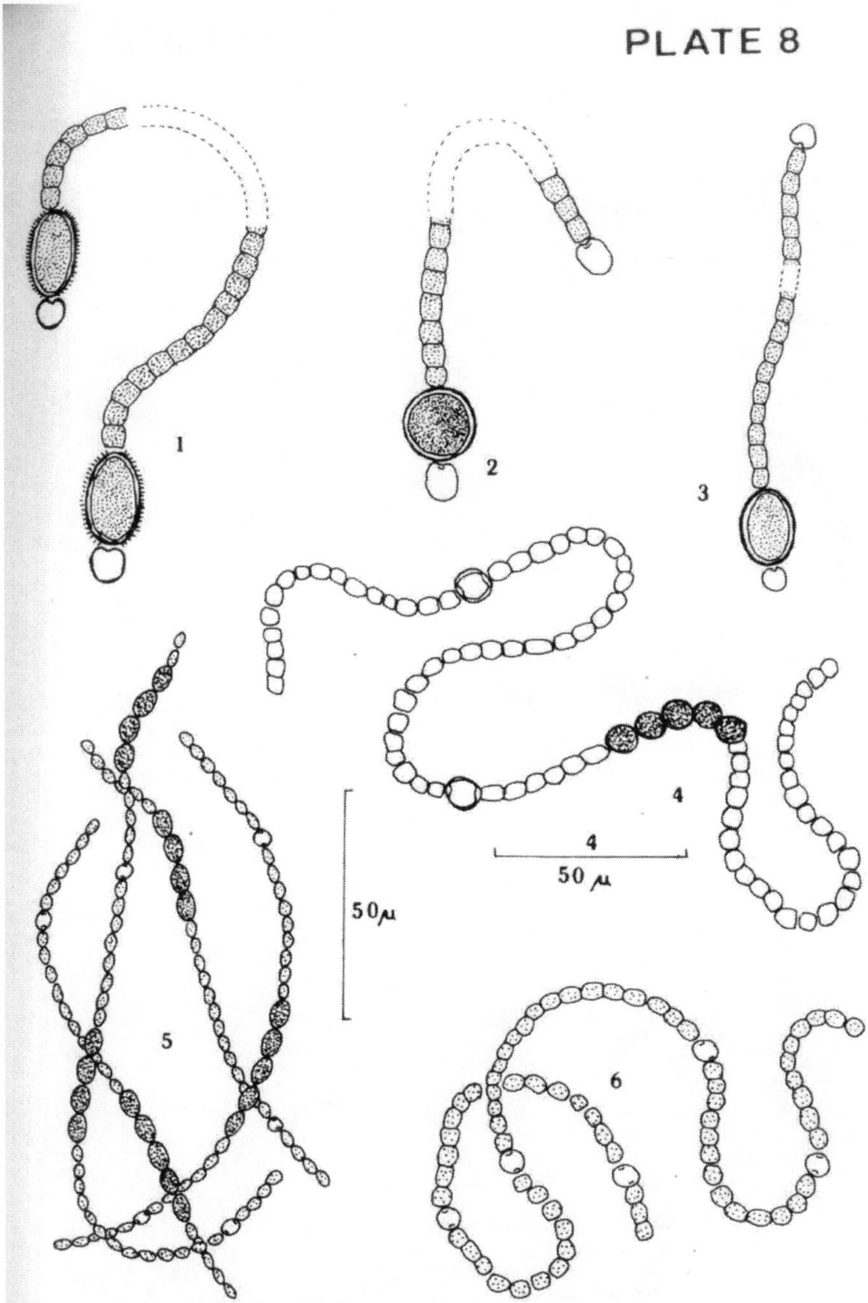


**Plate 8: 1. Cylindrospermum majus Kutz. ex Born. et Flah.; 2. C. sphaerica Prasad, B.N.; 3. C. musicola Kutz. ex Born. et Flah.; 4. Nostoc piscinale Kutz. ex Born. et Flah.; 5. N. humifusum Carm. ex Born. et Flah.; 6. N. sphaericum Vaucher ex Born. et Flah.**

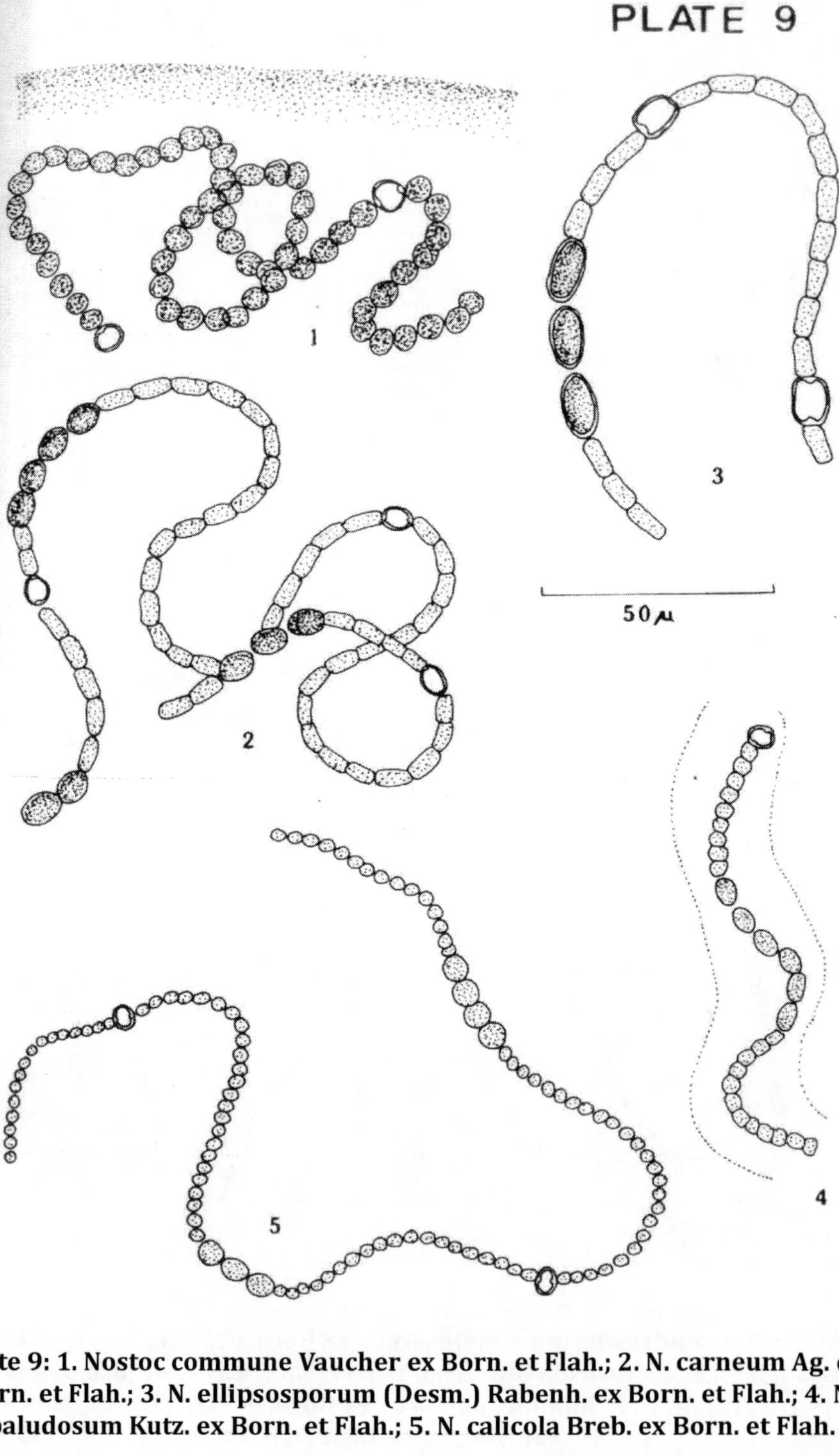


**Plate 9: 1. Nostoc commune Vaucher ex Born. et Flah.; 2. N. carneum Ag. ex Born. et Flah.; 3. N. ellipsosporum (Desm.) Rabenh. ex Born. et Flah.; 4. N. paludosum Kutz. ex Born. et Flah.; 5. N. calicola Breb. ex Born. et Flah.**

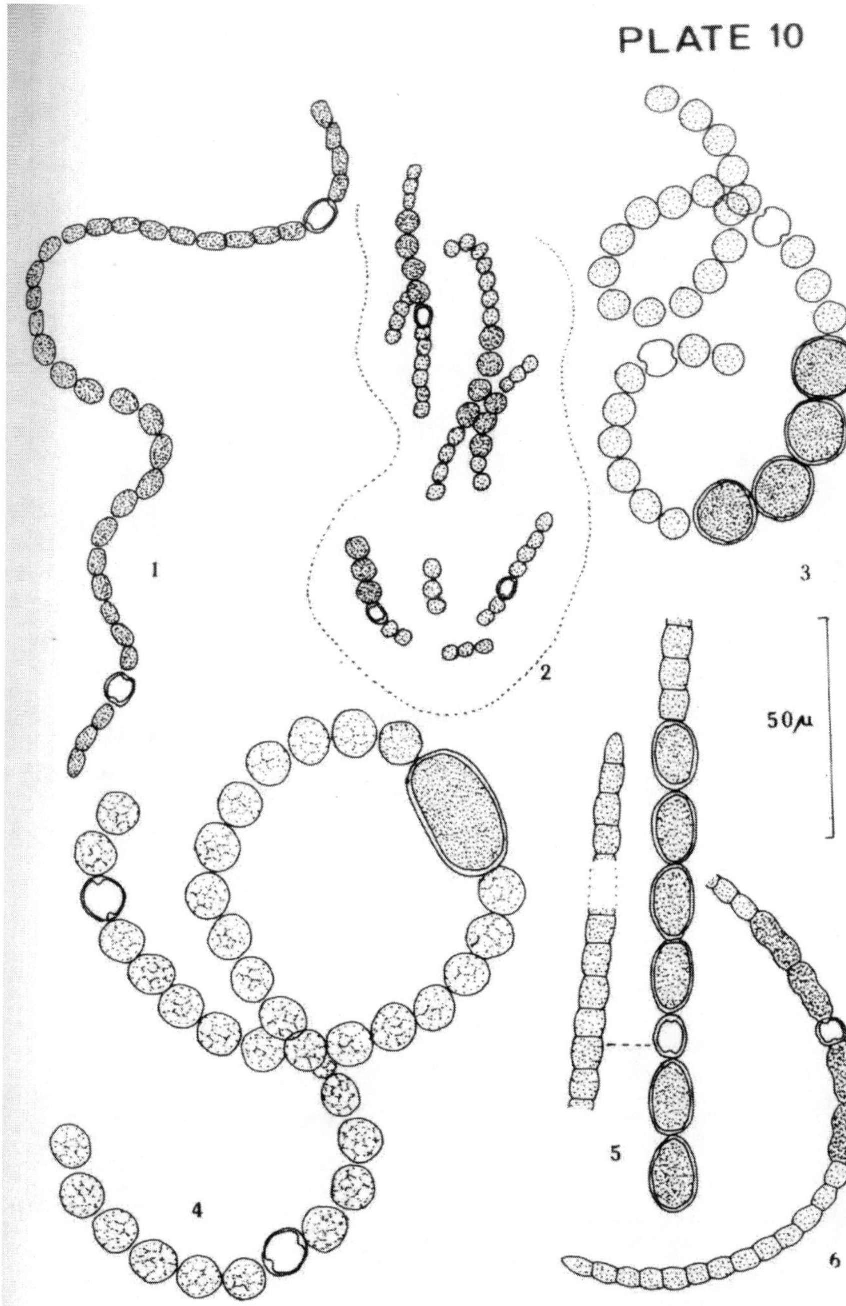


**Plate 10: 1. Nostoc muscorum Ag. ex Born. et Flah.; 2. N. punctiforme (Kutz.) Hariot; 3. Anabaena spiroides Klebahn.; 4. A. circinalis Rabenh. ex Born. et Flah.; 5. A. iyengarii Bharadwaja; 6. A. torulosa (Carm.) Lagerh. ex Born. et Flah.**

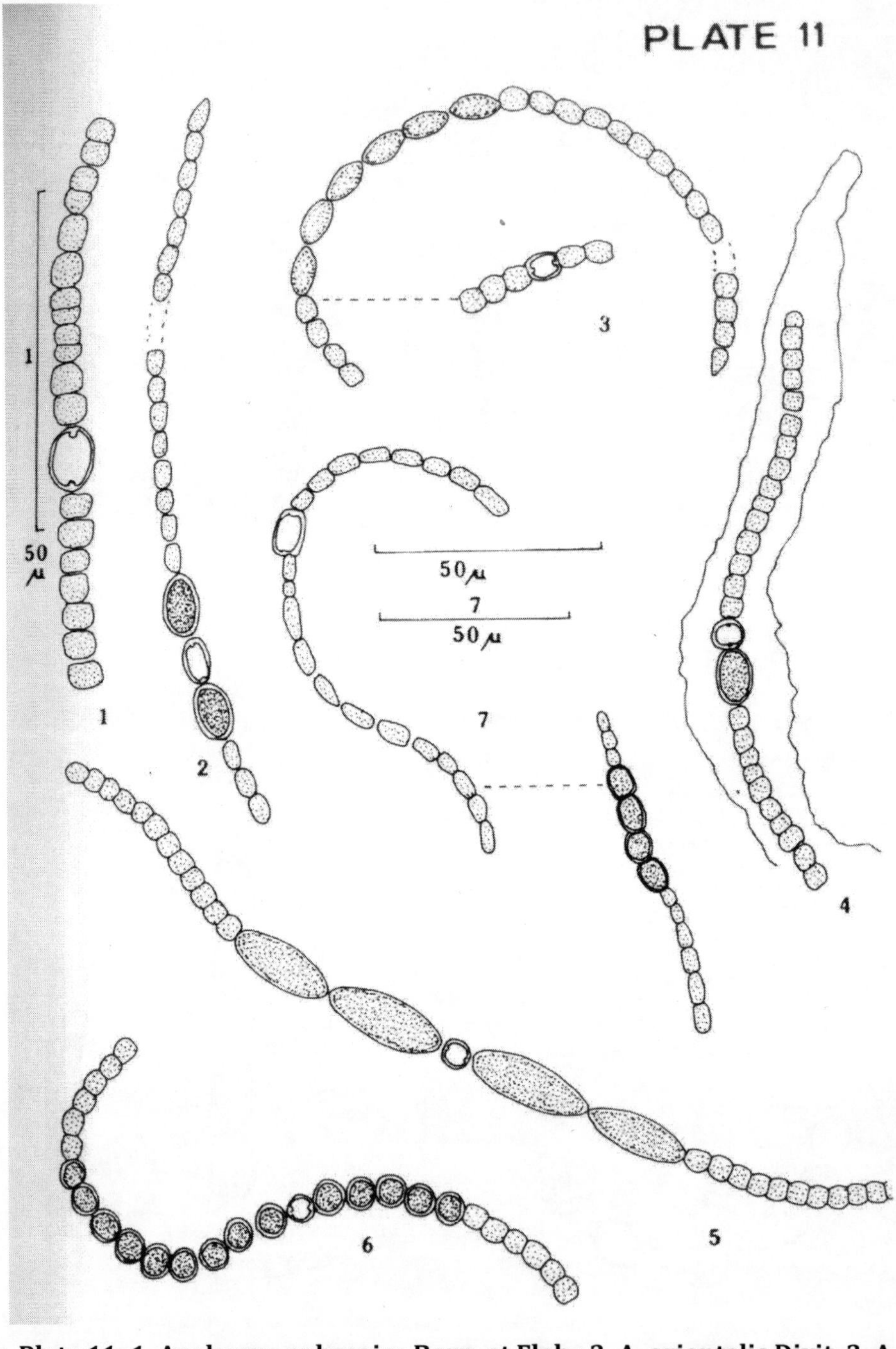


**Plate 11: 1. Anabaena sphaerica Born. et Flah.; 2. A. orientalis Dixit; 3. A. naviculoides Fritsch; 4. A. ambigua Rao, C.B.; 5. A. oscillarioides Bory ex Born. et Flah.; 6. A. fertilissima Rao, C.B.; 7. A. variabilis Kutz. ex Born. et Flah. var. ellipsospora Fritsch.**

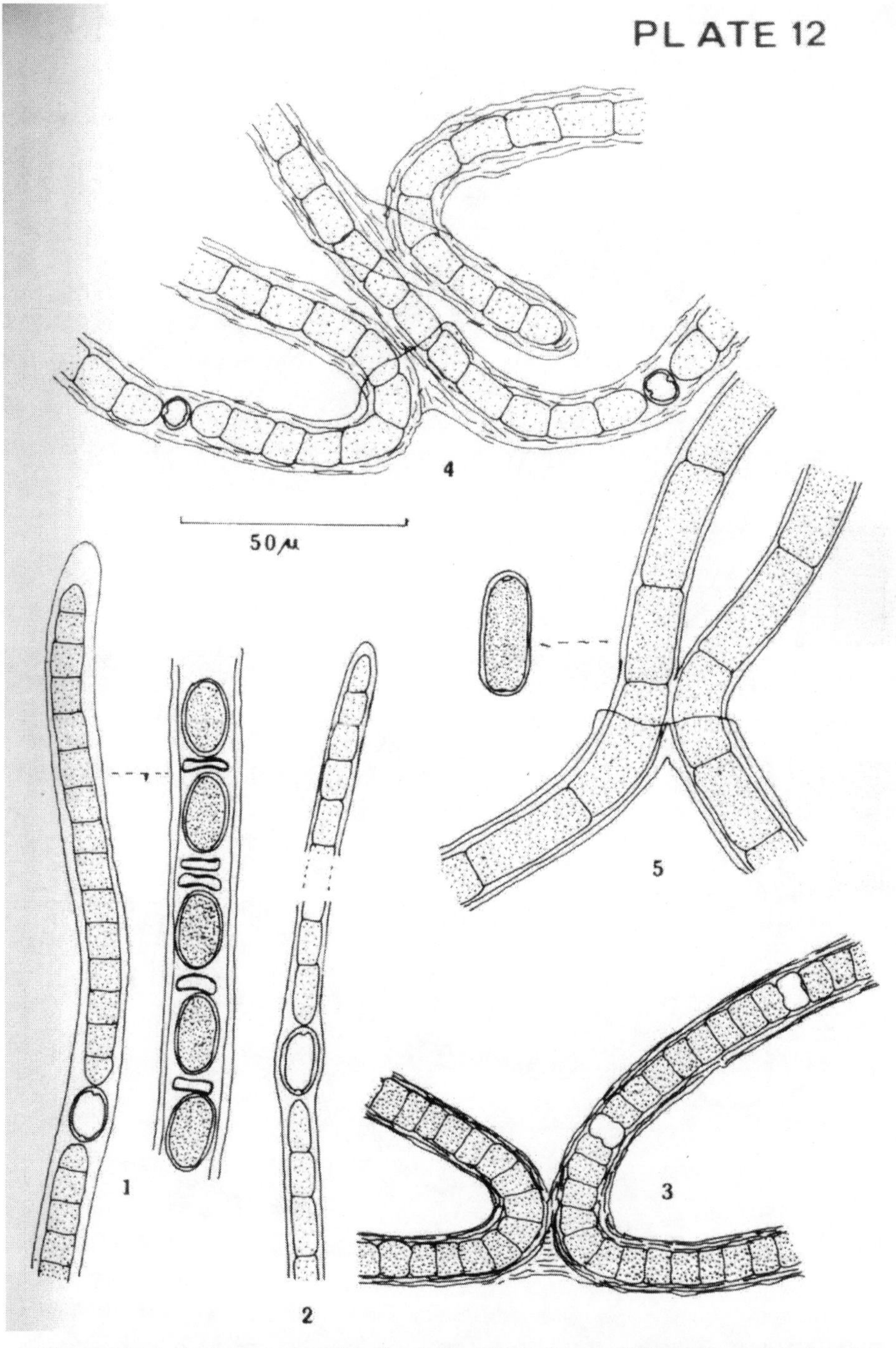


**Plate 12: 1. Aulosira fertilissima Ghosh; 2. A. prolifica Bharadwaja; 3. Scytonema ocellatum Lyngbye ex Born. et Flah.; 4. S. mirabile (Dillw.) Born.; 5. S. simplex Bharadwaja.**

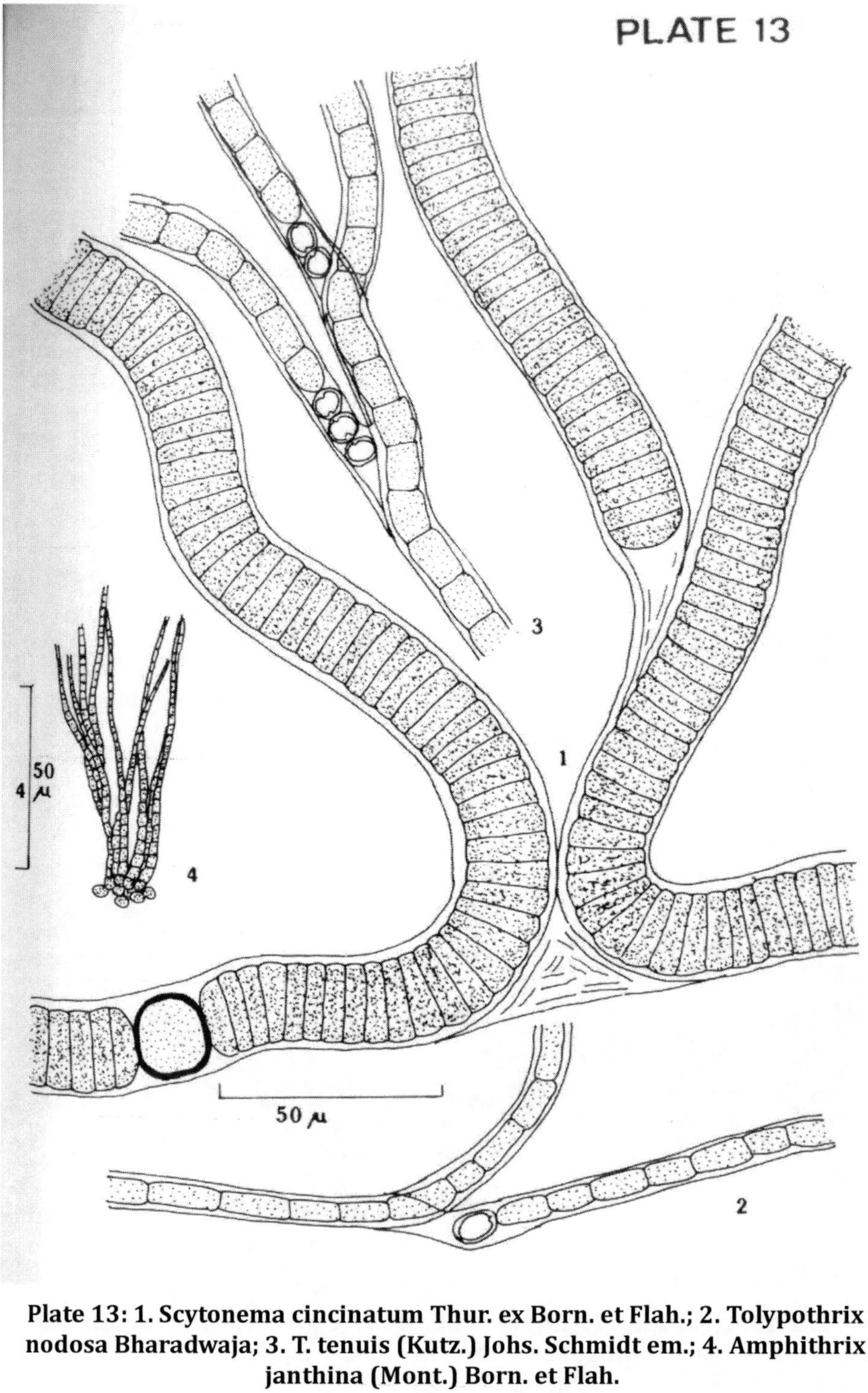


**Plate 13: 1. Scytonema cincinatum Thur. ex Born. et Flah.; 2. Tolypothrix nodosa Bharadwaja; 3. T. tenuis (Kutz.) Johs. Schmidt em.; 4. Amphithrix janthina (Mont.) Born. et Flah.**

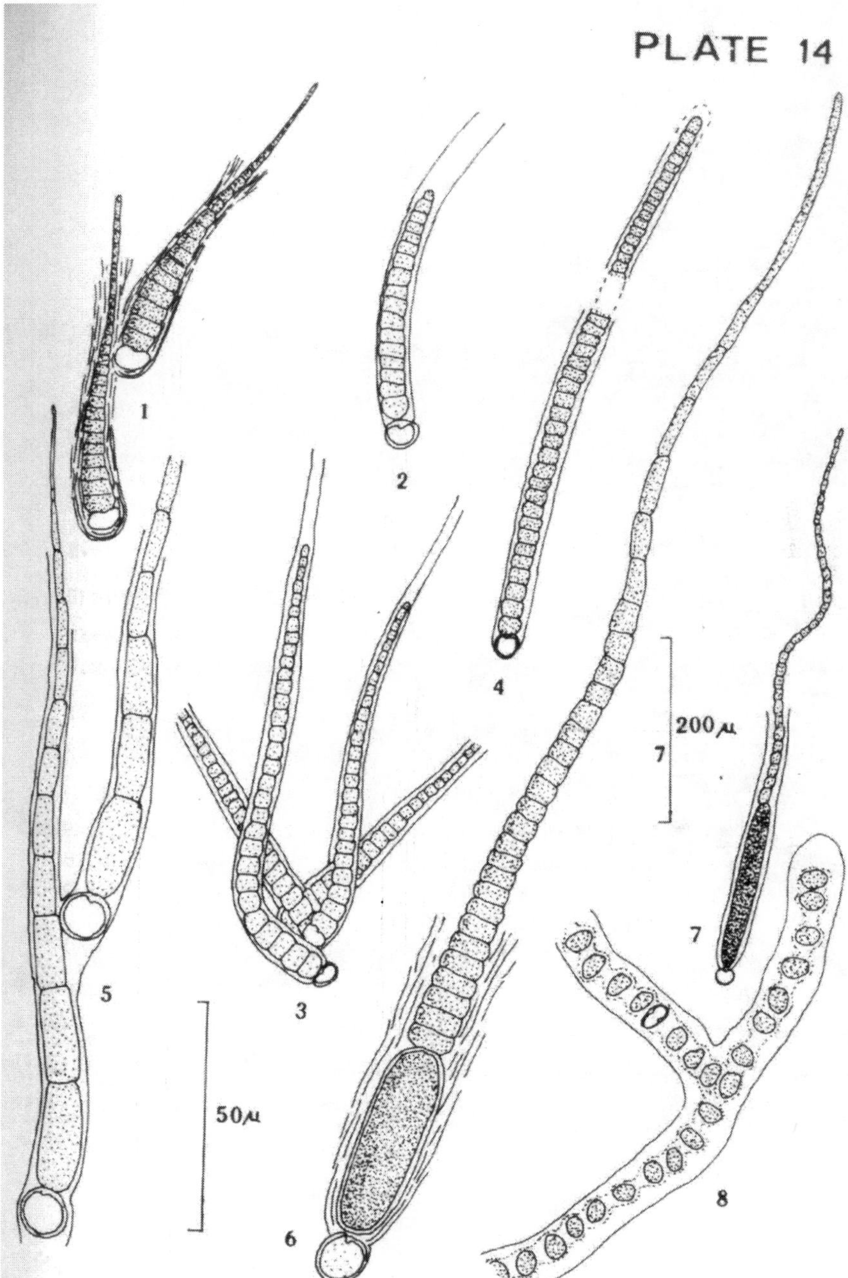


**Plate 14: 1. Calothrix parietina Thur. ex Born. et Flah.; 2. C. brevissima West. G.S.; 3. C. elenkinii Kossinskaja; 4. C. membranacea Schmidle; 5. Rivularia aquatica De Wilde.; 6. Gloeotrichia raciborskii Woloszynska var. kashiense Rao, C.B.; 7. Gl. pisum Thur. ex Born. et Flah.; 8. Stigonema hormoides (Kutz.) Born. et Flah.**

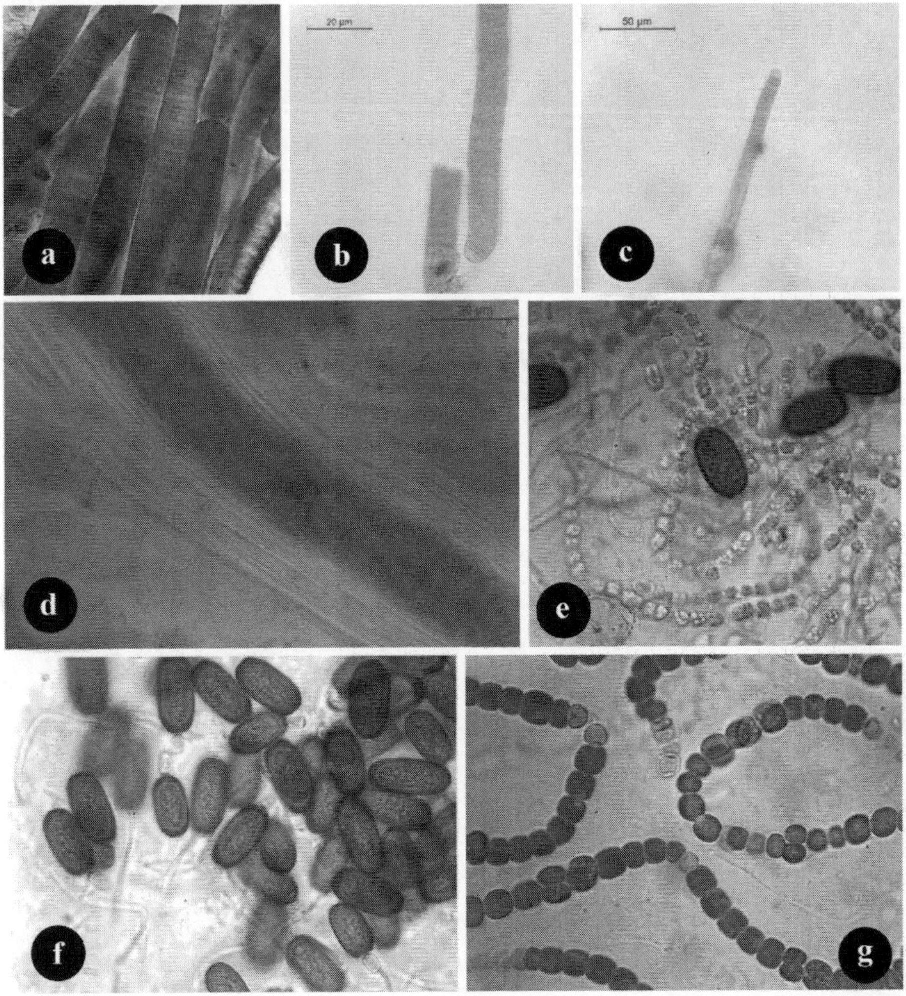


**Plate 15: a.** *Oscillatoria princeps*; **b.** *O. subbrevis*; **c.** *O. limosa*; **d.** *Lyngbya majuscula*; **e & f.** *Cylindrospermum muscicola*; **g.** *Nostoc commune*. [Courtesy: Geethu & Shamina, 2021]

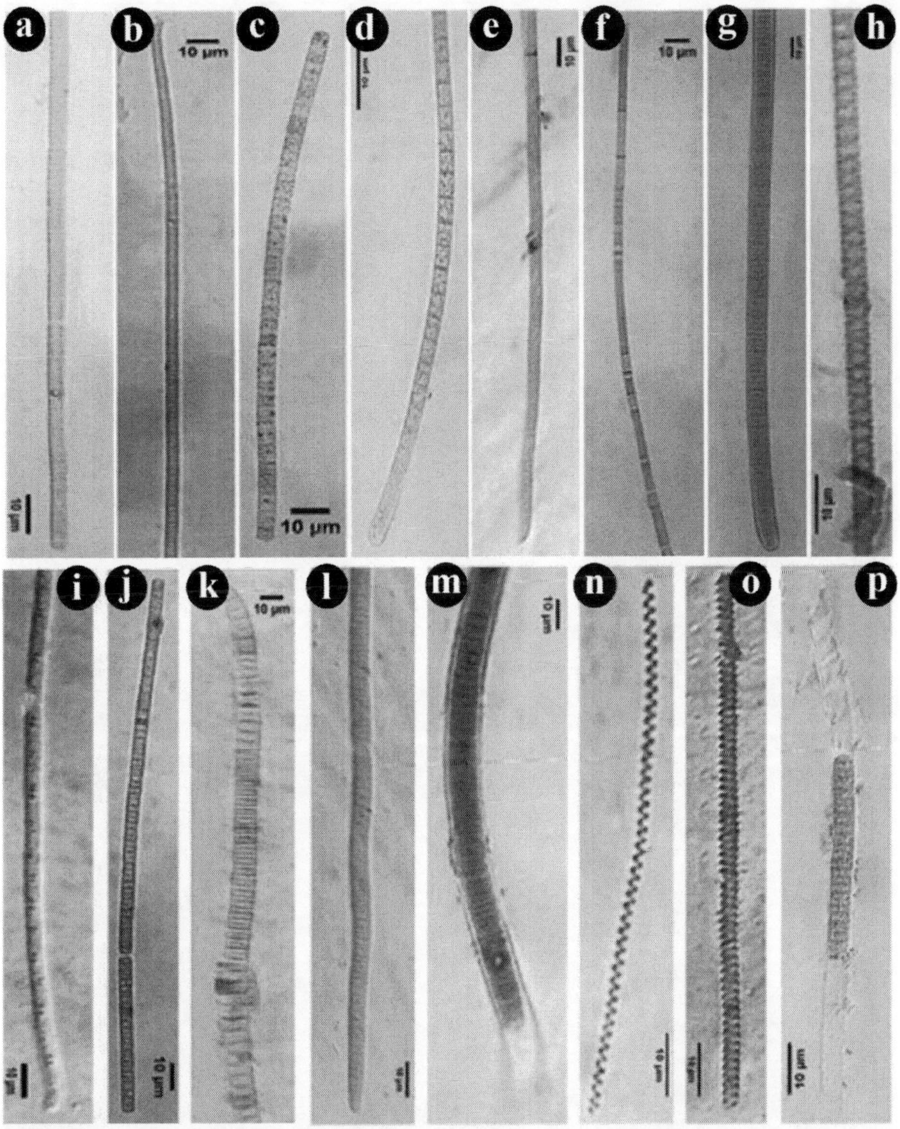

**Plate 16: Microphotographs of Freshwater Filamentous Cyanobacteria - a.** *Oscillatoria obscura;* **b.** *O. acuta;* **c.** *O. amphibia;* **d.** *O. irrigua;* **e.** *O. brevis;* **f.** *O. limosa;* **g.** *O. princeps;* **h.** *O. amoena;* **i.** *O. splendida;* **j.** *O. subbrevis;* **k.** *O. chalybea;* **l.** *O. terabiformis;* **m.** *Lyngbya major;* **n.** *Spirulina major;* **o.** *S. subsalsa;* **p.** *Phormidium anomala* [Courtesy: Dey & Chakraborty, 2020]

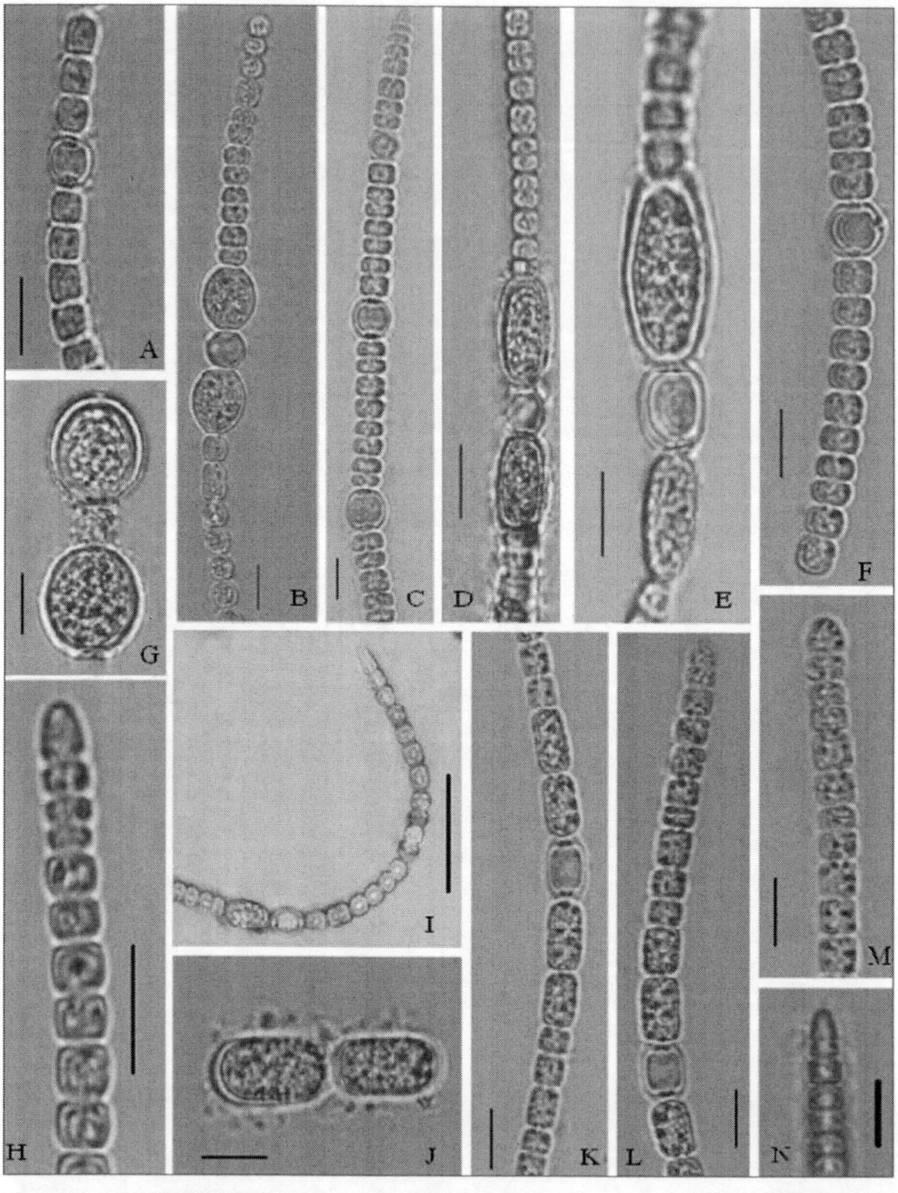


**Plate 17: A-I.** Trichomes of *Anabaena iyengarii* showing heterocytes with flanking akinetes of different shapes ranging from sphaerical to ellipsoidal; **J-M.** *Anabaena torulosa* trichomes with more than one akinetes flanking the heterocytes; **N.** Trichome of *Anabaena iyengarii* showing conical apical cell. Scale bar: 5 μm [Courtesy: Banerjee *et al.*, 2021]

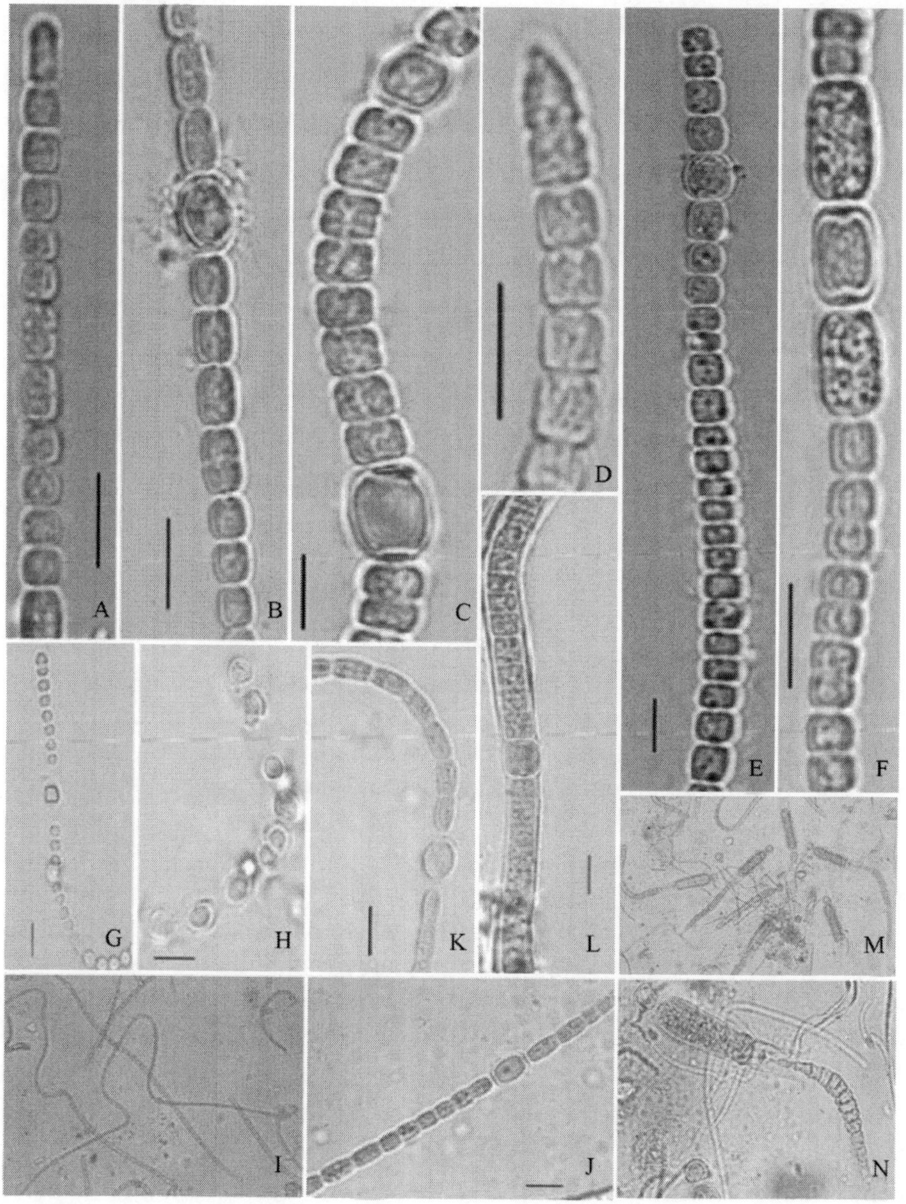


**Plate 18: A & B.** Trichomes of *Anabaena iyengarii* with elongated cells; **C & D.** Trichomes of *Anabaena iyengarii* with barrel shaped cells; **E & F.** *Anabaena torulosa* trichome; **G.** *Nostoc calcicola* trichome with intercalary heterocyte; **H.** *Nostoc calcicola* trichome with rows of akinetes; **I.** Association of *Nostoc carneum* trichomes with both intercalary and terminal heterocytes; **J & K.** *Nostoc carneum* trichomes with intercalary heterocytes; **l.** *Scytonema ocellatum*; **m.** Habit of *Gloeotrichia pisum*; **n.** Single filament of *Gloeotrichia pisum*. Scale bar: 5 µm.

[Courtesy: Banerjee *et al.*, 2021]

## SYSTEMATIC ACCOUNT

### GLOEOCAPSA Kutzing

Phyc.generalis, 174, 1843

**Key to the Species**

1. Sheath colourless ..........................................................................................2

1. Sheath coloured ..........................................................................................3

2. Cells more than 4.8 μ broad (5.4-7.8 μ) ........................................ 1. ***G. decorticans***

2. Cells less than 4.8 μ broad ........................................... 4

3. Sheath reddish, 3.6-4.2 μ broad ........................... 2. ***G. quaternata***

3. Sheath yellowish brown or yellow, 6.0-8.4 μ broad ....... 3. ***G. rupestris***

4. Cells 2.4-3.8 μ broad ........................................ 4. ***G. gelatinosa***

4. Cells 3.0-4.8 μbroad .................................... 5. ***G. polydermatica***

*Gloeocapsa decorticans* (A. Br.)Richter ex Wille, Nyt. Mag. Naturvid. 62: 186.1925;Desikachary, T.V. Cyanophyta.114, pl. 24, fig. 9, 1959.

Thallus pale yellowish in colour, with innumerable number of colonies, without calcium incrustation; colonies predominantly of 2 cells, often having 1, 3 or 4 cells in maximum, diameter of colony12.0-15.6 μ upto 18.0μ; sheath colourless; cells round or slightly oval, diameter without sheath between 5.4 μ and 7.8 μ, cell content yellowish blue-green, granular, individual sheath of the cell distinct, thick, stratified. (Pl. 2, fig. 3).

Formerly reported from Allahabad, U.P. (Mitra, 1951; Verma, 1965, Nainital, U.P. (Talapasayi, 1962); Kalcutta, West Bengal (Gupta, 1975); Ludhiana, Punjab (Pandhol and Grover, 1976); Karnataka (Bongale and Bharati, 1980); 24-Parganas and Howrah districts, West Bengal (Mukhopadhyay and Chatterjee, 1981); River Cauvery, Karnataka (Somashekar, 1983); Kalcutta, West Bengal (Santra, 1987;Sen and Gupta, 1987).

In a stagnant water in a pit near a pond at Hatiara near Kalcutta (Sen, June, 1984; Collection No. 2), in a drying small pool at Rahara near Khardaha, North 24-Parganas district (Collection No. 354).

Voucher No. 1 (Collection No. 354).

***Gloeocapsa quaternata*** (Breb.)Kutz.Tab. Phycologicae, 1: 15, pl.20,fig. 1, 1846; Desikachary, T.V. Cyanophyta.120, pl. 20, fig. 9, 1959.

Thallus pale brownish, expanded, with many colonies held together, upto 18.0μ diam.; colony predominantly 2 and 4 cells, 8 cells in maximum; cells mostly round, diameter without sheath between 3.6-4.2μ; sheath distinct, thick, reddish, lamellated, cell content pale blue-green, granular.(Pl. 2, fig. 4).

Formerly reported from Berhampur, Murshidabad district, West Bengal (Bruhl and Biswas, 1922); Kalcutta, West Bengal (Biswas, 1925; Banerjee, 1936);

Karnataka (Bongale and Bharati, 1980); Bhagabangola,Murshidabad District, West Bengal (Sen and Gupta. 1987); Brackish water wetland, North 24-Parganas, West Bengal (Naskar et al., 2006).

Growing in a tank at Bhagabangola, Murshidabad district (Collection No. 174).

Voucher No. 2 (Collection No. 174).

***Gloeocapsa rupestris*** Kutz. Tab. Phycologicae, 1 : 17, pl. 22, fig.2, 1846; Species Algarum, 223, 1849; Desikachary, T.V. Cyanophyta.117, 1959.

Thallus yellowish brown; with many colonies, sheath yellowish brown but faintly yellow in outer colonies, lamellated; colonies predominantly of 2 cells, 18.0-60.0 μ diam.; cells slightly oval, diameter without sheath 6.0-8.4μ, cell content yellowish blue-green, granular, individual sheath of the cell distinct, yellowish. (Pl. 2, fig. 5),

This species has been reported from Kalcutta, West Bengal (Prain, 1905; Banerji, 1936); Saharanpur district, U.P. (Pal and Yadav, 1974); Lucknow, U.P. (Prasad and Mehrotra, 1980); Karnataka (Bongale and Bharati, 1980); Rohilkhand Division, U.P. (Pandey,1982); Mathur, 24-Parganas district, West Bengal (Maity and Santra, 1985); Jalgaon region, Maharashtra (Bhoge and Ragothaman, 1986); Balidewanganj, Hooghly district, West Bengal (Sen and Gupta, 1987);Roychawk, South 24-Parganas district, West Bengal (Santra *etal.*,1988); Bokaro, Jharkhand (Sharma, 2010); Garia, West Bengal (Sen, 2006).

Growing on a small pull attached to aquatic plants at Balidewanganj, Hooghly district (Collection No. 69).

Voucher No.3(Collection No.69).

***Gloeocapsa gelatinosa*** Kutz.Phyc.generalis, 174, 1843; Desikachary, T.V. Cyanophyta. 114, pl. 27, fig. 6, 1959.

Thallus gelatinous, dense, with many colonies; colonies predominantly of 2 cells, may be four or more cells, diameter of colony 22.8-24.0 μ, sheath colourless; cells 2.4-3.8μ in diameter, without sheath, cell content faintly bluish-green or almost hyaline, homogeneous. (Pl. 2, fig. 12).

This species has been reported from Kalcutta, West Bengal (Banerji, 1936); Karnataka (Bongale and Bharati, 1980); River Cauvery, Karnataka (Somashekar, 1983); Jalgaon region, Maharashtra (Bhoge and Ragothaman, 1986); Tamilnadu (Anand and Revathi, 1987); Simurali, Nadia district, West Bengal (Sen and Gupta, 1987); Jadavpur, West Bengal (Sen, 2006).

Growing on moist soil beside a drain at Simurali near Kalyani, Nadia district (Collection No. 166).

Voucher No. 4 (Collection No. 166).

***Gloeocapsa polydermatica*** Kutz. Tab. Phycologicae, 1: 15, pl. 20.fig. 3, 1846; Species Algarum, 208, 1849; Desikachary, T.V.Cyanophyta. 114, pl. 25, fig. 1, 1959.

Thallus gelatinous, with many colonies, compact; colonies predominantly of 2 cells, often having 1, 3 or 4 cells; sheath colourless, fairly thick and lamellated, sheath around individual cell distinctly lamellated and thick; cells round or ovoid, without sheath 3.0-4.8 μ diam., cell content pale greenish blue, granular.(Pl. 2, fig. 2).

Formerly reported from Meerut, U.P. (Bendre and Kumar, 1975);Bankura district, West Bengal (Sinha and Mukherjee, 1975); Karnataka (Bongale and Bharati, 1980); Allahabad, U.P. (Chadha and Pandey,1982); Belghoria, West Bengal (Sen, 2006)

Adhering to the aquatic plants in a pond at Fakirtala near Nabadwip, Nadia district. (Collection No. 53).

Voucher No. 5 (Collection No. 53).

## CHROOCOCCUS Nag.

Gatt. ein zell. Algen, 45,1849

**Key to the Species**

1. Cells less than 7.2 μ diam. ........................................................ 1. ***C. minutus***

1. Cells up to 7.2 μ diam. or above ........................................................ 2

2. Lamellae of the sheath distinct, diam. above 7.2 μ.............................. 2. ***C. turgidus***

2. Lamellae of the sheath absent, diam. up to 7.2 μ................................ 3. ***C. pallidus***

***Chroococcus minutus*** (Kutz.) Nag. Gatt. ein Zell. Algen, 46, 1849;Desikachary, T.V. Cyanophyta.103, pl. 24, fig. 4& pl.26,figs. 4, 15, 1959.

Cells round, bluish olive, generally single and in groups of 2-4, sometimes 8, bound by a common gelatinous sheath, 4.2-6.6μ diam. without sheath; sheath of individual cell distinct, yellowish, gelatinous, indistinctly lamellar, inconspicuous in single cell which differentiates with growth; reproduction by cell division in three planes; nannocysts within the cells, 16 in number, round to slightly oval. (Pl. 1, fig. 12).

This species has been reported earlier from Kalcutta, West Bengal (Bruhl and Biswas, 1923; Banerjee, 1936); Benaras, U.P. (Rao, 1937; Singh, 1939); other places of U.P. (Singh, 1939) Tamil Nadu (Fremy, 1942); Allahabad, U.P. (Mitra, 1951); Saharanpur district, U.P. (Pal and Yadav, 1974); Meerut, U.P. (Bendre and Kumar, 1975); Ludhiana, Punjab (Pandhol and Grover, 1976); Patiala, Punjab (Sarnia et al., 1979); Gadigarh stream, Jammu (Anand, 1979);Lucknow, U.P. (Prasad and Mehrotra, 1980); Karnataka (Bongale and Bharati, 1980); 24-Parganas and Howrah districts, West Bengal (Mukhopadhyay and Chatterjee, 1981); Allahabad, U.P. (Chadha and Pandey, 1982); River Cauvery, Karnataka (Somashekar, 1983); Khandesh, Maharashtra (Barhate and Tarar, 1983); Bankura district, West Bengal (Sinha and Mukherjee, 1984); Bhubaneswar, Orissa (Mohanty,1984); Nagpur, Maharashtra (Tarar and Shewale, 1984); Diamond Harbour, South-24-Parganas district, West Bengal (Maity and Santra,1985); Kolaghat, Midnapore district, West

Bengal (Pal and Santra,1985); Vidarbha region, Maharashtra (Kolte and Goyal, 1985); Jalgaon region, Maharashtra (Bhoge and Ragothaman, 1986); Jabalpur,M.P. (Srivastava and Sen, 1987); Bhadrakali, Hooghly district, WestBengal (Sen and Gupta, 1987 ); College street area,Kolkata,WestBengal(Sen,2006) Bokaro, Jharkhand (Sharma, 2010).

Collected from trench water, associated with boat-shaped pennate diatom at Bhadrakali, Hooghly district (1985 Collection No. 164); from rice fields at Matiari near Debagram, Nadia district (Collection No. 248); Simurali, Nadia district (Collection No. 246), Dampal Bandh near Purbasthali, Burdwan district (Collection No. 313).

Voucher No. 6 (Collection No. 246).

***Chroococcus turgidus*** (Kutz.) Nag. Gatt. einzell. Algen, 46, 1849; Desikachary, T.V. Cyanophyta.101, pl. 26, fig. 6, 1959.

Planktonic, colony oblong; cells round or ovoid, in groups of 2-4, very rarely 8 together, blue-green, 7.2-15.6 μ diam. without sheath, 12.0-22.8 μ diam. with sheath; sheath hyaline, lamellated. (Pl. 1, fig. 13).

Formerly reported from Bombay (Schmidle, 1900; Gonzalves and Joshi, 1946); Berhampur, West Bengal (Bruhl and Biswas, 1922); Kalcutta (Bruhl and Biswas, 1923; Biswas, 1925; Banerjee, 1936);Benaras, U.P. (Rao, 1937); Andhra Pradesh (Rao, 1938); Saharanpur district, U.P. (Pal and Yadav, 1974); Marathawada, Maharashtra (Kamat, 1974); Patiala, Punjab (Sarma and Kanta, 1978); Gadigarh stream, Jammu (Anand, 1979); Karnataka (Bongale and Bharati, 1980); Lucknow, U.P. (Prasad and Mehrotra, 1980); Rohilkhand division, U.P. (Pandey, 1982); Jaipur, Rajasthan (Trivedy, 1982); Khandesh, Maharashtra (Barhate and Tarar, 1983); River Cauvery, Karnataka (Somashekar, 1983); Vidarbha region, Maharashtra (Kolte and Goyal, 1985); Jabbalpur, M.P. (Srivastava and Sen, 1987); Arambagh, Hooghly district, West Bengal (Gupta and Sen, 1987).

Free floating in a pond at Balidewanganj, Hooghly district (Collection No. 70); from the wall of a drain at Nandigram, Midnapore district. (Collection No. 375)

Voucher No. 7(Collection No. 375).

***Chroococcus pallidus*** Nag. Gatt.einzell. Algen, 46, pl. 1, fig. 2, 1849; Desikachary, T.V. Cyanophyta.108, pl. 26, fig. 5, 1959.

Stratum thick, slimy; colony oblong, having generally 2-4 cells, sometimes single, rarely 8 cells; cells round, blue-green, 5.4-7.2 μ diam. without sheath, 10.8-13.2 μ with sheath; sheath hyaline, unlamellated. (Pl. 2, fig. 1).

Formerly reported from Benaras, U.P. (Rao, 1937); Ludhiana, Punjab (Pandhol and Grover, 1976); Karnataka (Bongale and Bharati1980); Lucknow, U.P. (Prasad and Mehrotra, 1980); 24-Parganas andHowrah districts, West Bengal (Mukhopadhyay and Chatterjee, 1981); Allahabad, U.P. (Chadha and Pandey, 1982); Khandesh, Maharashtra(Berhate and Tarar, 1983); Bhubaneswar, Orissa (Mohanty,

1984);Jabbalpur, M.P. (Srivastava and Sen, 1987); Matiari, Nadia district,West Bengal (Sen and Gupta, 1987); Brackish water wetland, North 24 Pgs, West Bengal (Naskar *et al.*, 2006);Baghbazar area,Kolkata,WestBengal(Sen,2006); Bokaro, Jharkhand (Sharma, 2010).

Growing in paddy fields at Jagadanandapur near Dainhat, Burdwandistrict (Collection No. 19); Kalna, Burdwan district (Collection No. 44); Purbasthali, Burdwan district (Collection No. 39); In stagnant water near R.P.Sen Memorial High School at Matiari, Nadia district (Collection No. 177); in paddy fields at Matiari, Nadia district (Collection No. 228).

Voucher No. 8 (Collection No. 228).

**MICROCYSTIS Kutzing**

Linnaea: 8, 372 pp., 1833 em. Kutzing, Tab. Phycologicae, 1: 7, 1846; Species Algarum, 208, 1849

**Key to the Species**

1. Colony clathrate ........................................................................ 2

1. Colony not clathrate .................................................................... 3

2. Cells 3.6-4.8 μ diam. ........................................................ 1. ***M. aeruginosa***

2. Cells 6.0-8.4 μ diam. ........................................................ 2. ***M. robusta***

3. Colonies compound having partial colonies held together; cells more than 4.2 μ diam. ........................................................................................ 3. ***M. viridis***

3. Colonies simple, cells less than 4.2 μ diam. ................................... 4. ***M. flosaquae***

***Microcystis aeruginosa*** Kutz. Tab. Phycologicae, 1: 8, pl. 8, fig.1,1846; Desikachary, T.V, Cyanophyta. 93, pl. 17, figs. 1,2,6 & pl. 18, fig. 10, 1959.

Colonies free floating, mucilaginous, round or ellipsoidal, solid, three dimensional when young and clathrate when older, young colonies 74.0-220.0 μ in diam., larger when old and clathrate; mucilaginous matrix hyaline but distinct, not stratified, not extending beyond the colony; cells round, 3.6-4.8 μ diam., vacuoles present (Pl. 1, figs. 1-5).

Formerly reported from Kalcutta, West Bengal (Martens, 1870; Prain, 1905; Biswas, 1926, 1927, 1942; Banerji, 1936); Bombay, Maharashtra (Schmidle, 1900; Dixit, 1936; Gonzalves and Joshi, 1943);Benaras, U.P. (Bharadwaja, 1935); Berhampur, Orissa (Rao, 1938);Kerala (Parukutty, 1940); Delhi (Rao, 1940); Madras (Ganapathi, 1940); Marathawada, Maharashtra (Kamat, 1974); Allahabad, U.P. (Gupta andPandey, 1974); Meerut, U.P. (Bendre and Kumar, 1975); Bhagalpur,Bihar (Nasar and Datta Munshi, 1976); Ludhiana, Punjab (Pandhol andGrover, 1976); Patiala, Punjab (Sarma and Kanta, 1978); Gadigarh,Jammu (Anand, 1979); Karnataka (Bongale and Bharati, 1980); Lucknow,U.P. (Prasad and Mehrotra, 1980); Allahabad, U.P. (Pandey and Pandey,1982); Rohilkhand Division, U.P. (Pandey, 1982); Bhubaneswar, Orissa(Mohanty, 1984); Bankura district, West Bengal (Sinha and Mukherjee, 1984); Jalgaon region, Maharashtra (Bhoge and Ragothaman, 1986);Arunachal Pradesh, Assam and Meghalaya (Reddy *et al.*, 1986); Jabbalpur, M.P. (Srivastava and Sen, 1987); Pawagiri Oon, M.P. (Mahajan,1987); Bhagalpur, Bihar (Saha and Pandit, 1987); Kamarpukur, Hooghlydistrict, West Bengal (Gupta and Sen, 1987); Brackish water wetland, North 24 Pgs, West Bengal (Naskar *et al.*, 2006); Rice fields of Orissa (Dey *et al.*, 2010); Nager Bazar, Dumdum, West Bengal (Sen, 2006);Barasat,West Bengal(Talai-Mukhopadhyay and Naskar,2015).

Collected from Kamarpukur, Hooghly district (Collection No. 68); from rice fields at Bakulla near Guptipara,Hooghly district (Collection No. 300); Fakirtala, Nabadwip, Nadia district(collection No. 315); Rezinagar, Murshidabad district (Collection No.362).

Voucher No. 9 (Collection No. 68).

***Microcystis robusta*** (Clark) Nygaard, in Ostenfeld and Nygaard Phytoplankton Gatun Lake, Panama-Kanal, Dansk. Bot. Ark., 4(10): 8, figs. 1-4, 1925; Desikachary, T.V. Cyanophyta. 85, pl. 17,figs. 7-10, 1959.

Colonies planktonic, gelatinous, round when young, older colonies elongated, clathrate, mucilaginous matrix distinct but margins indefinite; cells dispersed in the mucilage, spherical, 6.0-8.4μ diam., gasvacuoles indistinct. (Pl. 1, figs. 6,7).

Earlier reported from Kalcutta, West Bengal (Biswas, 1927; Banerji, 1936); Madras (Ganapathi, 1940); Lucknow, U.P. (Prasad, and Mehrotra, 1980) ; River Cauvery, Karnataka (Somashekar, 1983); Midnapore district, West Bengal (Pal and Santra, 1985); Jalgaon, Region, Maharashtra (Bhoge and Ragothaman, 1986); Pusa, Bihar (Jha*et al.*, 1986); Kadamtala, Howrah district, West Bengal (Sen andGupta, 1987); Bokaro, Jharkhand (Sharma, 2010)|

Collected from stagnant water in a small pool at Kadamtala,Howrah district (Collection No. 25); from ricefield not under cultivation at Kali Park near Dum Dum airport, Kalcutta (Collection No. 346).

Voucher No. 10 (Collection No. 346).

***Microcystis viridis*** (A. Br.)Lemm.Zeitschr.fur fischerie, 11:101, 1903; Abh. Nat. Ver. Bremen, 17: 342, 1902; Desikachary,T.V. Cyanophyta.87, pl. 18, figs. 1-6, 1959.

Colonies planktonic, round, solid or variously shaped, havingdistinct partial colonies held together, colonial sheath mucilaginous, with distinct margins; daughter colonies having distinct limits; cells spherical or ellipsoidal, 4.2-6.6μ diam., gasvacuoles present. (Pl. 1, figs. 8,9).

Formerly reported from Kalcutta (Banerji, 1936); Saharanpurdistrict, U.P. (Pal and Yadav, 1974); Karnataka (Bongale and Bharati,1980); River Cauvery, Karnataka (Somashekar, 1983); Jalgaon region,Mabarashtra (Bhoge and Ragothaman, 1986); Jabbalpur, M.P. (Srivastavaand Sen, 1987); Khardah, North 24-Parganas district, West Bengal (Sen and Gupta, 1987); Brackish water wetland, North 24 Pgs., West Bengal (Naskar *et al.*, 2006);

Collected from a pond at Khardaha, North 24-Parganas district (Collection No. 23).

Voucher No. 11. (Collection No. 23).

***Microcystis flos-aquae*** (Wittr.) Kirchner, in Engler and Prantl, Naturlichen Pflanzenfam., I. (1a): 56, fig. 49(N), 1898; Desikachary, T.V. Cyanophyta. 94, pl. 17, fig. 11 & pl. 18, fig. 11, 1959.

Colonies free floating, planktonic, mucilaginous, round or ellipsoidal or elongated, solid, never clathrate, 74.0-185.0 μ in diam.; mucilaginous matrix not very distinct, diffluent; cells closely arranged, round, 3.0-4.2μ diam., gasvacuoles present.(Pl. 1, figs. 10, 111).

Formerly reported from Assam (Bruhl and Biswas, 1926); Benaras, U.P. (Bharadwaja, 1935); Bombay, Maharashtra (Dixit, 1936; Gonzalves and Joshi, 1946); Kalcutta, West Bengal (Banerji, 1936; Biswas, 1942);Andhra Pradesh (Rao, 1938); Madras, Tamilnadu (Fremy, 1942; Ganapathi,1940; 1943); Orissa (Rao, 1938); Delhi (Rao, 1940); Bihar (Rao, 1939); Bankura district, West Bengal (Sinha and Mukherjee, 1975); Ludhiana, Punjab (Pandhol and Grover, 1976); 24-Parganas and Howrah district,West Bengal (Mukhopadhyay and Chatterjee, 1981); Allahabad, U.P.(Pandey and Pandey, 1982); Rohilkhand Division, U.P. (Pandey, 1982);Jaipur, Rajasthan (Trivedy, 1982); River Cauvery, Karnataka (Somashekar, 1983); Khandesh, Maharashtra (Barhate and Tarar, 1983);Vidarbha region, Maharashtra (Kolte and Goyal, 1985); Jabbalpur, M.P. (Srivastave and Sen, 1987); Bansberia, Hooghly district, West Bengal (Gupta and Sen, 1987); Brackish water wetland, North 24 Pgs, West Bengal (Naskar *et al.*, 2006); Rice fields near Rajarhat, West Bengal (Sen, 2006).Barasat,West Bengäl(Talai-Mukhopadhyay and Naskar,2015).

Growing in drying small pool at Bansberia, Hooghly district (Collection No. 172); from rice fields at Fakirtala, Nabadwip, Nadia dist. (Collection No. 331), Kumirmara, Dankuni, Hooghly district (Collection No. 334).

Voucher No. 12 (Collection No. 172).

## APHANOCAPSA Nag

Gatt. einzell. Algen, 52, 1849

### Key to the Species

1. Cells single or in 2, never enclosed in common envelope .......................................... 2

1. Cells 2-4 in common envelope ................................................................1.***A.biformis***

2. Thallus brownish blue to brown,cells 6.0-6.6μ diam. ........................... 2.***A. grevillei***

2. Thallus light olive, cells 2.4-3.6μ diam. ........................................... 3. ***A. montana***

***Aphanocapsa biformis*** A. Br. in Rabenhorst, Fl. Eur. Alg., 2: 246, 1865; Desikachary, T.V. Cyanophyta. 134, pl. 21, figs. 3,4,1959.

Thallus more or less spherical, gelatinous, olive-green and more densely coloured towards the periphery; cells loosely arranged in a thallus, spherical, 4.8-

6.8μ diam.; 2-4 cells often enclosed in common envelope; contents dense, granular, greenish blue. (Pl.2,fig. 8).

Formerly reported from Valiyathura, Travancore, Kerala (Parukutty, 1940); Saharanpur district, U.P. (Pal and Yadav, 1974); Sagar, M.P. (Mishra and Purohit, 1979); Karnataka (Bongale and Bharati, 1980); Lucknow, U.P. (Prasad and Mehrotra, 1980); 24-Parganas and Howrah districts, West Bengal (Mukhopadhyay and Chatterjee, 1981); Bhubaneswar, Orissa (Mohanty, 1982); Allahabad, U.P. (Chadha and Pandey, 1982); River Cauvery, Karnataka (Somashekar, 1983); Gadchiroli and Bhandara districts, Maharashtra (Kolte and Goyal, 1985);Brackish water wetland, North 24 Pgs., West Bengal (Naskar *et al.*, 2006); Rice fields of Odisha (Dey *et al.*, 2010); Rice fields near Garia, West Bengal (Sen, 2006)

Collected, from rice field at Simurali, Nadia district (Collection No. 245).

Voucher No. 13 (Collection No. 245).

***Aphanocapsa grevillei***(Hass.)Rabenh. Fl. Eur. Alg., 2 : 50, 1865; Desikachary, T.V. Cyanophyta. 134, pl. 21, fig. 9, 1959.

Thallus spherical to elliptical, gelatinous, brownish blue to brown, more densely coloured towards the peripheral part of the colony, almost hollow, peripheral part more dense, with calcareous crystals, 5-15 mm diam.; cells closely arranged in a thallus, spherical, 6.0-6.6μ diam.; without distinct sheath; contents dense, distinctly granular, bluish yellow. (Pl. 2, fig. 6; Pl.15, fig. 2).

Formerly reported from Kalcutta (Bruhl and Biswas, 1923; Biswas,1925, 1942; Banerjee, 1936); Benaras, U.P. (Rao, 1937); Bereilly,U.P. (Chaturvedi and Pandey, 1976); Lucknow, U.P. (Prasad and Mehrotra, 1980); Allahabad, U.P. (Chadha and Pandey, 1982); Allahabad, U.P. (Pandey and Pandey, 1982); Bhubaneswar, Orissa (Mohanty, 1982),River Cauvery, Karnataka (Somashekar, 1983); Bhubaneswar, Orissa(Mohanty, 1984); Garhbeta, Midnapore district, West Bengal (Pal and Santra, 1985); Dainhat, Burdwan district, West Bengal(Sen and. Gupta, 1987); Brackish water wetland, North 24 Pgs., West Bengal (Naskar *et al.*, 2006);Baranagar near Kolkata ,West Bengal (Sen,2006);Rice fields of Sonitpur district, Assam (Dasgupta and Ahmed, 2013).

Collected from stagnant water in a ditch at Jagadanandapur near Dainhat, Burdwan district (Collection No.179); from rice field at Chiruti, Murshidabad district (Collection No. 348).

Voucher No. 14 (Collection No. 179).

***Aphanocapsa montana*** Cramer, in Wartmann and Schenk, Schweiz. Krypto. no. 134, 1862; Desikachary, T.V. Cyanophyta. 135, pl. 20, fig. 8, 1959.

Thallus gelatinous, variable in shape, irregular, attached to submerged plants, light olive in colour; cells spherical, single or in twos, 2.4-3.6μ diam.; sheath diffluent; contents dense, light blue-green. (Pl. 2, fig. 7).

This species has been reported earlier from Punjab (Randhawa,1936); Kalcutta, West Bengal (Banerji, 1936); Patiala, Punjab (Sarma and Kanta, 1978); Karnataka (Bongale and Bharati, 1980); Tejpur, U.P. (Pandey, 1982); Allahabad, U.P. (Pandey and Pandey, 1982);Rampara,Murshidabad district, West Bengal (Pal and Santra, 1984);Kalcutta, West Bengal (Santra, 1987); Debagram, Nadia district,West Bengal (Sen and Gupta, 1987); Sal Lake, Kolkata, West Bengal (Sen, 2006).

Collected from stagnant water in a field at Beldanga near Debagram, Nadia district (Collection No. 181); from paddy field Guptipara, Hooghly district (Collection No. 254).

Voucher No. 15 (Collection No. 181).

## APHANOTHECE Nag.

Gatt. einzell, Algen, 59, 1849.

### Key to the Species

1. Thallus often with calcareous crystals,cells 4.2-4.8μ broad, 8.4-10.2μ long ......... ............................................................ 1. ***A. stagnina***

1. Thallus without crystals, cells 6.0-6.6 μ broad, 12.0-13.2 μ long.......................... ............................................................ 2. ***A. pallida***

***Aphanothece stagnina*** (Spreng.) A. Br. in Rabenhorst, Fl. Eur. Alg., 2: 66, 1865; Desikachary, T.V. Cyanophyta.. 137, pl. 21, fig. 10, 1959.

Thallus thick, large, forming slimy, mucilaginous mass, yellowish green with bluish green surface, with calcareous crystals; cells ellipsoidal or cylindrical, arranged more densely in the peripheral part of gelatinous matrix, newly formed cells smaller and round, without individual envelope, 4.2-4.8μ broad, 8.4-10.2μ long; contents finely granular, yellowish blue-green, homogeneous. (Pl. 2, fig. 9).

Formerly reported from Kalcutta, West Bengal (Banerji, 1936); Gorakhpur, U.P. (Singh, 1939); Travancore, Kerala (Parukutty, 1940); Madurai, Tamilnadu (Fremy, 1942); Meerut, U.P. (Bendre and Kumar,1975); Ludhiana, Punjab (Grover and Pandhol, 1975); Bankura district, West Bengal (Sinha and Mukherjee, 1975); Ludhiana, Punjab (Pandhol and Grover, 1976); Patiala, Punjab (Sarnia *et al.*, 1979); Karnataka (Bongale and Bharati, 1980); River Cauvery, Karnataka (Somashekar, 1982); Pusa, Bihar (Jha *et al.*, 1986); Tamilnadu (Anand and Revathi, 1987); Bhandarhati, Hooghly district, West Bengal (Gupta and Sen, 1987); Rice fields of Nadia district, West Bengal (Sen, 2001); Rice fields near Garia, West Bengal (Sen, 2006);Ricefields of Sonitpur district,Assam(Dasgupta and Ahmed,2013);Barasat,West Bengal(Talai-Mukhopadhyay and Naskar,2015).

Collected from rice fields at Jagadanandapur near Dainhat, Burdwan district (Collection No. 33); Kalitala, Dankuni,Hooghly district (Collection No. 115); Beldanga near Debagram, Nadia district (Collection No.156); in stagnant water at Bhandarhati, Hooghly district (Collection No. 158); from rice fields at Samudragarh near Kalna, Burdwan district (Collection No. 236).

Voucher No. 16 (Collection No. 33).

***Aphanothece pallida*** (Kutz.)Rabenh.Fl. Eur. Alg., 2: 64, 1865; Desikachary, T.V. Cyanophyta.140, pl. 22, fig. 3, 1959.

Thallus soft, like mucilaginous ball, up to 5 mm diam., sheath yellowish blue-green, distinct near the periphery and diffluent inner; cells ellipsoidal or cylindrical, arranged densely in the peripheral region of the colony, 6.0-6.6μ broad, 12.0-13.2μ long; contents finely granular, blue-green, homogeneous. (Pl. 2, fig. 10).

Formerly reported from Bombay, Maharashtra (Schmidle, 1900; Gonzalves and Joshi, 1946); Benaras, U.P. (Rao, 1936); Panihati, 24-Parganas, West Bengal (Banerji, 1936); Andhra Pradesh (Rao,1938); Puri, Berhampur and Cuttack, Orissa (Rao, 1938); Bihar (Rao, 1939);Marathawada, Maharashtra (Kamat, 1974); Ludhiana, Punjab (Grover and Pandhol, 1975); Bankura district, West Bengal (Sinha and Mukher- jee, 1975); Cuttack, Orissa (Singh, 1975); Ludhiana, Punjab (Pandhol and Grover, 1976); Nagpur, Maharashtra (Kamat and Harankhedkar,1976); Bareilly, U.P. (Chaturvedi and Pandey, 1976); Patiala,Punjab (Sarna *et al.*, 1979); Lucknow, U.P. (Prasad and Mehrotra,1980); Karnataka (Bongale and Bharati, 1980); 24-Parganas and Howrah districts, West Bengal (Mukhopadhyay and Chatterjee, 19&1); Allahabad, U.P. (Chadha and Pandey, 1982); Bhubaneswar, Orissa (Mohanty, 1983); Jammu and Kashmir (Goyal *et al.*, 1984); Jhargram, Midnapore district, West Bengal (Pal and Santra, 1985); Vidarbha region, Maharashtra (Kolte and Goyal, 1985); Pusa, Bihar (Jha *et al.*,1986); Purbasthali, Burdwan district, West Bengal (Gupta and Sen, 1987);Tamilnadu (Anand and Revathi, 1987); Arambagh, Hooghly district,West Bengal (Sen and Gupta, 1987);Baghajatin Rail Station,Kolkata,West Bengal(Sen,2006); Bokaro, Jharkhand (Sharma, 2010).

Collected from paddy field at Purbasthali, Burdwan district(Collection No. 58); as mucilaginous balls in pond water at Nutanbazar, Arambagh, Hooghly district (Collection No. 75); Khajuti near Dainhat, Burdwan district (Collection No. 134); Palpara near Simurali, Nadia district (CollectionNo. 194); Kalatala near Nabadwip, Nadia district (Collection No. 230); Guptipara, Hooghly district (Collection No. 297).

Voucher No. 17 (Collection No. 194).

## CHLOROGLOEA Wille

Algol.Notizen, I-VI, Nyt Mag. Naturvid.,38: 5, 1900.

***Chlorogloea fritschii*** Mitra, Two new algae from Indian soils, Ann.Bot., Lond., n.s., 14: 457, figs. 25-43, 1950; Desikachary,T.V. Cyanophyta. 163, pl. 31, figs. 1-16, 1959.

Thallus irregular in size and shape, deep blue-green, outline definite, cells arranged in three dimensions, horizontal growth more predominant, thallus without any envelope; cells round or slightly ellipsoidal, some cells free from the groups in the thallus; cell sheath inconspicuous, 7.2-7.8 μ in diam.; endospore produced singly within the cell, naked, spherical or ovoid, 6.0-7.2μ in diam. (Pl. 1, fig. 14).

Formerly reported from Allahabad, U.P. (Mitra, 1950, 1951);Karnataka (Bongale and Bharati, 1980); 24-Parganas and Howrah districts (Mukhopadhyay and Chatterjee, 1981); River Cauvery, Karnataka (Somashekar, 1983); Khandesh, Maharashtra (Barhate and Tarar, 1983); Marathwada, Maharashtra (Chaporkar and Gangawane, 1984);Beldanga, Murshidabad districts, West Bengal (Sen and Gupta, 1987); Bokaro, Jharkhand (Sharma, 2010); Rajarhat, Kolkata, West Bengal (Sen, 2006).

Growing in crop field at Bhabta, Murshidabad district (Collection No. 365).

Voucher No. 18 (Collection No. 365).

**DERMOCARPA Crouan**

Note sur quelques algues marines nouvelles, Ann. Sci. nat., Bot.,ser. 4, 9: 70, 1858.

***Dermocarpa leibleiniae*** (Reinsch) Born.et Thur. Notes algologiques.2: 73.pl. 26. fig. 3-5. 1880; Desikachary, T.V, Cyanophyta.173, pl. 33, figs. 20-21, 1959.

Sporangium slightly elongate to pyriform, attached with *Lyngbva* filament, single, differentiated into small basal stalk like part, olive-green; 18-21.6μ broad; wall thick, stratified; endospores formed from the entire content by division in three planed, or asmall sterile basal part is left, endospores about 25 or so in number. (Pl. 2, fig. 11).

Formerly reported from Kalcutta, West Bengal (Gupta, 1965); Karnataka (Bongale and Bharati, 1980).

Growing on brick built reservoir at Sibpur, Howrah district (Collection No. 74).

Voucher No. 19 (Collection No. 74).

**MYXOSARCINA Printz.**

E. Norske Vidensk. Solsk.Skrift., 35; pl. 16, figs.342-357 (1920), 1921.

***Myxosarcina spectabilis*** Geitler Arch Hydrobiol.(u.Planktonk.)12(4):624, 1933;Desikachary, T.V. Cyanophyta. 178, pl. 30, figs. 1-5 & pl. 31, figs. 17-22, 1959.

Three dimensional colony, almost round in outline, attached with the substratum, sheath of the colonies thin, hyaline, very young colonies 4-celled, maximum diameter of colony 51μ; composed of regularly and closely arranged cells, 4.2-6.0μ broad; blue-green, with slightly angular corners, wall of individual cell very conspicuous, thick; some of cells of the colony enlarged forming endospores. (Pl. 3, fig. l).

Reported earlier from Kalcutta, West Bengal (Gupta, 1965); Marathawada, Maharashtra (Kamat, 1974); Patiala, Punjab (Sarma *et al.*, 1979); Lucknow, U.P. (Prasad and Mehrotra, 1980); Allahabad,U.P. (Pandey and Pandey, 1982); Allahabad, U.P. (Chadha and Pandey,1982); River Cauvery, Karnataka (Somashekar, 1983); Digha, Midnapore district, West Bengal (Santra *et al.*, 1988); Rajabazar, Kolkata, West Bengal (Sen, 2006).

Collected from constantly moist wall at Maheshganj near Nabadwip, Nadia district (Collection No . 52).

Voucher No. 20 (Collection No. 52).

**OSCILLATORIA Voucher**

Hist, conf., 165, 1803; Gomont, Monogr. Oscillariees,198, 1892.

**Key to the Species**

1. Trichomes prominently constricted .................................................................. 2

1. Trichomes not constricted or slightly so ........................................................... 3

2. Septa not granulated ........................................................................ 1. ***O. annae***

2. Septa granulated ........................................................................................... 4

3. Septa granulated ........................................................................................... 5

3. Septa not granulated........................................................................................ 6

4. Cells as long as broad or slightly longer ............................................ 2. ***O. schultzii***

4. Cells broader than long ..................................................................................... 7

5. Cross walls with single granule .......................................................................... 8

5. Cross walls with more than one granules ............................................................ 9

6. Trichomes bent at the apices ............................................................................ 10

6. Trichomes not bent at the apices ....................................................................... 11

7. Trichomes 10.8-16.8μ broad.......................................................... 3. ***O. sancta***

7. Trichomes 4.8-7.2μ broad .................................................................................. 12

8. Granule single on each side of the septa ........................................................... 13

8. Granule single on one side of the septa ............................................................. 14

9. Trichomes not narrow at the apical part ............................................ 4. ***O. limosa***

9. Trichomes narrow at the apical part ................................................................... 15

10. Trichomes slightly constricted at the cross-walls; cells over6.0 μ broad (7.2-7.8μ) ......... ............................................................ **5. *O. chalybea***

10. Trichomes not constricted at the cross-walls; cells up to 6.0μ broad .........16

11. Cells over 15.0μ broad, about 1/6-1/5 as long as broad .................... 6. ***O. princeps***

11. Cells up to 9.0 μbroad ....................................................................................... 17

12. Trichomes up to 6.0μbroad, cells slightly broader than long............7. ***O. formosa***

12.Trichomes over 6.0 μbroad, cells 1/2 as long as broad ................ **8. *O. proteus***

13. Trichomes more or less parallel in a colony, 2.2-2.4μbroad.................................. ........................................................... **9. *O pseudogeminata var. unigranulata.***

13. Trichomes entangled in a colony,3.0-3.6 μ broad.......................... 10. ***O. amphibia***

14. Trichomes spirally twisted at the apical part .......................... **11. *O. terebriformis***

14. Trichomes straight.............................................................................. **12. *O. salina***

15. Cells longer than broad, about twice as long as broad; end cells ending in a knob-like structure ........................................................................ **13. *O. splendida***

15. Cells as broad as or broader than long ....................................................... 18

16. End cell acute; cells 3.0-3.6μ broad, 2.4-3.0μ long ......................... **14. *O. animalis***

16. End cell narrow but not acute; cells 4.8-6.0μ broad, 3.0-4.8μ long......... **15. *O. acuta***

17. Cells over 7.2 μ broad; end cells flattened ....................................... **16. *O. irrigua***

17. Cells less than 7.2μ broad, end cells not flattened........................................ 19

18. Trichomes more or less parallel forming bundle, attached by base ........................................................................................................................ **17. *O. rubiscens***

18. Cells less than 5.5μ broad ........................................................................... 20

19. End cell with thickened outer membrane; cells 7.2-9.0 μ broad ........................................................................................................................ **18. *O. simplicissima***

19. End cell without thickened outer membrane; cells upto 6.0 μ broad .......... 21

20. Cells 0.8μ long; apical cell conical ............................................ **19. *O. obscura***

20. Cells over 3.0μ long ............................................................................. 22

21. Trichomes briefly attenuated at apex; cells 2.4-2.8μ long ................ **20. *O. brevis***

21. Trichomes not attenuated at apex; cells 1.8-2.4μ long ................. **21. *O. subbrevis***

22. Cells as long as or slightly longer than broad ................................ **22. *O. prolifica***

22. Cells mostly broader than long ..................................................................... **23**

23. Cells 2.4-3.6μ long; septa granulated on one side...................... **23. *O. agardhii***

23. Cells 4.2-5.0μ long; septa granulated on both sides ......................**24. *O. amoena***

***Oscillatoria annae*** van Goor, Zur Kenntnis des Oscillatoriaceen,Rec. Trav. bot. neerland., 15 : 289, pl. 2, fig. 1a-c, 1918; Desikachary, T.V. Cyanophyta. 203, pl. 38, fig. 13, 1959.

Thallus light blue-green, trichome straight, apical part slightly narrow from the rest of the trichome, distinctly constricted at the joints, ends not bent; apical cell rounded; cells about 1/3 as long as broad, 7.2-8.4μ broad, 2.4-3.0μ long, transverse wall not granulated; cell contents granular. (Pl. 3, fig. l).

Formerly reported from Allahabad, U.P. (Gupta, 1956); Patiala,Punjab (Sarma and Kanta, 1978); Marathwada, Maharashtra (Ashtekarand Kamat, 1980); Vidarbha region, Maharashtra (Kolte and Goyal,1985); Sultanpur, U.P. (Khan, 1985); Garhbeta , Midnapore district, West Bengal (Pal and Santra, 1985).

Collected from a wet muddy place at Mahanandapur, Midnaporedistrict (Collection No. 380).

Voucher No. 21. (Collection No. 380) .

***Oscillatoria schultzii*** *Lemm.*Forschb. Biol. Stat., Plon, 12: 145, pl. 4, figs.14-15, 1905; Desikachary, T.V. Cyanophyta. 232, pl. 41, fig. 11, 1959.

Thallus pale blue-green; trichome almost straight, short with apex distinctly and gradually tapering, distinctly constricted at the joints, ends not bent; apical cell capitate; cells as long as broad or slightly longer than broad, 2.2-2.8μbroad, 2.4-3.0 μ long; transverse walls granulated; cell contents homogeneous.(Pl. 3, fig. 9; Pl.).

This species has been reported from Bombay, Maharashtra (Gonzalves and Gangla, 1949); Kalcutta, West Bengal (Gupta, 1965); Saharanpur district, U.P. (Pal and Yadav, 1974); Ludhiana, Punjab (Grover and Pandhol, 1975, 1976); Karnataka (Bongale and Bharati, 1980); Ghatal, Midnapore district, West Bengal (Pal and Santra, 1985); Jalgaon region, Maharashtra (Bhoge and Ragothaman, 1986);Singhee, Burdwan district; Bhandarhati, Hooghly district; Manicktala and Kestopur, Kalcutta, West Bengal (Sen and Gupta, 1987); Belgharia, Kolkata, West Bengal (Sen, 2006).

Growing on roadside slime at Kestopur, Kalcutta (Collection No. 136); from rice fields at Singhee, Burdwan district (Collection No. BD-50); from crop field at Nandigram, Midnapore district (Collection No. 173).

Voucher No. 22 (Collection No. 173).

***Oscillatoria sancta*** (Kutz.)Gom.Monogr. Oscillariees, 209, pl. 6, fig. 12, 1892; Desikachary, T.V. Cyanophyta.203, pl. 42, fig. 10, 1959.

Bluish-green, free floating thallus; trichomes bluish-green, straight, slightly narrow towards the apical part, prominently constricted at the cross-walls; cells broader than long, about 1/3- 1/5 times as long as broad, 10.8-10.8μ broad, 2.4-5.4μ long, cross walls granulated on both sides, end cell hemispherical with a thick membrane. (Pl. 3, fig. 10),

This species has been previously reported from Bombay, Maharashtra (Schmidle, 1900; Gonzalves and Joshi, 1946); Benaras, U.P. (Rao, 1937); Cuttack (Rao, 1938); West Bengal (Banerji, 1938); Madhubani, Bihar (Rao, 1939); New Delhi (Rao, 1940); Allahabad,U.P. (Gupta, 1956); Meerut, U.P. (Bendre and Kumar, 1975); Ludhiana, Punjab (Pandhol and Grover, 1976); Patiala, Punjab (Sarma and Kanta, 1978); Lucknow, U.P. (Prasad and Mehrotra, 1980); Sohna river, Bihar (Sinha and Srivastava, 1980); Allahabad, U.P. (Pandey and Pandey, 1982); Bhubaneswar, Orissa (Mohanty, 1982); Nagpur, Maharashtra (Nandkar *etal.*, 1983); Contai, Midnapore district, West Bengal (Pal and Santra, 1985); Meghalaya and Tripura (Reddy *etal.*, 1986); Tamilnadu (Anand and Revathi, 1987); Pawagiri Oon, M.P. (Mahajan, 1987);Behala, Kalcutta, West Bengal (Sen and Gupta, 1987); Sakherbazar, Behala, West Bengal (Sen, 2006).

Free floating in stagnant water associated with *Langbya digueti* Gom.and some species of *Anabaena* at Sakherbazar, Behala, Kalcutta (Collection No. 21); growing beside a roadside drain at Nababganj near Ichapur, North 24-Parganas district (Collection No. 351).

Voucher No. 23 (Collection No. 351).

***Oscillatoria limosa*** Ag.ex Gom. Agardh, Dispositio Algarum Sueciae, 35, 1812; Gomont, Monogr.Oscillariees, 210, pl. 6, fig. 13,1892; Desikachary, T.V. Cyanophyta.206, pl. 42, fig. 11,1959.

Thallus dark blue-green; trichome almost straight, not tapering, not prominently constricted at the cross-walls; apical cell convex, more or less flat, outer wall slightly thickened; cells much shorter than broad, 12.0-l4.4μ broad, 2.4-3.6μ long, cross walls granulated; cell contents granular dark blue-green. (Pl. 3, fig. 11).

This species has been formerly recorded from Howrah, Kalcutta and Hooghly river in West Bengal (Prain, 1905; Biswas, 1926, 1942); Berhampur, Orissa (Rao, 1938); Madras, Tamilnadu (Ganapathi, 1940); Saharanpur district, U.P. (Pal and Yadav, 1974); Meerut, U.P. (Bendre and Kumar, 1975); Patiala, Punjab (Sarma and Kanta, 1978); Gadigarh stream, Jammu (Anand, 1979); Lucknow, U.P. (Prasad and Mehrotra,1980); Marathwada, Maharashtra (Ashtekar and Kamat, 1980); Karnataka (Bongale and Bharati, 1980); Allahabad, U.P. (Pandey and Pandey,1982); Rohilkhand division, U.P. (Pandey, 1982); Bhubaneswar, Orissa(Mohanty, 1982); River Cauvery, Karnataka (Somashekar, 1983); Bankura district, West Bengal (Sinha and Mukherjee, 1984); Kupwara, Pulwama, Srinagar, Jammu and Kashmir (Goyal *etal.*, 1984); Maharashtra (Patil and Satav, 1986); Pusa, Bihar (Jha *etal.*, 1986); Meghalaya, Nagaland (Reddy *etal.*, 1986); Bagbazar and Kaikhali, Kalcutta, West Bengal (Sen and Gupta, 1987); Junput, Midnapore district, West Bengal (Santra *etal.*, 1988); Bidhanhagar, North 24 Pgs., West Bengal (Dey and Chakraborty, 20220);Baguihati,near Dum Dum,West Bengal(Sen,2006); North 24 Pgs., West Bengal (Naskar *et* al., 2008).

On road-side slime at Kaikhali, nearDum Dum Airport (Collection No. 145); from rice fields at Sankarpur, near Nabadwip, Nadia district (Collection No. 189); Kumirmara near Dankuni, Hooghly district (Sen, June, 1988; Collection No. 338).

Voucher No. 24 (Collection No. 145).

***Oscillatoria chalybea*** (Mertens) Gom. Mertens in Jurgens, Alg.aquaticae, Decas., 12, nr. 4, 1822; Gomont, Monogr.Oscillariees, 232, pl. 7, fig. 19, 1892; Desikachary, T.V. Cyanophyta.218, pl. 38, fig. 3, 1959.

Trichome straight and slightly bent and narrow at the apex, slightly constricted at the cross walls; apical cell round, not capitate, calyptra absent, outer wall not thickened; cells 1/2-1/3 times as long as broad, 7.2-7.8μ broad, 3.0-4.2μ long; transverse walls not granulated; cell contents dark blue-green, homogeneous. (Pl. 3, fig. 12).

This species was collected from Kalcutta, West Bengal (Biswas, 1926); Mussorie, U.P.; Narsapur, A.P.; Berhampur, Orissa (Rao, 1937,1938); Travancore, Kerala (Parukutty, 1940); Delhi (Rao, 1940);Bombay, Maharashtra (Gonzalves and Joshi, 1946); Rohilkhand division,U.P. (Singh and Chaturvedi, 1970); Meerut, U.P. (Bendre and Kumar,1975); Bankura district, West Bengal (Sinha and Mukherjee,

1975);Rohilkhand division, U.P. (Chaturvedi and Pandey, 1976); Karnataka(Bongale and Bharati, 1980); 24-Parganas and Howrah districts, WestBengal (Mukhopadhyay and Chatterjee, 1981); Allahabad, U.P. (Pandeyand Pandey, 1982); Bhubaneswar, Orissa (Mohanty, 1982); River Cauvery,Karnataka (Somashekar, 1983); Ghatal, Midnapore district (Pal andSantra, 1985); Jalgaon region, Maharashtra (Bhoge and Ragothaman,1986); Bhagalpur, Bihar (Saha and Pandit, 1987); Katwa, Burdwan district; Dunlop Bridge, Kalcutta; Mogra, Hooghly district, West Bengal (Sen and Gupta, 1987); Nurpur, South 24-Parganas district,West Bengal (Santra *etal.*, 1988); Paddy fields of Bardhaman, Nadia and Hooghly district (Sen and Gupta, 1998); Hooghly and Bardhaman district, West Bengal (Sen, 2001); Rice fields in Orissa (Dey *et al.*, 2010;); Dhakuria, Kolkata, West Bengal (Sen, 2006); North 24 Pgs., West Bengal (Naskar *et al.*, 2008).Bidhannagar,WestBengal(Dey and Chakrobarty,2020).

Growing on constsantly moist brick built wall at Narendranagar near Dunlop Bridge, Kolkata (Collection No. 61); from paddy fields at Kalna, Burdwan district (Collection No. 63); Jagadanandapur near Dainhat, Burdwan district (Collection No. 30); Khajuti near Katwa, Burdwan district (Collection No. BD-18); in a flower pot at Gurupur Nursery, Mogra, Hooghly district (Collection No. 155); from paddy fields at Simurali, Nadia district (Collection No. 199)

Voucher No. 25 (Collection No. 61).

***Oscillatoria princeps*** Vauch.ex Gom. Vaucher, Histoire des confervesd'eau douce, 190, pl. 15, fig. 2, 1803; Gomont, Monogr.Oscillariees, 206, pl. 6, fig. 9, 1892; Desikachary, T.V. Cyanophyta.210, pl. 37, figs. 1, 10, 11, 13 and 14, 1959.

Thallus dirty green, free floating; trichomes dark blue-green with a slight brownish tinge, straight, slightly narrow at the apex;cells at the apical part much shorter than those in the mature region,without any constriction at the cross-walls; cells much broader than long, about 1/6-1/5 as long as broad, 15.5-18.5μ broad, 3.0μ or less long, cross walls not granulated, end cells more or less flat,capitate with a slightly thick membrane; cell contents granular, dense near the centre, dark greenish blue.

Cells of collection from Arambagh measure 45.5-48.0μ broad and 4.2-4.6μ long. (Pl. 4, fig. 1).

Cosmopolitan in distribution in India; reported from Saharanpur district, U.P. (Pal and Yadav, 1974); Marathawada, Maharashtra (Kamat, 1974); Ludhiana, Punjab (Grover and Pandhol, 1975); Bankura district, West Bengal (Sinha and Mukherjee, 1975); Meerut, U.P. (Bendre andKumar, 1975); Ludhiana, Punjab (Pandhol and Grover, 1976); Gadigarh stream, Jammu (Anand, 1979); Karnataka (Bongale and Bharati, 1980); Rohilkhand division, U.P. (Pandey, 1982); Bhubaneswar, Orissa (Mohanty,1982); River Cauvery, Karnataka (Somashekar, 1983);Jammu and Kashmir(Goyal *etal.*, 1984); Garhbeta, Midnapore district, West Bengal (Pal and Santra, 1985); Vidarbha region, Maharashtra (Kolte and Goyal, 1985); Sultanpur, U.P. (Khan, 1985); Meghalaya, Arunachal Pradesh, Tripura, Nagaland and Assam (Reddy *etal.*, 1986); Jalgaon region, Maharashtra (Bhoge and Ragothaman, 1986);

Bhagalpur, Bihar (Saha and Pandit, 1987); Tamilnadu (Anand and Revathi, 1987); Kalcutta, West Bengal (Santra, 1987); Kalcutta, Arambagh, Hooghly district,West Bengal (Sen and Gupta, 1987); Diamond Harbour, South 24-Parganasdistrict; Haldia, Midnapore district, West Bengal (Santra *etal.*,1988); Ricefields of Hooghly, Nadia and Bardhaman district, West Bengal (Sen, 2001); Belgharia, Kolkata, West Bengal (Sen, 2006); North 24 Pgs., West Bengal (Naskar *et al.*, 2008); Rice fields in Orissa (Dey *et* al., 2010); Sonitpur district, Assam (Dasgupta and Ahmed, 2013); ); Barasat, West Bengal (Talai-Mukhopadhyay and Naskar, 2015); Bakkhali, Chheramatha, South 24 Pgs, West Bengal (Barman *et al.*, 2015); Bidhannagar, North 24 Pgs, West Bengal (Dey and Chakraborty, 2020).

Collected from rice fields at Naihati near Dainhat, Burdwan district(Collection No. 27); in ponds at Baghajatin area near Garia, Kalcutta (Collection No.67); from soil water enrichment culture of vegetable field at Narayanpur near Dum Dum Airport (Collection No. 55); Simurali, Nadia district (Collection No. 147); Guptipara, Hooghly district (Collection No. 288); Dankuni, Hooghly district (Collection No.316).

Voucher No. 26 (Collection No. 316).

***Oscillatoria formosa*** Bory ex Gom. Bory, Diet, class, d'hist. nat., 12: 474, 1827; Gomont, Monogr.Oscillariees, 230, pl. 7, fig.16, 1892; Desikachary, T.V. Cyanophyta.232, pl. 40, fig. 15, 1959.

Blue-green thallus; trichomes straight and apex gradually tapering and bent, constricted at the cross-walls; apical cell round, not capitate, calyptra absent; cells slightly broader than long, 4.8- 6.0μ broad, 4.2-4.8μ long, slightly granulated at the septum.(Pl. 4, fig. 2).

Formerly reported from Kalcutta, West Bengal (Martens, 1870; Biswas, 1926); Chinsurah, Hooghly district, West Bengal (Bruhl and Biswas, 1922); Bombay, Maharashtra (Dixit, 1936); Mussorie, Benaras, U.P.; Cuttack, Orissa (Rao, 1937, 1938); Darbhanga, Bihar (Rao, 1939); Travancore, Kerala (Parukutty, 1940); Delhi (Rao, 1940); Allahabad, U.P. (Gupta, 1956); Marathawada, Maharashtra (Kamat, 1974); Meerut, U.P. (Bendre and Kumar, 1975); Bankura district, West Bengal (Sinha and Mukherjee, 1975); Patiala, Punjab (Sarma *et al.*, 1979); Karnataka (Bongale and Bharati, 1980); 24-Parganas and Howrah districts, West Bengal (Mukhopadhyay and Chatterjee, 1981); Allahabad, U.P. (Pandey and Pandey, 1982); Nagpur, Maharashtra (Nandkar *et al.*, 1983); River Cauvery, Karnataka (Somashekar, 1983); Haldia, Midnapore district; West Bengal (Pal and Santra, 1985); Bhagalpur, Bihar (Saha and Pandit, 1987); Kalighat, Kalcutta; Deulpara, Hooghly district, West Bengal (Sen and Gupta. 1987); Haldia, Midnapore district, West Bengal (Santra *et al.*, 1988); Mintu Park, Kolkata, West Bengal (Sen, 2006).

Growing in waste water drain near Kalighat temple, Kalcutta (Collection No. 85); paddy fields at Adisaptagram, near Mogra, Hooghly district (Collection No. 114); on moist soil at Deulpara, Hooghly district (Collection No. 167); on paddy fields at Brahmapara, Simurali, Nadia district (Collection No. 201); Paglachandi near Debagram, Nadia district (Collection No. 239).

Voucher No. 27 (Collection No. 239).

***Oscillatoria proteus*** Skuja, Zur Susswasseralgenflora Burmas, Nov. Acta Reg. Soc. Upsal., ser. 4, 14: 48, pl. 8, figs. 11-13, 1949; Desikachary, T.V. Cyanophyta. 221, pl. 41, figs. 15, 16 and 18, 1959.

Thallus bluish-green, free floating; trichomes straight or sometimes slightly curved, having same width throughout except the slightly narrow apical cell, cross-walls prominently constricted, cell joint prominent, septa granulated; cells 1/2 as long as broad,6.0-7.2µ broad, 2.7-3.5µ long, end cell not capitate, round, outer wall not thickened, calyptra absent; cell contents blue-green, finely granular. (Pl. 4, fig. 3).

Formerly recorded from Saharanpur district, U.P. (Pal and Yadav, 1974); Bankura district, West Bengal (Sinha and Mukherjee, 1975); Meerut, U.P. (Bendre and Kumar, 1975); Patiala, Punjab (Sarma and Kanta, 1978); Sagar, M.P. (Mishra and Purohit, 1979); Jaipur,Rajasthan (Trivedy, 1982); Khandesh, Maharashtra (Barhate and Tarar,1983); Nagpur, Maharashtra (Nandkar *et al.*, 1983); Salbani, Midnapore district, West Bengal (Pal and Santra, 1985).

Collected from a rice field near Dampal bandh, Purbasthali, Burdwan district (Collection No. 10).

Voucher No. 28 (Collection No. 10).

***Oscillatoria pseudogeminata*** G. Schmid, var. **unigranulata** Biswas, Malayan aquatic Biology, J. Fed. Malay Stat. Mus., 14: 409, pl. 9, fig. 7, 1929; Desikachary, T.V. Cyanophyta. 229, pl. 39, fig. 19 and pl. 41, fig. 17, 1959.

Greenish blue colony clinged to the soil; trichomes arranged more or less parallely in a colony, tenuous, straight or somewhat bent at the apical part, not constricted at the cross-walls, apex truncate and slightly tapering, calyptra absent; cells 4.0-4.2 µ long, 2.2-2.4µ broad; septa thick, prominent, with one granule on each side at the centre of the septum; cell contents blue-green, granular. (Pl. 4, fig. 4).

***O.pseudogeminata*** described by Rao, (1940) from Delhi appears to be same as var.***unigranulata*** Biswas.

Formerly reported from Meerut, U.P. (Bendre and Kumar, 1975);24Parganas and Howrah districts, West Bengal (Mukhopadhyay and Chatterjee, 1981); Bhubaneswar, Orissa (Mohanty, 1982); Khandesh,Maharashtra (Barhate and Tarar, 1983); Pusa, Bihar (Jha *et al.*, 1986);Arunachal Pradesh and Tripura (Reddy *et al.*, 1986); Hatiara near Dum Dum, Kolkata, west Bengal (sen, 2006).

Matiari near Debogram,Nadia district(Collection No. 131); Somrabazar near Guptipara, Hooghly district (Collection No. 292).

Voucher No. 29 (Collection No. 131).

***Oscillatoria amphibia*** Ag.ex Gom. Agardh, Aufzahlung etc., Flora,10: 632, 1827; Gomont, Monogr.Oscillariees, 221, pl. 7, figs. 4 and 5, 1892; Desikachary, T.V. Cyanophyta.229, pl.37, fig. 6, 1959.

Trichome straight, not constricted at the cross-walls; not attenuated at the apices, obtusely rounded, not capitate, calyptra absent; cells 4.8-6.0μ long, 3.0-3.6μ broad, septa granulated,with two granules at each septum; cell contents finely granular,blue-green. (Pl. 4, fig. 5).

This species was formerly collected from Howrah, West Bengal (Prain,1905) Chinsura, Berhampur, Burdwan, West Bengal (Bruhl and Biswas, 1922); Kalcutta, West Bengal (Biswas, 1925, 1926; Banerji, 1938); Hooghly river, West Bengal (Biswas, 1942); Loktak Lake, Assam (Bruhl and Biswas, 1926); Saharanpur district, U.P. (Pal and Yadav,1974); Bankura district, West Bengal (Sinha and Mukherjee, 1975);Meerut, U.P. (Bendre and Kumar, 1975); 24-Parganas and Howrah districts, West Bengal (Mukhopadhyay and Chatterjee, 1981); River Cauvery, Karnataka (Somashekar, 1983); Jadavpur, Kalcutta, West Bengal (Gupta and Sen, 1987);Sakuntlapark,Behala,West Bëngal(Sen,2006); Bidhannagar, North 24 Pgs., West Bengal (Dey and Chakraborty, 2020).

Growing in shallow water at Santoshpur, Jadavpur, Kalcutta(Collection No. 9); on rice fields at Swarupganj near Nabadwip, Nadia district (Collection No. 193); Somrabazar near Guptipara, Hooghly district (Collection No.294).

Voucher No. 30 (Collection No. 294).

***Oscillatoria terebriformis*** Ag.ex Gom. Agardh. Aufzahlung, Flora,10: 634, 1827; Gomont, Monogr.Oscillariees, 234, pl. 7,fig. 24, 1892; Desikachary, T.V. Cyanophyta.217, pl. 38, fig. 16, 1959.

Thallus blue-green; apex of trichome not tapering, apical part spirally twisted, not prominently constricted at the cross-wall, slightly constricted at older part; apical cell round, not capitate,outer wall not thickened; cells shorter than broad, 3.0-3.6μ long,5.0- 5.4μ broad; partition walls marked by prominent single granule on one side; cell contents uniformly granular blue-green. (Pl. 4,fig. 6).

This species has been recorded from Assam (Biswas, 1934);Benaras, U.P.; Khurda Road, Orissa (Rao, 1936, 1938); Patna, Bihar(Rao, 1939); Basti, Gorakhpur and Mirzapur, U.P. (Singh, 1939);New Delhi (Rao, 1940); Kalcutta, West Bengal (Gupta, 1965); Ludhiana,Punjab (Grover and Pandhol, 1975); Rohilkhand division, U.P. (Chaturvedi and Pandey, 1976); Lucknow, U.P. (Prasad and Mehrotra, 1980); Karnataka (Bongale and Bharati, 1980); Bhubaneswar, Orissa (Mohanty, 1982); Khandesh, Maharashtra (Barhate and Tarar, 1983); Nagpur,Maharashtra (Tarar and Shewale, 1984); Dum Dum, Kalcutta; Purbasthali, Burdwan district, West Bengal (Sen and Gupta, 1987); Rice fields of Bardhaman and Nadia district, West Bengal (Sen and Gupta, 1998);Bhawbanipur, Kolkata,West Bengal (Sen,2006); Bidhannagar, North 24 Pgs., West Bengal (Dey and Chakraborty, 2020);

Growing in shallow stagnant muddy water near Kazipara, DumDum, Kolkata (Collection No. 7), on the moistdripping wall of a building in Purba Sinthee, Dum Dum, Kolkata (Collection No. 60); collected from a rice field at Purbasthali, Burdwan district(Collection No. BD-30); adhering to the sides of cement cisterns at Police Garden, Kalyani, Nadia district (Collection No. 176).

Voucher No. 31. (Collection No. 60).

***Oscillatoria salina*** Biswas, Flora of Salt Lakes, J. Dept. Sci., Kalcutta Univ., 8: 21, pl. 6, fig. 6a-d, 1926; Desikachary,T.V. Cyanophyta. 239, 1959.

Thallus deep blue-green, adpressed with the substratum, inter-mingled with other blue-green algae, trichomes scattered but parallely disposed in some cases; trichome straight, having same width throughout, ends tapering, 0.2-1.2 mm long; apical cell hyaline, elongated, tapering and bent or curved; cells about 1.5 to 2 times broader than long, 3.6-4.2μ broad, 1.8-2.4μ long, some of the intercalary cells shorter with dense contents, cross-walls inconspicuous, single granule on one side; cell contents granular, homogeneous. (Pl. 4, fig. 7).

Formerly reported from Benaras, U.P. (Rao, 1937); Calcutta, West Bengal (Banerji, 1938; Biswas, 1942); Lucknow, U.P. (Prasad snd Mehrotra, 1980); Marathwada, Maharashtra (Ashtekar and Kamat, 1980); River Cauvery, Karnataka (Somashekar, 1983); Nagpur, Maharashtra (Tarar and Shewale, 1984).

Growing on brick submerged in shallow water at Phulbagan Central Govt. Housing, Narkeldanga, Kalcutta (Collection No. 355).

Voucher No. 32. (Collection No. 355).

***Oscillatoria splendida*** Grev.ex Gom. Greville, Flora Edinensis, 305, 1824; Gomont, Monogr.Oscillariees, 224, pl. 7, figs. 7 and 8, 1892; Desikachary, T.V. Cyanophyta. 234, pl. 37, figs. 7 and 8,pl. 38, fig. 10 and pl. 40, fig. 11, 1959.

Thallus brownish blue-green; trichome almost straight but slightly bent towards the gradually attenuated apical part involving3-4 cells, not constricted at the cross-walls; cells almost twice as long as broad, 2.4-3.0μ broad, 3.6-4.8μ long, transverse walls granulated, end cell narrow, bent and terminated in a knob-like structure, capitate, calyptra absent; cell contents homogeneous. (Pl. 4, fig. 8).

This species has been reported from Palta, Jadavpur, Hooghly river near Kalcutta, West Bengal (Bruhl and Biswas, 1922; Banerji, 1938; Biswas, 1942); Bankura district, West Bengal (Sinha and Mukherjee, 1975); Bhagalpur, Bihar (Nasar and Datta Munshi, 1976); Marathwada, Maharashtra (Ashtekar and Kamat, 1980); River Cauvery, Karnataka (Somashekar, 1983); Ramgarh, Kalcutta, West Bengal (Sen and Gupta, 1987); Rice fields of Murshidabad and Nadia districts, West Bengal (Sen and Gupta, 1998); Bidhannagar, North 24 Pgs., Kalighat, Kolkata, West Bengal (Sen, 2006).West Bengal (Dey and Chakraborty, 2020);

Growing on submerged decaying plant at Ramgarh near Garia, Kolkata (Collection No. 112); from rice fields at Sikarpur, Simurali, Nadia district (Collection No. 198); Fakirtala near Nabadwip, Nadia district (Collection No. 326); on moist soil near a tube well at Kalyanpur near Baruipur, South 24 Pgs. district (Collection No. 372).

Voucher No. 33. (Collection No. 326).

***Oscillatoria animalis*** Ag.ex Gom. Agardh, Aufzahlung, Flora, 10 : 632, 1827; Gomont, Monogr. Oscillariees, 227, pl. 7, fig. 13, 1892; Desikachary, T.V. Cyanophyta.239, pl. 40, fig. 14, 1959.

Trichomes forming dark blue-green thallus, straight, gradually attenuating and slightly bent at the apices, attenuation involving 4-5 cells, not constricted at the cross-walls; cells 2.4- 3.0μ long and 3.0-3.6μ broad, not granulated at the cross walls, end cell acute, not capitate, calyptra absent, cell contents deep blue-green. (Pl. 4, fig. 9).

Formerly reported from Calcutta, West Bengal (Biswas, 1925,1942; Banerji, 1938); Benaras, U.P.; Puri, Orissa (Rao, 1937, 1938); Barielly, U.P. (Singh and Chaturvedi, 1970); Saharanpur district,U.P. (Pal and Yadav, 1974); Meerut, U.P. (Bendre and Kumar, 1975); Marathwada, Maharashtra (Ashtekar and Kamat, 1980); Bhubaneswar,Orissa (Mohanty, 1982); Nagpur, Maharashtra (Nandkar, 1983); River Cauvery, Karnataka (Somashekar, 1983); Marathwada, Maharashtra(Chaporkar and Gangawane, 1984); Bhagalpur, Bihar (Saha and Pandit, 1987); Dum Dum, Kalcutta; Dainhat, Katwa and Bishnupur, Burdwan district, West Bengal (Sen and Gupta, 1987); Gorabazar, Dum Dum, West Bengal (Sen 2006).

Growing in a pond at Purba Sinthee at Dum Dum, Kalcutta (Collection No. 64); on rice fields at Gopkhanji near Dainhat, Burdwan district (Collection No. BD-12); Khajuti near Katwa,Burdwan district (Collection No.BD-16).

Voucher No. 34. (Collection No. 64).

***Oscillatoria acuta*** Bruhl et Biswas, orth.mut. Geitler, Bruhl andBiswas, Indian Bark algae, J. Dept. Sci., Kalcutta Univ., 5: 3, pl. 1, figs. 6a and b, 1923; Geitler, Kryptogamenflora, 978, 1932; Desikachary, T.V. Cyanophyta. 240, pl.39, figs. 5 and 8, 1959.

Trichomes mostly aggregated in a bluish-green colony, not constricted at the cross-walls, straight but narrow and prominently bent towards the apices; apical cell not capitate, without calyptra; cells 4.8-6.0μ broad, 3.0-4.8μ long; cell contents bluish-green granular with some large granules. (Pl. 4, fig. 10).

This species has been reported from Calcutta,West Bengal(Bruhl and Biswas, 1923; Biswas, 1925); Saharanpur district, U.P. (Pal and Yadav, 1974); Meerut, U.P. (Bendre and Kumar, 1975); Sagar, M.P. (Mishra and Purohit, 1979); Lucknow, U.P. (Prasad and Mehrotra, 1980); Karnataka (Bongale and Bharati, 1980); River Cauvery, Karnataka (Somashekar, 1983); Khandesh, Maharashtra (Barhate and Tarar,1983); Nagpur, Maharashtra (Tarar and Shewale, 1984); Arambagh, Hooghly district; Lake Town, Calcutta, West Bengal (Gupta and Sen, 1987);Lketown,Kolkata , Wet Bengal(Sen,2006); Rice fields of Orissa (Dey *et al.*, 2010); Bidhannagar, North 24 Pgs., West Bengal (Sen, 2020); Jharkhali, South 24 Pgs. district, West Bengal (Barman *et al.*, 2015);

Gourhati near Arambagh, Hooghly district (Collection No. 76); from muddy water at Jessore Road near Lake Town, Kalcutta (Collection No. 47).

Voucher No. 35 (Collection No. 76).

***Oscillatoria irrigua*** (Kutz.)Gom.Monogr.Oscillariees, 218, pl. 6, figs. 22-23, 1892; Desikachary, T.V. Cyanophyta.224, pl. 42, figs. 7 and 9, 1959.

Thallus dirty blue-green; trichome straight, slightly attenuated towards the apex, not constricted at the cross-walls; cells about 1/2 to almost as long as broad, 7.2-8.4μ broad, 4.8-6.0μ long, septa prominent, not granulated, end cell flattened convex, sub-capitate with thickened outer wall; cell contents light blue, granular. (Pl. 4, fig. 11).

Formerly recorded from Saharanpur district, U.P. (Pal and Yadav, 1974); Meerut, U.P. (Bendre andKumar, 1975); Karnataka (Bongale and Bharati, 1930); River Cauvery,Karnataka (Somashekar, 1983); Khandesh, Maharashtra (Barhate and Tarar, 1983); Silabati, Ghatal, Midnapore district, West Bengal (Pal and Santra, 1985); Jalgaon region, Maharashtra (Bhoge and Ragothaman, 1986); North 24 Pgs., West Bengal (Naskar *et al.*, 2008); Bidhanagar, North 24 Pgs., West Bengal (Dey and Chakraborty, 2020).

Floating in pools and puddles along the railway track at Janai Road, Hooghly district (Collection No. 6); growing on rice field at Jatia near Halisahar, North 24-Parganas district (Collection No. 345).

Voucher No. 36 (Collection No. 345)

***Oscillatoria rubiscens*** DC.ex Gom. D.C., Notice sur la Matiere qui colora en rouge la lac de Morat au printemps de 1825, Mem.Soc. Phys. Sc. Hist. Nat. Geneve, 3(2) : 29, fig. 1825;Gomont, Monogr. Oscillariees, 204, pl. 6, figs. 6, 7, 1892; Desikachary, T.V. Cyanophyta.235, pl. 42, fig. 12, 1959.

Thallus dirty bluish-green, having more or less parallelly arranged trichomes forming bundle and attached by the basal part with the substratum; trichome straight, not constricted at the cross-walls; cells about 1/2-1/3 as long as broad, 6.6-7.0 μ broad, 4.0-4.8 μ long, septa prominent, granulated, end cell hemispherical, capitate with massive, translucent yellowish, convex calyptra; cell contents finely granular, yellowish green. (Pl. 4, fig. 12)

Formerly reported from Chunar, U.P. (Rao, 1937); and Polamur in Andhra Pradesh (Rao, 1938); Rohilkhand Division, U.P. (Singh and Chaturvedi, 1970); Saharanpur district, U.P. (Pal and Yadav,1974); Meerut, U.P. (Bendre and Kumar, 1975); Jammu (Anand, 1979);Lucknow, U.P. (Prasad and Mehrotra, 1980); River Cauvery, Karnataka(Somashekar, 1983); Khandesh, Maharashtra (Barhate and Tarar,1983);Meghalaya (Reddy etal., 1986).

Growing on muddy moist soil at Bantra, Howrah district (Collection No. 88).

Voucher No. 37. (Collection No. 88).

***Oscillatoria simplicissima*** Gom.Monogr.Oscillariees, 219, pl. 7,fig. 1, 1892; Desikachary, T.V. Cyanophyta.224, 1959.

Thallus blue-green, free floating in stagnant water; trichome straight, not constricted at the cross-walls, septa not granulated,breadth uniform throughout; cells about 1/3 as long as broad, 7.2-9.0μ broad, 2.4-3.6μ long, apical cell not capitate with a rounded apex and slightly thick membrane; cell contents finely granular,faint blue-green. (Pl. 4, fig. 13).

Formerly collected from Calcutta, West Bengal (Biswas, 1926);Saharanpur district, U.P. (Pal and Yadav, 1974); Meerut, U.P. (Bendre and Kumar, 1975); Nagpur, Maharashtra (Kamat and Harankhedkar, 1976);Patiala, Punjab (Sarma and Kanta, 1978); Jaipur, Rajasthan (Trivedy,1982); Midnapore, West Bengal (Pal and Santra, 1985); Kolkata, WestBengal (Sen and Gupta, 1987); Chetla, Kolkata, West Bengal (Sen, 2006).

Growing in stagnant water at Zoo garden, Alipore, Kalcutta (Collection No. 40); in a muddy pit with little water at Gocharan, South 24 Pgs. district (Collection No. 371).

Voucher No. 38 (Collection No. 40).

***Oscillatoria obscura*** Bruhl et Biswas, Algae of the Bengal Filter Beds, J. Dept. Sci., Kalcutta Univ., 4: 6, pl. 2, fig. 9, 1922; Desikachary, T.V. Cyanophyta. 207, 1959.

Thallus deep blue-green; trichome almost straight, tapering at the apical part, not constricted at the cross-walls; apical cell rounded assuming a conical shape; cells much shorter than broad, 3.6-4.2μ broad, 0.8 μ long; cross walls granulated. (Pl. 4, fig. 14).

This species has been formerly collected from Bengal (Bruhl and Biswas, 1922); Benaras, U.P. (Rao, 1937); Berhampur, Orissa (Rao, 1938); Travancore, Kerala (Parukutty, 1940); Ludhiana, Punjab (Grover and Pandhol, 1975, 1976); Bankura district, West Bengal (Sinha and Mukherjee, 1975); Meerut, U.P. (Bendre and Kumar, 1975); Lucknow, U.P. (Prasad and Mehrotra, 1980); Karnataka (Bongale and Bharati, 1980); Bhubaneswar, Orissa (Mohanty, 1982, 1984); Jammu and Kashmir (Goyal *et al.*, 1984); Pusa, Bihar (Jha *et al.*, 1986); Tamilnadu (Anand and Revathi, 1987); Kalcutta, West Bengal (Sen and Gupta, 1987); Hooghly district, West Bengal (Gupta and Sen, 1987); Garia, Kokata, West Bengal (Sen, 2006);Bokaro, Jharkhand (Sharma, 2010); Barasat, West Bengal (Talai-Mukhopadhyay and Naskar, 2015); Basanti, Herobhanga, South 24 Pgs. district, West Bengal (Barman et al., 2015); Bidhannagar, North 24 Pgs., West Bengal (Dey and Chakraborty, 2020).

From water filled ditch near Khanakul Post Office, Hooghly district (Collection No. 105); from rice fields at Chanditala, Dankuni, Hooghly district (Collection No. 125); from shallow puddle at Mominpur, Kalcutta(Collection No. 197); from rice fields at Khajuti near Dainhat, Burdwan district (Collection No. 204); Bhandartikuri, Purbasthali, Burdwan district (Collection No. 279); Bakulla near Guptipara, Hooghly district (Collection No. 286).

Voucher No. 39. (Collection No. 197).

***Oscillatoria brevis*** (Kutz.)Gom.Monogr.Oscillariees, 229, pl. 7, figs. 14 and 15, 1892; Desikachary, T.V. Cyanophyta.241, 1959.

Trichomes straight, briefly attenuated and slightly bent at the apices, attenuation involves 1-2 cells, not constricted at the cross-walls; cells 2.4-2.8μ long and 4.8-6.0μ broad, not granulated at the cross walls; end cell convex, not capitate, calyptra absent; cell contents blue-green finely granular. (Pl. 4, fig. 15).

This species has been reported from Kalcutta, West Bengal (Banerji, 1938); Madhubani, Bihar (Rao, 1939); New Delhi (Rao, 1940); Bareilly, U.P. (Singh and Chaturvedi, 1970); Bankura district, West Bengal (Sinha and Mukherjee, 1975); Marathawada, Maharashtra(Ashtekar and Kamat, 1980); Lucknow, U.P. (Prasad and Mehrotra, 1980); Allahabad, U.P. (Pandey and Pandey, 1982); Khandesh, Maharashtra (Barhate and Tarar, 1983); Calcutta, West Bengal (Sen and Gupta, 1987); Bidhannagar, North 24 Pgs., West Bengal (Sen and Gupta, 1987); Bidhannagar, North 24 Pgs., West Bengal (Dey and Chakraborty, 2020).

Growing in a pond near Bankra Housing Society at Birati, Dum Dum, Kolkata (Collection N0. 130); on paddy fields at Kestampur, Dankuni, Hooghly district (Collection No. 260).

Voucher No. 40. (Collection No. 260).

***Oscillatoria subbrevis*** Schmidle, Engler's Bot. Jahrb., 30; 243, pl. 4, fig. 7, 1901; Desikachary, T.V. Cyanophyta. 207, pl.37, fig. 2, pl. 40, fig. 1, 1959.

Trichome almost straight, having same width throughout, not constricted at the joints, ends not bent, apical cell rounded; cells shorter than the length, 4.8-6.0μ broad, 1.8-2.4μ long; cross-walls not granulated, calyptra absent; cell contents finely granular. (Pl. 4, fig. 16) .

Former records are from Punjab (Singh, 1941); Bombay, Maharashtra (Gonzalves and Joshi, 1943); Bareilly, U.P. (Singh and Chaturvedi,1970); Saharanpur district, U.P. (Pal and Yadav, 1974); Meerut, U.P.(Bendre and Kumar, 1975); Ludhiana, Punjab (Grover and Pandhol,1975);Patiala, Punjab (Sarma and Kanta, 1978); Karnataka (Bongale and Bharati, 1980); Rohilkhand Division, U.P. (Pandey, 1982); Nagpur,Maharashtra (Nandkar et al., 1983); Khandesh, Maharashtra (Barhate and Tarar, 1983); Nagpur, Maharashtra (Tarar and Shewale, 1984);Jammu and Kashmir (Goyal *et al.*, 1984); Vidarbha region, Maharashtra (Kolte and Goyal, 1985); 24-Parganas district, West Bengal (Maity and Santra, 1985); Midnapore district, West Bengal (Pal and Santra, 1985); Pusa, Bihar (Jha *et al.*, 1986); Tamilnadu (Anand and Revathi, 1987); Bhagalpur, Bihar (Saha and Pandit, 1987); Diamond Harbour, South 24 Parganas district, West Bengal (Santra *et al.*, 1988); Rice fields in Murshidabad distrct, West Bengal (Sen and Gupta, 1998); Baranagar, Kolkata, West Bengal (Sen, 2006); Baduria, North 24 Pgs, West Bengal (Naskar *et al.*, 2008); Ricefields in Orisa (Dey *et al.*, 2010); Amrabani, Basanti, South 24 Pgs. district, West Bengal (Barman, 2015); Bidhannagar, North 24 Pgs. district, West Bengal (Dey and Chakraborty, 2020).

Collected from rice field at Bhabta, Murshidabad district (Collection No. 365).

Voucher No. 41 (Collection No. 365).

***Oscillatoria prolifica*** (Grev.)Gom., Monogr. Oscillariees, 205, pl. 6, fig. 8, 1892; Desikachary, T.V. Cyanophyta.234, 1959.

Thallus having irregularly arranged trichomes forming bundles; trichomes almost straight, gradually attenuated at the apices, not constricted at the cross-walls, end cell capitate; cells as long as broad or slightly longer, 2.8-5.0μ broad, 3.0-5.4μ long; cross walls granulated, calyptra present. (Pl. 4, fig. 17).

Formerly reported from A.P. (Rao, 1938); Saharanpur district, U.P, (Pal and Yadav, 1974); Murshidabad district, West Bengal (Pal and Santra, 1984); South-Sinthee, Kolkata, West Bengal (Sen, 2006).

Growing in water filled pit at Nalpur near Sankrail, Howrah district (Sen, September, 1984; Collection No. 82); collected from soil of flower pot at Khagra, Berhampur, Murshidabad district (Collection No. 347).

Voucher No. 42. (Collection No. 82).

***Oscillatoria agardhii*** Gom.Monogr.Oscillariees, 205, 1892; Desikachary, T.V. Cyanophyta.235, 1959.

Trichomes forming entangled mass, slightly curved, gradually narrow towards the apex; not constricted at the cross-walls; septa granulated on one side, breadth of the trichome-uniform except the apical part, attenuation involve the apical 4 or 5 cells; cells 2.4-3.6μlong, 3.6-5.4μbroad, mostly broader than long; end a convex, with cell convex calyptra; cell content finely granular (Pl.4, fig. 18).

Formerly reported from Benaras, U.P. (Rao, 1936); Madras,Tamilnadu (Ganapathi, 1940); Calcutta, West Bengal (Biswas, 1942);Chamba, Punjab (Singh, 1941); Saharanpur district, U.P. (Pal and Yadav, 1974); River Cauvery, Karnataka (Somashekar, 1983); Pusa, Bihar (Jha *et al.*, 1986); Kalcutta, West Bengal (Sen and Gupta, 1987); S.S.K.M. Hospital campus, Kolkata, West Bengal (Sen, 2006).

Growing in stagnant water near Thakurpukar, Behala, Kalcutta (Collection No. 79); From a stagnant pond at Ramgarh near Garia, Kalcutta (Collection No. 350).

Voucher No. 43. (Collection No. 350).

***Oscillatoria amoena*** (Kutz.) Gom.Monogr. Oscillariees, Ann. Sci.nat. Bot. Ser. 7, 16: 91-264, pl. 7, fig. 9, 1892; Desikachary, T.V. Cyanophyta. 230, pl. 40, fig. 12, 1959.

Thallus blue-green, forming felty strata on muddy soil; trichomes straight, having very slight constrictions at the cross-wall, septa granulated and not very distinct, breadth of the trichome uniform except at the apical part, attenuation involves the apical 2 or 3 cells; cells slightly shorter than broad, rarely as long, 4.2-5.0μ long, 4.8-5.5μ broad; cell content finely granular; apical cell capitate with a round nearly straight apex with calyptra.(Pl. 4, fig. 19).

This species has been reported from Ghatchatti (Srivastava,1967); Manikaran, H.P. (Vasishta, 1968); Chirraiya Bagh (Prasad and Srivastava, 1968); Sardhana

(Kumar, 1970); Uttar Pradesh (Singh *et al.*, 1970); Saharanpur district, U.P. (Pal and Yadav, 1974);Bankura district, West Bengal (Sinha and Mukherjee, 1975); Kalcutta,West Bengal (Gupta, 1975); Meerut, U.P. (Bendre and Kumar, 1975);Bareilly, U.P. (Chaturvedi and Pandey, 1976); Patiala, Punjab (Sarma *et al.*, 1979); Sagar, M.P. (Mishra and Purohit, 1979); Lucknow, U.P. (Prasad and Mehrotra, 1980); Karnataka (Bongale and Bharati, 1980);Marathswada, Maharashtra (Ashtekar and Kamat, 1980); 24-Parganas and Howrah districts, West Bengal (Mukhopadhyay and Chatterjee, 1981);Rohilkhand division, U.P. (Pandey, 1982); River Cauvery, Karnataka(Somashekar, 1983); Khandesh, Maharashtra (Barhate and Tarar, 1983);Jammu and Kashmir (Goyal *et al.*, 1984); Haripal,Hooghly district and Kalcutta, West Bengal (Gupta and Sen, 1987); Rice fields of Bardhaman and Hooghly district, West Bengal (Sen and Gupta, 1998); Rice field at Dainhat, Bardhaman district, West Bengal (Sen, 2001); Paddy field near Dum Dum airport, Kolkata, West Bengal (Sen, 2006); Bidhannagar, North 24 Pgs., West Bengal (Dey & Chakraborty, 2020).

Collected fnom rice fields at Khajuti near Dainhat, Burdwan district (Collection No. 24); Osmanpur near Kalna, Burdwan district (Collection No. 49); from moist soil near Haripal station near Tarakeswar, Hooghly district (Collection No. 101); from the side of canal near Bidhan Nagar Road Railway Station, Kalcutta (Collection No. 113); Agradip near Purbasthali, Burdwan district (Collection No. 235); Bakulla near Guptipara,Hooghly district (Collection No. 293).

Voucher No. 44 (Collection No. 113).

## ARTHROSPIRA Stizenberger

*Spirulina* and *Arthrospira*, Hedwigia, 1: 32, 1852; Gomont, Monogr.Oscill., 246, 1892.

### Key to the Species

1. Trichomes Densely Spirally Coiled ........................................... 1. ***A. spirulinoides***

1. Trichomes Loosely Coiled ........................................................................ 2

2. Spirals 9.3-12.0μ Broad, 18.0-19.5μ Distant ...................................... 2. ***A.jenneri***

2. Spirals 24.0-30.0 μ Broad, 36.0-45.0μ Distant ................................ 3. ***A. platensis***

***Arthrospira spirulinoides*** Ghose, A. systematic and ecological account of a collection of blue-green algae from Lahore and Simla, J. Linn. Soc. Bot., 56: 338, pl. 31, fig. 4, 1923; Desikachary, T.V. Cyanophyta. 189, pl. 35, fig. 1, 1959.

Thallus greenish blue; trichomes flexuous, regularly and densely spirally coiled, 48.0-54.0μ long, breadth 4.8μ, septa prominent, not constricted at cross-walls; spirals very close to each other, diameter about 10.2μ(Pl. 3, fig. 2).

A collection from Jagatballavpur in Howrah district shows spirals about 1.0 μ distant from each other.

Formerly reported from 24-Parganas and Howrah districts, West Bengal (Mukhopadhyay and Chatterjee, 1981).

Collected from rice fields at Bhandartikuri near Purbasthali, Burdwan district (Collection No. 66), Andulberia, Murshidabad district (Collection No. 364).

Voucher No. 45. (Collection No. 364).

***Arthrospira jenneri*** Stizenb.ex Gom. Stizenberger, *Spirulina* and *Arthrospira*. Hedwigia, 1: 32, 1852; Gomont, Monogr.Oscillariees, 247, pl. 7, fig. 26, 1892; Desikachary, T.V. Cyanophyta.192, pl. 35, fig. 3, 1959.

Thallus blue-green; trichomes flexuous, more or less regularly and loosely coiled, at the ends slightly and gradually narrower than the other part, 6.0-6.6μbroad; septa distinct, slightly constricted at the cross-walls; spirals 9.3-12.0 μ broad, distances between the spirals 18.0-19.5μ; cells shorter than broad, 4.0-4.2μlong, septa granulated, end cells rounded; cell contents homogeneous, blue-green. (Pl. 3, fig. 3).

Formerly reported from Ludhiana, Punjab (Grover and Pandhol, 1975); Ludhiana, Punjab (Pandhol and Grover, 1976); Allahabad, U.P. (Pandey and Pandey, 1982); Pusa, Bihar (Jha *etal.*, 1986).

Growing in a ditch at Chiruti, Murshidabad district (Collection No. 361).

Voucher No. 46 (Collection No. 361).

***Arthrospira platensis*** (Nordst.)Gom.Monogr.Oscillariees, 247, pl. 7, fig. 27, 1892; Desikachary, T.V. Cyanophyta.190, pl. 35, fig. 2, 1959.

Thallus blue-green; trichomes flexuous, regularly and loosely coiled, 144.0μ or more long, 6.6-7.2μ broad, septa distinct, constricted at cross walls; spirals 24.0-30.0μbroad, 36.0-45.0μ distant; cells shorter than broad, 3.0-4.2μ long, septa granulated, end cells rounded; cell contents granular, homogeneous, blue-green. (Pl. 3, fig. 4).

Formerly reported from Calcutta (Biswas, 1927; 1942); Rohilkhand division, U.P. (Singh and Chaturvedi, 1970); Meerut, U.P. (Bendre and Kumar, 1975); Patiala, Punjab (Sarma and Kanta, 1978); River Cauvery, Karnataka (Somashekar, 1983); Jalgaon region, Maharashtra (Bhoge and Ragothaman, 1986); Bakkhali, South 24 Pgs. district, West Bengal (Barman *et* al., 2015).

Growing on a field not under cultivation at Kalipark near Dum Dum Airport (Collection No. 346).

Voucher No. 47 (Collection No. 346).

**SPIRULINA Turpin em. Gardner**

Turpin in Dic.Sci. nat. de Levrault, 50: 309, 1827; em.Gardner, New Pacific Coast Marine Algae I, Univ. Calif. Publ. Bot., 6: 379, 1917; Gomont Monogr.Oscillariees, 249, 1892.

**Key to the Species**

1. Spirals less than 3.6μ diam. ........................................................**1. *Sp. major***

1. Spirals more than 3.6μ diam. ............................................................... 2

2. Spirals close to each other, about 0.9μ distant ................................ **2. *Sp. subsalsa***

2. Spirals distant, 3.0-4.8μ from each other ................................ **3. *Sp. meneghiniana***

***Spirulina major*** Kutz.ex Gom. Kutzing, Phyc. gene., 183, 1843; Gomont, Monogr. Oscillariees, 251, pl. 7, fig. 29, 1892; Desikachary, T.V. Cyanophyta.196, pl. 36, fig. 13, 1959.

Thallus blue-green; trichomes entangled, flexuous, spirally coiled, breadth 1.2-1.8μ, septa obscure; spirals 3.0-3.6μ diameter and 3.0-4.8μ distant from each other (Pl. 3, fig. 5).

Formerly recorded from Loktak Lake, Manipur (Bruhl and Biswas,1926); Cherrapunji, Meghalaya (Biswas, 1934); Assam (Parukutty, 1939); Calcutta, West Bengal (Biswas, 1926; Banerji, 1938); Bombay, Maharashtra (Dixit, 1936; Gonzalves and Joshi, 1946); Benaras, U.P.; A.P. (Rao, 1937, 1938); Allahabad, U.P. (Gupta, 1956); Travancore, Kerala (Parukutty, 1940); Marathawada, Maharashtra (Kamat, 1974); Saharanpur district, U.P. (Pal and Yadav, 1974); Bankura district, West Bengal (Sinha and Mukherjee, 1975); Meerut, U.P. (Bendre and Kumar, 1975); Nagpur, Maharashtra (Nandkar *et al.*,1983); River Cauvery, Karnataka (Somashekar, 1983); Midnapore district,West Bengal (Pal and Santra, 1985); Jalgaon region, Maharashtra (Bhoge and Ragothaman, 1986); Pawagiri Oon, M.P. (Mahajan, 1987); Bhagalpur, Bihar (Saha and Pandit, 1987); Haldia, Midnapore district and Sagar Island, South 24-Parganas, West Bengal (Santra *et al.*, 1988); North 24 Pgs. district, West Bengal (Naskar *et al.*, 2008).

Collected from paddy field soil at Burikhali near Bauria, Howrah district (Collection No. 81).

Voucher No. 48 (Collection No. 81).

***Spirulina subsalsa*** Oerst.ex Gom. Oersted., Beretn. Exc. tilTrindelen etc., Nat. Tidskr., 17, pl. 7, fig. 4, 1842; Gornont, Momgr. Oscillariees, 253, pl. 7, fig. 32, 1892; Desikachary, T.V. Cyanophyta.193, pl. 36, figs. 3, 9, 1959.

Thallus greenish blue, trichomes entangled and mixed with fine trichomes of ***Oscillatoria salina***, flexuous, more or less regularly and densely spirally coiled, breadth 1.8-2.0μ, septa of cells obscure; spirals close to each other, diameter 3.6-4.2μ and about 0.9μ distant from each other (Pl. 3, fig. 6).

Formerly recorded from Calcutta, West Bengal (Turner, 1892; Banerji, 1938); A.P. (Rao, 1938); Orissa (Rao, 1938); Assam (Parukutty, 1939); Delhi (Rao, 1940); Bombay, Maharashtra (Gonzalves,1947; Dixit, 1936); Saharanpur district, U.P. (Pal and Yadav, 1974); Ludhiana, Punjab (Pandhol and Grover, 1976); Balmikinagar, Bihar (Sinha and Srivastava, 1980); Karnataka (Bongale and Bharati, 1980); Pusa, Bihar (Jha *et al.*, 1986); Assam (Reddy *et al.*, 1986); Bhagalpur, Bihar (Saha and Pandit, 1987); Tarakeswar, Hooghly district,West Bengal (Gupta and Sen, 1987); Jharkhali, Zambudweep, South 24 Pgs. district, West Bengal (Barman *et* al., 2015).

Growing on paddy field at Purbasthali, Burdwan district (Collection No. 42); from paddy fields at Guptipara, Hooghly district (Collection No. 259); from a wet muddy place at Mahanandapur, Midnapore district (Collection No. 379).

Voucher No. 49. (Collection No. 42).

***Spirulina meneghiniana*** Zanard.ex Gom. Zanardini, Not. intorno alleCellul. marine ed elittor. di Venezia, Atti. R. Inst. Veneto,6: 80, 1847; Gomont, Monogr. Oscillariees, 250, pl. 7, fig. 28, 1892; Desikachary, T.V. Cyanophyta.195, pl. 36, fig. 8,1959.

Thallus more or less thick, deep blue-green, trichomes entangled, flexuous, loosely and somewhat irregularly spirally coiled,1.2-1.8μ broad, spirals distant, diameter of spirals 3.6-4.8μ distant; septa of cells inconspicuous; cell contents finely granular,homogeneous, blue-green. (Pl. 3, fig. 7).

Formerly reported from Calcutta, West Bengal (Banerji, 1938);Bankura district, West Bengal (Sinha and Mukherjee, 1975); Balmikinagar, Bihar (Sinha and Srivastava, 1980); Allahabad, U.P. (Pandey and Pandey, 1982); River Cauvery, Karnataka (Somashekar, 1983);Nagpur, Maharashtra (Nandakar *et al.*, 1983); Tamilnadu (Anand and Revathi, 1987); North 24 Pgs. district, West Bengal (Naskar *et al.*, 2008); Gangasagar, Zambudweep, South 24 Pgs. district, West Bengal (Barman *et al.*, 2015).

Collected from rice fields at Palpara near Simurali, Nadia district (Collection No. 175), Amghata nearNabadwip, Nadia district (Collection No. 336).

Voucher No. 50 (Collection No. 175).

## LYNGBYA Ag.

Syst. Alg., 25, 1824: Gomont, Monogr, Oscillariees, 118, 1892.

### Key to the Species

1. Trichomes unconstricted or slightly so .................... 2

1. Trichomes constricted .................... 3

2. Sheath prominently stratified .................... 4

2. Sheath not stratified or feebly so .................... 5

3. Sheath 5.0-5.7μ thick; cells nearly quadrate or slightly shorter, septa not granulated.................... ***1. L. baculum***

3. Sheath 2.5μ thick; cells about 1/2 as long as broad, septa granulated.................... ***2. L. putealis***

4. Sheath 10.0 μ thick; filaments curved or bent, cross-walls not granulated; breadth of filaments 34.8-37.2 μ .................... ***3. L. majuscula***

4. Sheath up to 3.0μthick; filaments straight; cross-walls granulated; breadth of filaments 14.4-15.0μ.................... ***4. L. major***

5. Cells shorter than broad .................... 6

5. Cells more or less quadrate or longer than broad ........................................ 7

6. Filaments spirally coiled for major part........................................... **5. *L. spiralis***

6. Filaments not spirally coiled ........................................................................ 8

7. Sheath pink; cross-walls not granulated; cells more than 3.0 μ broad ................... ........................................................................................................ **6. *L. rubida***

7. Sheath hyaline; cross-walls granulated; cells less than 3.0μ broad ............... 9

8. Sheath more than 2.0μ broad ............................................... **7. *L. confervoides***

8. Sheath up to 2.0μ broad ............................................................................... 10

9. Filaments not straight; septa with single granule on each side; cells 1.5-1.8μ broad, 1.5μ long ........................................................................ **8. *L. lagerheimii***

9. Filaments tortuous and straight towards ends; cells 2.4-3.0μ broad, 2.5-2.7μ long .......... ............................................................................... **9. *L. digueti***

10. Septa not granulated ........................................................................... 11

10. Septa granulated ................................................................................ 12

11. Cells much broader than long, about 1/5 as long as broad, 18.0-23.0μ broad ................ ............................................................................ **10. *L. birgei***

11. Cells about 1/2 as long as broad, up to 12.6μ broad .................................. 13

12. Sheath pale yellow, feebly stratified, coloured violet by chlor-zinc-iodide; calyptra present ........................................................................ **11. *L. stagnina***

12. Sheath hyaline, not stratified, not coloured violet by chlor-zinc-iodide; calyptra absent.. ........................................................................................................14

13. End cells without calyptra ................................................ **12. *L. ceylanica***

13. End cells with slightly thickened outer membrane .................................... 15

14. Sheath upto 0.75μ thick; cells 10.4-12.2μ broad, 2.6-4.8μ long; end cell broadly rounded ..............................................................**13. *L. hieronymusii***

14. Sheath upto 2.0μ thick; cells 6.0-7.2μ broad, 1.8-3.0μ long; end cell rounded............. ........................................................ **14. *L. martensiana***

15. Trichomes not constricted at the crosswalls; cells 6.0-7.2μ broad, 2.4-3.6μ long ......... ................................................................. **15. *L. aerugineocoerulea***

15. Trichomes constricted at the cross-walls; cells 7.2-8.4μ broad, 3.6-4.2μ long .............. ....................................................................................**16. *L. gracilis***

***Lyngbya baculum*** Gom.Essai class.Nost., J. Bot., 4: 354, 1890;Monogr. Oscillariees, 123, pl. 2, fig. 17, 1892; Desikachary,T.V. Cyanophyta.285, pl. 49, fig. 5 and pl. 52, fig. 3, 1959.

Thallus dense, pinkish blue-green, filaments not attached by base; sheath uniform, moderately thick, generally not stratified, sometimes inconspicuously stratified, 5.0-5.75μ thick, stained violet with chlor-zinc-iodide; trichome uniformly broad, straight or slightly bent, constricted at the cross-walls, cells nearly quadrate or slightly shorter, 7.4-8.6μ broad, 6.2-9.8μ (mostly 6.2-7.4μ) long, septa prominent, not granulated, end cell swollen and enlarged, without calyptra; cell contents finely granular, homogeneous, blue-green (Pl. 5, fig. 10).

Formerly recorded from Pamban, South Indian Coast (Iyengar and Desikachary, 1944); Khandesh, Maharashtra (Barhate and Tarar, 1983).

Growing on exposed damp ground at Belghoria, Kolkata (Collection No. 344), wet soil under the bridge at Bager-Khal near Kanchrapara, North 24-Parganas district(Collection No. 357).

Voucher No. 51 (Collection No. 344).

***Lyngbya putealis*** Mont. ex Gom. Montagne, 20 centurie de Plantes cellulaires exotiques, Ann. Sci. nat., 2 ser., Bot., 13: 200, 1840; Gomont, Monogr. Oscillariees, 143, pl. 3, fig. 14, 1892; Desikachary, T.V. Cyanophyta.317, pl. 52, fig. 12, 1959.

Thallus compact, formed of closely tufted filaments attached with the substratum and free, more or less parallely disposed above, pale blue-green, 10.8-12.0 μ broad; sheath more or less uniform, projecting beyond the trichome, thin, not stratified, hyaline, about 2.5μ thick, stained violet with chlor-zinc-iodide; trichome uniformly broad, straight or slightly and irregularly curved, constricted at the cross-walls; cells about 1/2 as long as broad, 6.0- 12.0μ broad, 3.6-5.0μ long, septa prominent, cross-walls granulated, end cell rounded, without calyptra; cell contents granular, dull blue-green (Pl. 4, fig. 11).

Formerly recorded from Benaras, Uttar Pradesh (Rao, 1937); Berhampur, Orissa (Rao, 1938); Bombay, Maharashtra (Gonzalves, 1947); Saharanpur district, Uttar Pradesh (Pal and Yadav, 1974); Meerut, Uttar Pradesh (Bendre and Kumar, 1975); Sagar, Madhya Pradesh (Mishra and Purohit, 1979); Lucknow, U.P. (Prasad and Mehrotra, 1980); Karnataka (Bongale and Bharati, 1980); Bhubaneswar, Orissa (Mahanty, 1982); Bhagalpur, Bihar (Saha and Pandit, 1987); Behala, Kolkata, West Bengal (Sen 2006).

Growing on moist soil at Chakdaha, Nadia district(Collection No. 141), constantly moist wall of an old building at Sakher-bazar, Behala, Kalcutta (Collection No. 360).

Voucher No. 52 (Collection No. 141).

***Lyngbya majuscula*** Harvey ex Gom. Harvey in Hooker, English Flora,5 (part 1): 370, 1833; Gomont, Monogr. Oscillariees, 151,pl. 3, figs. 3, 4, 1892; Desikachary, T.V. Cyanophyta.313, pl. 48, fig. 7; pl. 49, fig. 12 and pl. 52, fig. 10, 1959.

Thallus compact, well developed, formed by entangled mass of curved, bent and somewhat coiled filaments; sheath up to 10μ thick, hyaline, stratified with rough outer surface, not stained violet by chlor-zinc-iodide; trichome not attenuated at apex, not constricted at the cross-walls; cells about 1/6 as long as broad, or shorter, 34.8-37.2μ broad, 4.8-7.2μ long, septa prominent, not granulated,end cell hemispherical, not capitate, calyptra absent; cell contents finely granular, brownish blue-green (Pl. 5, fig. 12).

The alga described here corresponds well with the diagnosis of its variety var. ***chakiaense*** (Rao, 1936). But variations in the thickness of sheath and in the measurement of cells of trichomes are so frequent among the thalli growing in neighbourhood and in distant localities that it seems undesirable to call it a variety. Study on the collection (Voucher No. 53) records the thickness of sheath about 5.0-6.0μ, cells 22.8μ broad, 3.6μ long, sometimes up to 5.4μ. In some collections in the vicinity, the thickness of sheath is measured up to 9.0μ.

Formerly reported from Uttar Pradesh (Rao, 1936); Bihar (Rao, 1939); Rohilkhand division, Uttar Pradesh (Singh and Chaturvedi, 1970); Saharanpur district, Uttar Pradesh (Pal and Yadav, 1974); Meerut, Uttar Pradesh (Bendre and Kumar, 1975); Lucknow, Uttar Pradesh (Prasad and Mehrotra, 1980); Allahabad, Uttar Pradesh (Pandey and Pandey, 1982); Karnataka (Somashekar, 1983); Nagpur, Maharashtra (Tarar and Shewale, 1984); Jalgaon region, Maharashtra (Bhoge and Ragothaman, 1986); Bhandarhati, Hooghly district, West Bengal (Gupta and Sen, 1987); Diamondharbour, South 24-Parganas district, WestBengal (Santra *et al.*, 1988); Sarberia, South 24Pgs. district, West Bengal (Barman, 2015).

Growing as free floating masses at Sibpur, Howrah district (Collection No. 74);on the side of a pond at Kansari Bazar near Berhampur, Murshidabad district (Collection No. 363).

Voucher No. 53 (Collection No. 74).

***Lungbya major*** Menegh.ex Gom. Meneghini, Conspectus Algologiae euganeae, 12, 1837; Gomont, Monogr. Oscillariees, 144, pl. 3, fig. 15, 1892; Desikachary, T.V. Cyanophyta,320, pl. 52,fig. 11, 1959.

Thallus formed by straight, long filaments forming bundles, pale blue-green, filaments not attached, 20.4-21.6μ broad; sheath uniform, thick, smooth, stratified, upto about 3.0μ thick, not coloured violet with chlor-zinc-iodide; trichome not attenuated at the apex, slightly constricted at the cross-walls; cells about 1/5-1/6 as long as broad, 14.4-15.0μ broad, 2.4-3.0μ long, septa prominent, cross-walls granulated, end cell rounded with slightly-thickened outer wall; cell contents finely granular, pale blue-green (Pl. 6, fig. 1).

Formerly recorded from Benaras, Uttar Pradesh (Rao, 1937);Andhra Pradesh (Rao, 1938); Calcutta, West Bengal (Banerji, 1938);Saharanpur district, Uttar Pradesh (Pal and Yadav, 1974); Meerut,Uttar Pradesh (Bendre and Kumar, 1975); Bankura district, West Bengal (Sinha and Mukherjee, 1975); Sagar, Madhya Pradesh (Mishra and Purohit, 1979); Lucknow, Uttar Pradesh (Prasad and Mehrotra, 1980); 24-Parganas and Howrah districts, West Bengal (Mukhapadhyay and Chatterjee, 1981); Bhubaneswar, Orissa (Mohanty, 1982; 1984); Karnataka (Somashekar,1983).; Pusa, Bihar (Jha *et al.*, 1986); North 24 Pgs. district, West Bengal (Naskar *et al.*, 2008); Varanasi, Uttar Pradesh (singh *et al.*, 2015).

Collected from rice fields at Tribeni near Mogra, Hooghly district (Collection No. 108); Paglachandi near Debagram, Nadia district (Collection No. 233).

Voucher No. 54 (Collection No. 233).

***Lyngbya spiralis*** Geitler, Kryptogamenflora, 1042, fig. 659, 1932;Desikachary, T.V. Cyanophyta.289, pl. 48, fig. 1, 1959.

Thallus thin, dense and compact, dirty blue-green, formed by entangled mass of spirally coiled filaments, spirals present for the most part of the filament; sheath smooth, moderately thick, hyaline, not stratified, not stained violet by chlor-zinc-iodide; trichomes not attenuated at apex, not constricted at the cross-walls; cells about 1/2-1/4 as long as broad, 4.8-5.4μ broad, 1.8-2.4μ long, septa prominent, not granulated, end cell rounded, not capitate, calyptra absent; cell contents granular, dark blue-green (Pl. 6, fig. 2).

Formerly reported from Bankura district, West Bengal (Sinha andMukherjee, 1975); Karnataka (Bongale and Bharati, 1980); Marathwada,Maharashtra (Ashtekar and Kamat, 1980); 24-Parganas and Howrah districts, West Bengal (Mukhapadhyay and Chatterjee, 1981); Nagpur,Maharashtra (Tarar and Shewale, 1984); Pusa, Bihar (Jha *et al.*, 1986); Barasat, West Bengal (Talai-Mukhopadhyay and Naskar, 2015).

Growing on muddy place near a tube-well at Moradighee, Murshidabad district (Collection No. 360).

Voucher No. 55 (Collection No. 360).

***Lyngbya rubida*** Fremy, Myxo.d'Afr.equat. franc., 187, fig. 155,1929; Desikachary, T.V. Cyanophyta. 298, pl. 53, fig. 10, 1959.

Thallus expanded, caespitose, pinkish blue-green to pale brown, formed by loosely entangled long filaments; filaments 6.5-7.2μ broad, straight; sheath slightly above 1.0 μ thick, not stratified, pink, not coloured violet with chlor-zinc-iodide; trichome not attenuated at the apex, not constricted at the cross-walls; cells about 1½ as long as broad, 4.2-4.8μ broad, 6.6-7.2μ long, septa not granulated, end cell rounded, calyptra absent; cell contents finely granular (Pl. 6, fig. 3).

Formerly recorded from Allahabad, Uttar Pradesh (Mitra, 1951).Some forms have been reported from Benaras and Berhampur (Rao, 1937, 1938); Rohilkhand division, Uttar Pradesh (Singh and Chaturvedi, 1970); Karnataka (Bongale and Bharati, 1980); Bhubaneswar, Orissa (Mohanty,1982); Khandesh, Maharashtra (Barhate and Tarar, 1983); Bakkhali,South 24-Parganas district, West Bengal (Santra *et al.*, 1988); Dakshineswar, Kolkata, West Bengal (Sen, 2006).Basanti, Bhagabatpur, South 24 Pgs., West Bengal (Barman, 2015);

Collected from rice fields at Ramchandrapur, Howrah district (Collection No. 77); Bhabta, Murshidabad dist.Collection No. 365).

Voucher No. 56 (Collection No. 77).

***Lyngbya confervoides*** C. Ag.ex Gom. C. Agardh, Syst. Alg., 73, 1824;Gomont, Monogr. Oscillariees, 156, pl. 3, figs. 5, 6, 1892;Desikachary, T.V. Cyanophyta.314, pl. 49, fig. 9 & pl. 52, fig. 13, 1959.

Thallus thick, compact, having irregularly arranged filaments forming bundles, yellowish green; filaments long, flexible but straight; sheath distinct, colourless, sometimes stratified, rough on the outer side, 4.2-6.0μ thick, not coloured violet by chlor-zinc-iodide; trichome not attenuated at the apex, not constricted at the cross-walls; cells nearly 1/4 as long as broad, 12.0-15.6μ broad, 3.0μ long, septa granulated, end cell rounded, calyptra absent; cell contents finely granular, homogeneous, olive-green (Pl. 6, fig. 4).

Formerly recorded from Bombay, Maharashtra (Boergesen, 1935;Dixit, 1936); Krusudai island, South Indian coasts (Iyengar and Desikachary, 1944); Chilka, Orissa (Biswas, 1932); Benaras, Uttar Pradesh (Rao, 1937); Cuttack, Orissa (Rao, 1938); Polamur, Namdur and Chittavaram, Andhra Pradesh (Rao, 1938); Khajiar, Punjab (Gupta, 1950); Karnataka (Bongale and Bharati, 1980); Bhubaneswar, Orissa (Mohanty, 1982); River Cauvery, Karnataka (Somashekar, 1983);Bakkhaii, 24-Parganas district, West Bengal (Maity and Santra, 1985); Haldia and Bakkhali, West Bengal (Santra *et al.*, 1988); North 24 Pgs. district, West Bnegal (Naskar et al., 2008); Basanti, Bakkhali, South 24 Pgs. district, West Bengal (Barman, 2015).

Growing in shallow stagnant water in a field at Ghola near Sodepur, North 24-Parganas district (Collection No. 352).

Voucher No. 57 (Collection No. 352).

***Lyngbya lagerheimii*** (Mob.)Gom. Essai Class. Homo., J. de. Bot., 4:354, 1890; Monogr.Oscillariees, 147, pl. 4, figs. 6, 7, 1892;Desikachary, T.V. Cyanophyta.290, pl. 48, fig. 6 and pl. 53.fig. 2, 1959.

Thallus formed by entangled mass of long filaments, spirally coiled, coiling frequently irregular assuming serpentine, occasionally straight, not attenuated at the apex, pale blue-green, 2.0-2.2μ broad; sheath uniform, thin, smooth, not stratified; trichome not constricted at the cross-walls; cells more or less quadrate,1.5-1.8μ broad, 1.5μ long, septa prominent, cross-walls with singlegranule at each side, end cell rounded;cell contents homogeneouspale blue-green(Pl.6, fig.5).

Formerly recorded from Delhi(Rao,1940);Rohilkhand division,U P (Singh and Chaturvedi, 1970);Bankura district,West Bengal (Sinha and Mukherjee,1975); Lucknow, UP (Prasad and Mehotra,1980); Marathwada, Maharashtra (Ashtekar and Kamat,1980); Karnataka (Bongale and Bharati, 1980); Rohilkhand division,U P (Pandey, 1982); River Cauvery, Karnataka (Somashekar,1983); Jalgaon region, Maharashtra (Bhoge and Ragothaman, 1986); Varanasi, U P(Singh et al., 2015); Bhawanipur area, Kolkata, West Bengal (Šen, 2006).

Growing on paddy fields at Senpur near Mogra, Hooghly district (Collection No. 87), Siakhala near Dankuni, Hooghly district (Collection No. 117), Jatia near Halisahar, North 24-Parganas district (Collection No. 345).

Voucher No. 58 (Collection No. 117).

***Lyngbya digueti*** Gom. in Hariot, Algues du Golfe de Californie recuilles, J. de Bot., 9: 169, 1895; Desikachary, T.V. Cyanophyta. 310, pl. 53, fig. 8, 1959.

Thallus formed by entangled mass of tortuous filaments, straight towards ends, attached at the base; filaments long; sheath up to 1.8 μ thick, entire, hyaline, thin, non-stratified, coloured violet by chlor-zinc-iodide; trichome not attenuated at apex, not constricted at the cross-walls; cells slightly shorter than broad, 2.4-3.0μ broad, 2.5-2.7μ long, septa prominent, cross-walls not granulated, end cell rounded, not capitate, calyptra absent; cell contents greenish blue, homogeneous (Pl. 6, fig. 6).

Formerly reported from Benaras, Uttar Pradesh (Rao, 1937); West Bengal (Banerji, 1938); Saharanpur district, Uttar Pradesh (Pal and Yadav, 1974); Meerut, Uttar Pradesh (Bendre and Kumar, 1975); Patiala, Punjab (Sarma *et al.*, 1979); Karnataka (Bongale and Bharati,1980); River Cauvery, Karnataka (Somashekar, 1983); Khandesh, Maharashtra (Barhate and Tarar, 1983); Gondalpara, Hooghly district, West Bengal (Gupta and Sen, 1987).

Growing on exposed damp ground at Gondalpara, Hooghly district (Collection No. 168), moist cemented place at R.G.Kar Medical College, Kolkata (Collection No. 358).

Voucher No. 59 (Collection No. 358).

***Lyngbya birgei*** Smith, G.M.New or interesting Algae from the Lakes of Wisconsin, Bull.Torrey bot. 61; 43: 482, pl. 26, fig. 28,1916; Desikachary, T.V. Cyanophyta.296, pl. 50, figs. 7, 8,1959.

Thallus greenish blue, free floating; filament straight or curved; sheath very thin, uniform, smooth, not stratified, about 0.5μ in thickness, hyaline; trichome not attenuated at the apex, without any constriction at the cross-walls; cells much broader than long, about 1/5 as long as broad, 18.0-23.0 μ broad, 4.2-4.8μ long, cross-walls not granulated, end cells more or less flat, not capitate; cell contents granular, dense near the centre, dark greenish blue (Pl. 6, fig. 7).

Formerly recorded from Alleppey and Travancore, Kerala (Parukutty, 1940); Bombay, Maharashtra (Gonzalves and Joshi, 1946); Allahabad, Uttar Pradesh (Gupta, 1956); Ludhiana, Punjab (Pandhol and Grover, 1976); Lucknow, Uttar Pradesh (Prasad and Mehrotra, 1980);Karnataka (Bongale and Bharati, 1980); Nprth 24 Pgs. district, West Bengal (Naskar *et al.*, 2008); Basanti, Bakkhali, South 24 Pgs. district, West Bengal (Barman, 2015).

Growing on moist muddy place at Jangalpara near Dankuni, Hooghly district (Collection No. 5).

Voucher No. 60 (Collection No. 5).

***Lyngbya stagnina*** Kutzing, Species Algarum, 281, 1849; Tab. Phycologicae, 1: 48, pl. 87, fig. 5, 1849; Desikachary, T.V. Cyanophyta. 317, 1959.

Thallus dull blue-green, formed by loosely interwoven flexuous filaments, 10.2-11.4μ broad; sheath pale yellow, thin, feebly stratified, about 2.0 μ thick, coloured violet with chlor-zinc-iodide; trichome 8.4-9.6μ broad, not attenuated, not constricted at the cross-walls; septa prominent, cells very short, less than 1/4 as long as broad, 2.0-2.4μ long, cross-walls granulated, end cell broadly rounded, capitate and calyptra present; cell contents homogeneous, granular, greenish blue (Pl. 6, fig. 8).

This specimen corresponds in measurement with the narrower trichome recorded by Rao (1937). Formerly reported from Benaras, Uttar Pradesh (Rao, 1937); Meerut, Uttar Pradesh (Bendre and Kumar, 1975); Bankura district, West Bengal (Saha and Mulherjee, 1975); Belghoria, Kolkata, West Bengal (Sen, 2006).

Collected from rice fields at Swarupganj near Nabadwip, Nadia district (Collection No. 329), Chiruti, Murshidabad district (Collection No. 348).

Voucher No. 61 (Collection No. 329).

***Lyngbya ceylanica*** Wille, Denkschr.Akad.Wiss. Wien. Math. nat. Kl.,91 : 161, pl. 3, figs. 27-29, 1914; Desikachary, T.V. Cyanophyta.299, pl. 54, fig. 4, 1959.

Thallus dense, compact, reddish brown, filaments anastomosing;sheath uniform, thin, smooth, not stratified, about 1.8μ in thickness, reddish, not stained violet with chlor-zinc-iodide; trichom not attenuated at the apex, not constricted at the cross-walls but insome very little constrictions present; cells about 1/2 as long as broad, 8.4-12.6μ broad, 4.5-6.6μ long, septa prominent, cross-walls not granulated, end cell rounded, without calyptra; cell contents homogeneous, finely granular (Pl. 6, fig. 9).

Formerly recorded from Cherrapunji, Assam (Biswas, 1934);Benaras, Uttar Pradesh (Rao, 1936); Saharanpur district, Uttar Pradesh (Pal and Yadav, 1974); Nagpur, Maharashtra (Kamat and Haran- khedkar, 1976); Karnataka (Bongale and Bharati, 1980); Lucknow,Uttar Pradesh (Prasad and Mehrotra, 1980); Allahabad, Uttar Pradesh (Pandey and Pandey, 1982); Bankura district, West Bengal (Sinha and Mukherjee, 1984); Diamond harbour, 24-Parganas district, West Bengal(Maity and Santra, 1985); South 24-Parganas district, West Bengal (Santra *et al.*, 1988).

Collected from paddy field at Sikarpur, Simurali, Nadia district (Collection No. 200); from moist soil at Hotar, South 24-Parganas district (Collection No. 374).

Voucher No. 62 (Collection No. 200).

***Lyngbya hieronymusii*** Lemm. Forschbers., Plon, 12: 146, pl. 4, figs.12, 13, 1905; Desikachary, T.V. Cyanophyta. 297, pl. 48, fig. 4, 1959.

Thallus formed by entangled mass of free-floating filaments, filaments straight, 12.0-13.8μ broad; sheath distinct, colourless, not stratified, homogeneous, up to 0.75μ thick, not stained violet by chlor-zinc-iodide; trichomes of uniform breadth, not narrow at the apex, not constricted at the joints; cells about 1/2.5-1/4 as long as broad, 10.4-12.2μ broad, 2.6-4.8μ long, septa granulated, end cell broadly rounded, calyptra absent; cell contents granular, blue-green (Pl. 6, fig. 10).

Formerly reported from Allahabad, Uttar Pradesh (Mitra, 1951; Gupta, 1956); Meerut, Uttar Pradesh (Bendre and Kumar, 1975); Rohilkhand division, Uttar Pradesh (Chaturvedi and Pandey, 1976); Nagpur, Maharashtra (Kamat and Harankhedkar, 1976); Ludhiana, Punjab (Pandhol and Grover, 1976); Patiala, Punjab (Sarma and Kanta, 1978); Karnataka (Bongale and Bharati, 1980); 24-Parganas and

Howrah districts, West Bengal (Mukhopadhyay and Chatterjee, 1981); Allahabad, Uttar Pradesh (Pandey and Pandey, 1982); Baramulla, Jammu and Kashmir (Goyal *et al.*, 1984); Haldia, Midnapore district, West Bengal (Paland Santra, 1985); Jalgaon region, Maharashtra (Bhoge and Ragothaman, 1986); Sagar Island, South 24-Parganas district, West Bengal (Santra *et al.*, 1988); North 24 Pgs. district, West Bengal (Naskar *et al.*, 2008).

Growing on paddy fields at Osmanpur near Kalna, Burdwan district (Collection No. 8), Sikarpur, Simurali, Nadia district (Collection No. 242), Kalachara near Dankuni, Hooghly district (Collection No. 305).

Voucher No. 63 (Collection No. 242).

***Lyngbya martensiana*** Menegh.ex Gom. Meneghini, Conspectus Algologiae euganeae, 12, 1837; Gomont, Monogr.Oscillariees, 145, fig. 17, pl. 3, 1892; Desikachary, T.V. Cyanophyta.318, pl. 52, fig. 6, 1959.

Thallus compact, pinkish blue-green, filaments long, flexible; sheath distinct, colourless, not stratified, rough on the outer side, upto 2μ thick, not stained violet by chlor-zinc-iodide; trichome not attenuated at the apex, not constricted at the cross-walls; cells about 1/2-1/4 as long as broad, 6.0-7.2 μ broad, 1.8-3.0μ long,septa granulated, end cell rounded, calyptra absent; cell contents finely granular, light blue-green (Pl. 6, fig. 11).

Formerly recorded from Manipur, Assam (Biswas, 1930); Travan- core, Kerala (Parukutty, 1940); Rohilkhand division, Uttar Pradesh (Singh and Chaturvedi, 1970); Saharanpur district, Uttar Pradesh (Pal and Yadav, 1974) ; Meerut, Uttar Pradesh (Bendre and Kumar,1975); Patiala, Punjab (Sarma and Kanta, 1978); Karnataka (Bongale and Bharati, 1980); Lucknow, Uttar Pradesh (Prasad and Mehrotra,1980) ; Marathwada, Maharashtra (Ashtekar and Kamat, 1980); Allahabad,Uttar Pradesh (Pandey and Pandey, 1982); River Cauvery, Kamataka(Somashekar, 1983); Nagpur, Maharashtra (Nandkar *et al.*, 1983); Haldia,Midnapore district, West Bengal (Pal and Santra, 1985); Diamondharbour, South 24-Parganas district, West Bengal (Maity and Santra,1985); Jalgaon region, Maharashtra (Bhoge and Ragothaman, 1986); Pusa,Bihar (Jha *et al.*, 1986); Meghalaya and Tripura (Reddy *et al.*, 1986);Kalcutta, West Bengal (Santra, 1987); Diamond-Harbour, South 24-Parganas district, West Bengal (Santra *et al.*, 1988); North 24 Pgs., West Bengal (Naskar *et al*, 2008); Basanti, Lothian Island, South 24 Pgs. district, West Bengal (Barman, 2015).

Collected from roadside ditch at Kestampur near Dankuni, Hooghlydistrict (Collection No. 3).

Voucher No. 64 (Collection No. 3).

***Lyngbya aerugineo-coerulea*** (Kutz.)Gom.Monogr.Oscillariees, 146,pl. 4, figs. 1-3, 1892; Desikachary, T.V. Cyanopbyta.315, pl.48, figs. 9, 10, 1959.

Thallus expanded, blue-green, flexuous; sheath uniform, thin,smooth, not stratified, 1.2-1.4μ thickness, not stained violet with chlor-zinc-iodide; trichome straight, not attenuated at the apex, not constricted at the cross-walls; cells 1/2 as long as broad or slightly shorter, 6.0-7.2μ broad, 2.4-3.6μ long, septa prominent,not granulated, end cell round or sometimes conical, capitate with slightly thickened outer membrane; cell contents light blue-green, homogeneous (Pl. 6, fig. 12).

This species has been reported from Kalcutta (Biswas, 1925); Chilka Lake, Orissa (Biswas, 1932); Meerut, Uttar Pradesh (Bendre and Kumar, 1975); Nagpur, Maharashtra (Kamat and Harankhedkar, 1976); Ludhiana, Punjab (Pandbol and Grover, 1976); Patiala, Punjab (Sarma and Kanta, 1978); Kamataka (Bongale and Bharati, 1980); Lucknow,Uttar Pradesh (Prasad and Mehrotra, 1980); Khandesh, Maharashtra(Barhate and Tarar, 1983); Nagpur, Maharashtra (Tarar and Shewale,1984); Bankura district, West Bengal (Sinha and Mukherjee, 1984); Pusa, Bihar (Jha *et al.*, 1986); Patuli, Burdwan district, West Bengal (Gupta and Sen, 1987).

Collected from rice fields at Simlon near Kalna, Burdwan district (Collection No. 14), Chanditala, Dankuni, Hooghly district (Collection No. 262).

Voucher No. 65 (Collection No. 262).

***Lyngbya gracilis*** (Menegh.)Rabenh.Fl. Eur. Alg., 2: 145, 1865;Desikachary, T.V. Cyanophyta.285, pl. 52, fig. 2, 1959.

Thallus dense, velvetty, light yellowish violet, filaments attached by the middle with the ends free; sheath uniform, thin, smooth, not stratified, up to 1.75μ thick, not coloured violet with chlor-zinc-iodide; trichome not attenuated at the apex, slightly constricted at the cross-walls; cells about 1/2 as long as broad, 7.2-8.4μ broad, 3.6-4.2μ long, septa prominent, cross-walls not granulated, end cell rounded with slightly thickened outer wall; cell contents homogeneous, light pinkish blue (Pl. 6, fig. 13).

Formerly recorded from Benaras, Uttar Pradesh (Rao, 1937); Kalcutta (Banerji, 1938); Saharanpur district, Uttar Pradesh (Pal and Yadav, 1974); Ludhiana, Punjab (Grover and Pandhol, 1975); Meerut,Uttar Pradesh (Bendre and Kumar, 1975); Marathwada, Maharashtra (Ashtekar and Kamat, 1980); Karnataka (Bongale and Bharati, 1980);Tamilnadu (Anand and Revathi, 1987); Dasghara, Hooghly district,West Bengal (Gupta and Sen, 1987); North 24 Pgs., West Bengal (Naskar *et al.*, 2008).

Growing on rice field at Dashghara, Hooghly district (Collection No. 96).

Voucher No. 66 (Collection No. 96).

## PHORMIDIUM Kutz.

Kutzing, Phyc. gene. 190, 1843; Monogr.Oscillariees, 156, 1892.

### Key to the Species

1. Trichomes constricted, cross-walls not granulated .................................. 2

1. Trichomes not constricted ................................................................ 3

2. End cells rounded ........................................................................ **1. *P. molle***

2. End cells conical ......................................................................... 4

3. Cross-walls granulated ........................................................................ 5

3. Cross-walls not granulated........................................................................ 6

4. Cells nearly quadrate; sheath not stained by chlor-zinc-iodide ............ **2. *P. fragile***

4. Cells upto 3 times longer than broad, sheath stained by chlor-zinc- iodide ........ ........................................................................................................**3. *P. tenue***

5. Sheath lamellated, stained by chlor-zinc-iodide ............................ **4. *P. ambiguum***

5. Sheath diffluent, not stained by chlor-zinc-iodide ......................... **5. *P. subfuscum***

6. Cells much broader than long; cells more than 5 times broader than long ......... ..........................................................................................................7

6. Cells nearly quadrate or slightly longer than broad ................................... 8

7. Sheath thin; end cells without calyptra........................................... **6. *P. anomala***

7. Sheath 2.0-2.5μ thick; end cells with calyptra................................ **7. *P. stagnina***

8. Breadth of cells less than 5.4μ; end cells obtuse conical; sheath stained by chlor-zinc-iodide ................................................................ **8. *P.corium var.capitatum***

8. Breadth of cells more than 5.4μ; end cells truncate; sheath not stained by chlor-zinc-iodide ..........................................................................................***9. P. retzii***

***Phormidium molle*** (Kutz.)Gom.Monogr.Oscillariees, 163, pl. 4, fig. 12, 1892; Desikachary, T.V. Cyanophyta.255, pl. 59, fig. 8, 1959.

Thallus blue-green, mucilaginous; sheath diffluent, hyaline, indistinct, overarching the apex, not coloured by chlor-zinc-iodide trichome more or less straight or slightly bent towards the apical part, not attenuated, uniformly broad throughout, constricted at the cross-walls, septa very prominent, assuming a gap between the cells; cells slightly longer than broad, but in some filaments 2.25 times longer, 2.2μ broad, 3.0-3.5μ long, in some filaments 5.0μ long, end cells rounded, calyptra absent; cell contents light blue-green, homogeneous (Pl. 5, fig. 1).

The thallus shows a mixture of trichomes differing in measurements of the cells. Accordingly majority of the filaments are to be identified as P. molle f. tenuior :W. & G.S. West (The Alg. Fl. Yorkshire, Bot. Trans. York nat. Union, 5: 146, 1902), while others correspond with ***P. molle***. The validity of ***f. tenuior*** is doubtful.

Formerly recorded from Calcutta (Martens, 1871; Prain, 1905); Allahabad, Uttar Pradesh and Madras (Mitra, 1951); Rohilkhand division, Uttar Pradesh (Singh and Chaturvedi, 1970); Saharanpur district, Uttar Pradesh (Pal and Yadav, 1974); Patiala, Punjab (Sarma and Kanta, 1978); Karnataka (Bongale and Bharati, 1980); Lucknow,Uttar Pradesh (Prasad and Mehrotra, 1980); 24-Parganas and Howrah districts, West Bengal (Mukhopadhyay and Chatterjee, 1981); Rohilkhand division, Uttar Pradesh (Pandey, 1982); Jaipur, Rajasthan(Trivedy, 1982); Vidarbha region, Maharashtra (Kolte and Goyal,1985);Burdwan district, West Bengal (Gupta and Sen, 1987); Hooghly district,West Bengal (Gupta and

Sen,1987);Bardhaman,Nadia,Hooghly districts(Sen,2005);Jadavpur,Kolkata,West Bengal(Sen,2006); Barasat, West Bengal (Talai-Mukhopadhyay and Naskar, 2015).

Collected from paddy fields at Jagadanandapur near Dainhat,Burdwan district (Collection No. 28), Putsuri near Purbasthali, Burdwan district (CollectionNo. 65), Siakhala, Dankuni, Hooghly district (Collection No. 90); from paddy fields at Adisaptagram near Mogra, Hooghly district (Collection No. 119), Jagadanandapur near Dainhat, Burdwan district(Collection No. BD-44), Purbasthali, Burdwan district (Collection No.BD-36), Jatrapur, Simurali,Nadia district (Collection No. 138), Osmanpur near Kalna, Burdwan district (Collection No. 184).

Voucher No. 67 (Collection No. 184).

***Phormidium fragile*** (Menegh.)Gom.Monogr.Oscillariees, 163, pl. 4, figs. 13-15, 1892; Desikachary, T.V. Cyanophyta.253, pl. 44, figs. 1-3, 1959.

Thallus deep blue-green with a brownish tinge, mucilaginous, thin, having more or less parallel arranged filaments; sheath diffluent hyaline, indistinct,not coloured violet by chlor-zinc-iodide; trichome flexuous, more or less straight and gradually attenuated towards the apex, constricted at the cross-walls; cells 1.8-2.4μ broad, 1.8-3.0μ long, septa not granulated, end cell not bent, more or less conical, not capitate; cell contents blue-green, homogeneous (Pl. 5, fig. 2).

Formerly recorded from Chilka Lake, Rambha Bay, Orissa (Biswas,1932); Bombay, Maharashtra (Schmidle, 1900); Calcutta (Banerji,1938); Delhi (Rao, 1940); Trivandrum, Kerala (Parukutty, 1940); Saharanpur district, Uttar Pradesh (Pal and Yadav, 1974); Bankura district,West Bengal (Sinha and Mukherjee, 1975); Patiala, Punjab (Sarma and Kanta, 1978); Lucknow, Uttar Pradesh (Prasad and Mehrotra, 1980); 24-Parganas and Howrah districts, West Bengal (Mukhopadhyay and Chatterjee, 1981); Nagpur, Maharashtra (Tarar and Shewale, 1984); Jalgaon region, Maharashtra (Bhoge and Ragothaman, 1986); Sagar Island, South 24-Parganas district, West Bengal (Santra *et al.,* 1988); Murshidabad district, West Bengal (Sen, 2005);Barasat, West Bengal (Talai-Mukhopadhyay and Naskar, 2015);).

Growing on crop field at Rezinagar, Murshidabad district(Collection No. 349).

Voucher No. 68 (Collection No. 349).

***Phormidium tenue*** (Menegh.)Gom.Monogr.Oscillariees, 169, pl. 4,figs. 23-25, 1892; Desikachary, T.V. Cyanophyta.259, pl.43, figs. 13-15 and pl. 44, figs. 7-9, 1959.

Thallus thin, not so compact, expanded, dull blue-green; trichomes long, more or less straight, sometimes slightly bent towards the apices, anastomosing, septa prominent, constricted at the cross¬walls; sheath thin, diffluent, stained violet by chlor-zinc-iodide; cells 1.2-1.8μ broad, 3.0-4.8 μ long, septa not granulated; end cell somewhat conical and more or less pointed, calyptra absent(Pl. 5, fig. 3).

Formerly reported from Calcutta (Martens, 1870; Prain, 1905;Bruhl and Biswas, 1922; Biswas, 1925, 1926, 1942); Delhi (Rao,1940);Saharanpur district, Uttar Pradesh (Pal and Yadav, 1974); Bankura district, West Bengal (Sinha and Mukherjee, 1975); Nagpur, Maharashtra (Kamat and Harankhedkar, 1976); Sagar,

M.P. (Mishra and Purohit,1979); Lucknow, Uttar Pradesh (Prasad and Mehrotra, 1980); Karnataka(Bongale and Bharati, 1980); 24-Parganas and Howrah districts, WestBengal (Mukhopadhyay and Chatterjee, 1981); Rohilkhand division,Uttar Pradesh (Pandey, 1982); River Cauvery, Karnataka (Somashekar,1983); Khandesh, Maharashtra (Barhate and Tarar, 1983); Tamilnadu(Anand and Revathi, 1987); Kalcutta, West Bengal (Santra,1987); Agradwip, West Bengal (Gupta and Sen, 1987); Bardhanman and Nadia district, West Bengal (Sen, 2005).

Growing on rice fields at Agradwip, Burdwan district (Collection No. BD-27), Kaliganj near Debagram, Nadia district (Collection No. 149), Ranibandh near Kalna, Burdwan district (Collection No. 215),Siakhala near Dankuni, Hooghly district (Collection No. 317); on a dripping wall at Kalyanpur near Baruipur, South 24-Parganas district (Collection No. 373).

Voucher No. 69 (Collection No. 222).

***Phormidium ambiguum*** Gom.Monogr.Oscillariees, 178, pl. 5, fig. 10, 1892; Desikachary, T.V. Cyanophyta.266, pl. 44, fig. 16 and pl. 45, figs. 5-8, 1959.

Thallus expanded, thin, deep greenish blue, compact, having anastomosing, long, flexible, more or less straight filaments; sheath distinct, thick, stratified, stained violet by chlor-zinc-iodide, trichome not attenuated at the apex, not constricted at the cross- walls; cells 4.2-4.8µ broad, 1.8-2.0µ long, septa granulated, end cell more or less rounded, calyptra absent; cell contents granular, homogeneous, deep blue-green (Pl. 5, fig. 4).

Formerly recorded from Benaras, Uttar Pradesh (Rao, 1936); Calcutta, West Bengal (Banerji, 1938); Kamrup, Assam (Parukutty,1939); Travancore, Kerala (Parukutty, 1940); Allahabad, Uttar Pradesh (Mitra, 1951); Bankura district, West Bengal (Sinha and Mukherjee, 1975); Karnataka (Bongale and Bharati, 1980); Lucknow, Uttar Pradesh (Prasad and Mehrotra, 1980); River Cauvery, Karnataka (Somashekar, 1983); Midnapore district, West Bengal (Pal and Santra, 1985); Nagaland and Tripura (Reddy *et al.*, 1986); Howrah and Midnapore district, West Bengal (Sen, 2005); North 24 Pgs., West Bengal (Naskar *et al.*, 2008); Varanasi, U.P. (Singh *et al.*, 2015).

Collected from moist muddy place at Swarupganj near Nabadwip, Nadia district (Collection No. 50), from the side of a canal at Mohanandapur, Midnapore district (Collection No. 378).

Voucher No. 70 (Collection No. 378).

***Phormidium subfuscum*** Kutz.ex Gom. Kutzing, Phyc. gene., 195, 1843;Gomont, Monogr. Oscillariees, 182, pl. 5, figs. 17 to 20, 1892;Desikachary, T.V. Cyanophyta.273, pl. 44, figs. 22 and 23,1959.

Thallus dark olive-green, thin, expanded having straight, short, more or less parallely arranged filaments; sheath diffluent into a mucous matrix, not coloured violet by chlor-zinc-iodide; trichome straight, slightly attenuated at the apex, not constricted at the cross-walls; cells about 1/2.5 as long as broad, 9.0-9.6µ broad,

3.6μ long, septa prominent, granulated, end cell conical, pointed, capitate with a massive conical calyptra; cell contents granular, olive-green (Pl. 5, fig. 5).

Formerly recorded from Benaras, Uttar Pradesh (Rao, 1937);Bankura district, West Bengal (Sinha and Mukherjee, 1975); Marathwada, Maharashtra (Ashtekar and Kamat, 1980); 24-Parganas and Howrah districts, West Bengal (Mukhopadhyay and Chatterjee, 1981); Howrah and Bardhaman districts, West Bengal (Sen, 2005).

Growing on rice fields at Gopalpur near Kalna, Burdwan district (Collection No. 34), Burikhali near Bauria, Howrah district (Collection No. 81).

Voucher No. 71 (Collection No. 81).

***Phormidium anomala*** Rao, C.B. The Myxophyceae of the United Provinces, III, Proc. Indian Acad. Sci., B, 6: 371, fig. 7F-I, 1937; Desikachary, T.V. Cyanophyta. 286, pl. 45, figs. 11-13, 1959.

Thallus compact, deep blue-green, formed by entangled bundle like mass of more or less flexible filaments; sheath thin, hyaline,not stratified, sometimes with rough outer surface, not stained by chlor-zinc-iodide; trichome not attenuated at apex, not constricted at cross-walls; cells about 1/8 as long as broad, 8.4-9.6μ broad,1.0-1.2μ long, septa prominent, not granulated, end cell rounded,not capitate, calyptras absent; cell contents finely granular, deep blue-green (Pl. 5, fig. 6).

Formerly recorded from Benaras, Uttar Pradesh (Rao, 1937);Berhampore, Orissa (Rao, 1938); Marathawada, Maharashtra (Kamat,1974); Rohilkhand division, Uttar Pradesh (Chaturvedi and Pandey,1976); Patiala, Punjab (Sarma and Kanta, 1978); Lucknow, Uttar Pradesh (Prasad and Mehrotra, 1980); Karnataka (Bongale and Bharati, 1980); Bankura district, West Bengal (Sinha and Mukherjee, 1984);Vidarbha region, Maharashtra (Kolte and Goyal, 1985); Sultanpur,Uttar Pradesh (Khan, 1985); Pusa, Bihar (Jha *et al.*, 1986); Tamilnadu(Anand and Revathi, I987); Murshidabad, Nadia and Bardhaman districts (Sen, 2005); North 24 Pgs. district, West Bengal (Naskar *et al.*, 2008).

Collected from rice fields at Manosapota, Simurali, Nadia district (Collection No. 140), Belpukur, Nabadwip, Nadia district (Collection No. 224), Jagadanandapur near Dainhat, Burdwan district(Collection No. 264).

Voucher No. 72 (Collection No. 224)

***Phormidium stagnina*** Rao, C.B. The Myxophyceae of the Madras Presidency, I, J. Indian bot. Soc., 17: 93, figs. 4-7, 1938; Desikachary, T.V. Cyanophyta. 265, pl. 45, figs. 16-18, 1959.

Thallus thin, membranous, blue-green, compact, having anastomosing, long, flexible, straight sometimes bent filaments; sheath distinct, colourless, outside not always smooth, thick, firm, often diffluent, 2.0-2.5μ thick, not coloured violet by chlor-zinc-iodidetrichome not attenuated at the apex, not constricted at cross-walls;cells 8.4-9.0μ broad, 1.2-1.8μ long, septa not granulated, end cell rounded, calyptra present; cell contents finely granular, homogeneous, blue-green (Pl. 5, fig. 7)

Formerly reported from Andhra Pradesh (Rao, 1938); Bankura district, West Bengal (Sinha and Mukherjee, 1975); Nagpur, Maharashtra(Kamat and Harankhedkar, 1976); 24-Parganas and Howrah districts,West Bengal (Mukhopadhyay and Chatterjee, 1981); Gosaba, South 24 Parganas district, West Bengal (Santra *et al.*, 1988); Nadia and Bardhaman districts, West Bengal (Sen, 2005), North 24 Pgs., West Bengal (Naskar *et al.,* 2008).

Collected from paddy fields at Khajuti near Dainhat, Burdwandistrict (Collection No. 31), Matiari near Debagram, Nadia district (Collection No. 150),Swarupganj near Nabadwip, Nadia district (Collection No. 225), Nirmalpalli, Simurali, Nadia district(Collection No. 226).

Voucher No. 73 (Collection No. 150).

***Phormidium corium var. capitatum*** Gardner, Univ. Calif. Publ. Bot.,14: 4, pl. 1, fig. 1, 1927; Desikachary, T.V. Cyanophyta. 271, pl. 43, figs. 11, 12, 1959.

Thallus expanded, moderately thick, coriaceous, blue-green,compact, having anastomosing, more or less straight filaments; sheath distinct, thin, not stratified, firm, smooth, stained violet by chlor-zinc-iodide, trichome not attenuated at the apex, not constricted at the cross-walls; cells 4.8-5.4 μ broad, 5.4-6.6 μ long, septa not granulated, end cell obtuse conical, with thickened outer membrane; cell contents granular, homogeneous, blue-green (Pl. 5, fig.8).

Formerly recorded from Berhampore, Orissa (Rao, 1938); Meerut,Uttar Pradesh (Bendre and Kumar, 1975); Allahabad, Uttar Pradesh(Pandey and Pandey, 1982); Bankura district. West Bengal (Sinhaand Mukherjee, 1984); South 24 Pgs. district, West Bengal (Sen, 2005); North 24-Pgs. district, West Bengal (Naskar *et al.*, 2008).

Growing on rice fields at Balidanga-Faridpur near Debagram,Nadia district (Collection No. 192),Sankarpur near Nabadwip, Nadia district (Collection No. 231),in shallow water in a pond at Usumpur near Ghola, North 24-Parganas district (Collection No. 353).

Voucher No. 74 (Collection No. 353).

***Phormidium retzii*** (Ag.)Gom.Monogr.Oscillariees, 175, pl. 5, figs. 6-9, 1892; Desikachary, T.V. Cyanophyta.268, pl. 44, figs. 13-15, 1959.

Thallus moderately thick, compact, attached with the substratum at the base, greyish blue; filaments clustered together, straight, slightly constricted at the cross-walls; not attenuated at the apices; sheath thin, diffluent, not stained violet by chlor-zinc-iodide,1.0μ thick; cells shorter than broad, 7.8-9.6μ broad, 5.4-7.5μ long, septa not granulated; end cell not bent, truncated with slightly thickened outer wall; cell contents homogeneous, granular,brownish blue-green (Pl. 5, fig. 9).

Formerly recorded from Parel near Bombay, Maharashtra, (Schmidle, 1900); Benaras, Gorakhpur and Mirzapur, Uttar Pradesh (Rao,1937; and Singh, 1939); Cuttack, Orissa (Rao, 1938); Patna, Bihar(Rao, 1939); Gauhati, Assam (Parukutty,

1939); Travancore, Kerala(Parukutty, 1940); Rohilkhand division, Uttar Pradesh (Singh andChatruvedi, 1970); Lucknow, Uttar Pradesh (Prasad and Mehrotra,1980); Khandesh, Maharashtra (Barhate and Tarar, 1983); Bankura district, West Bengal (Sinha and Mukherjee, 1984); Mathur, 24Parganas district, West Bengal (Maity and Santra, 1985); Kulpi,24-Parganas (South) district, West Bengal (Santra et al., 1988); Rice fields of Nadia, Hooghly and North 24 Pgs. districts, West Bengal (Sen, 2005).

Collected from rice fields at Ranibandh near Kalna, Burdwan district (Collection No. 36), Keota near Mogra, Hooghly district (Collection No. 123), Jagadanandapur near Dainhat, Burdwan district (Collection No. 206), Charmaizdia near Nabadwip, Nadia district (Collection No. 283); Bhandartikuri near Purbasthali, Burdwan district (Collection No. 295).

Voucher No. 75 (Collection No. 206).

**MICROCOLEUS Desmazieres**

Catal.pl. Botanagr belg., 7, 1823; Gomont, Monogr. Oscillariees, 350, 1892.

**Key to the Species**

1. Trichomes narrower than 4.8μ .................................................................. 2

1. Trichomes broader than 4.8μ .................................................................. 3

2. Trichomes not constricted at the cross-walls, septa granulated ........................................................................................................ **1. *M. acutissimus***

2. Trichomes constricted at the cross-walls, septa not granulated ........................................................................................................ **2. *M. chthonoplastes***

3. Trichomes narrower than 6.0μ, not coloured violet by chlor-zinc-iodi de........................................................................................ **3. *M. paludosus***

3. Trichomes broader than 6.0 μ, coloured violet by chlor-zinc-iodide ........................................................................................................ **4. *M. subtorulosus***

***Microcoleus acutissimus*** Gardner, New Myxophyceae from Porto Rico, Nem. N.Y. Bot. Gdn., 7: 55, pl. 11, fig. 2, 1927; Desikachary, T.V. Cyanophyta. 344, pl. 60, fig. 1, 1959.

Thallus light blue-green; filaments unbranched, straight, with 12-22 parallely arranged trichomes; sheath hyaline, gelatinising, with rough outer surface, not stained violet by chlor-zinc-iodide; trichomes sharply attenuated at the apices, not constricted at the cross-walls; cells 3-4 times longer than broad, 1.8-2.1μbroad, 4.8-7.2μlong, septa prominent, granulated, end cell acute, sharply conical, not capitate, calyptra absent; cell contents finely granular, light blue-green (Pl, 7, fig. 4).

Formerly recorded from Meerut, Uttar Pradesh (Bendre and Kumar, 1975); Ludhiana, Punjab (Grover and Pandhol, 1975, 1976);Sagar, Madhya Pradesh (Mishra and Purohit, 1979); Lucknow, UttarPradesh (Prasad and Mehrotra, 1980); Karnataka (Bongale and Bharati,1980); Rohilkhand division, Uttar Pradesh (Pandey, 1982); Tamilnadu(Anand and Revathi, 1987).

Collected from moist soil at Charmazdia near Nabadwip, Nadia district (Collection No. 51), cultivated field at Madanpur, Nadia district (Collection No. 142).

Voucher No. 76 (collection No. 51).

***Microcoleus chthonoplastes*** Thur. ex Gom.Thuret, Ann. Sci. nat.Bot. 6th ser., 1: 378, 1875; Gomont, Monogr.Oscillariees, 353, pl. 14, figs. 5-8, 1892; Desikachary, T.V. Cyanophyta.343, pl. 60, figs. 7- 9, 1959.

Thallus dense, expanded, dense blue-green; filaments flexuous, long, twisted, 42-60μ in breadth; sheath thick with variable thickness, not stratified, hyaline, outer surface rough, covering many closely aggregated trichomes, not coloured violet by chlor-zinc-iodide; trichomes densely aggregated and twisted, sometimes assuming a cord-like structure, attenuated at the apices, constricted at the cross-walls; cells longer than broad, 3.6-4.2μbroad, 5.2-7.2μlong, septa prominent, transverse walls not granulated, end cell acute conical, not capitate; cell contents granular, blue-green (Pl. 7, fig. 5).

Formerly recorded from Calcutta, West Bengal (Biswas, 1926 ;Banerji, 1938); Hooghly river, West Bengal (Biswas, 1942); near Bombay, Maharashtra (Schmidle, 1900; Iyengar and Desikachary, 1944); Chilka lake, Orissa (Biswas, 1932); Shillong, Assam (Biswas, 1934); Uttar Pradesh (Bharadwaja, 1935; Rao, 1937; Singh, 1939; Mitra,195l); Andhra Pradesh (Rao, 1938); Bihar (Rao, 1939); Punjab (Singh, 1941); Tamilnadu (Iyengar and Desikachary, 1944; Venkataraman, 1957); Marathawada, Maharashtra (Kamat, 1974); Meerut, Uttar Pradesh(Bendre and Kumar, 1975); Lucknow, Uttar Pradesh (Prasad and Mehrotra, 1980); Karnataka (Bongale and Bharati, 1980); 24-Parganas and Howrah districts, West Bengal (Mukhopadhyay and Chatterjee, 1981); Allahabad, Uttar Pradesh (Chadha and Pandey, 1982); Jammu and Kashmir (Goyal *et al.*, 1984); Vidarbha region, Maharashtra (Kolte and Goyal, 1985); Pusa, Bihar (Jha *et al.*, 1986); South 24-Parganas district, West Bengal (Santra *et al.*, 1988); Basanti, Hammambere, South 24 Pgs. district, West Bengal (Barman *et al.*, 2015).

Growing on paddy fields at Kestampur near Dankuni, Hooghlydistrict (Collection No. 118), Jagadanandapur near Dainhat, Burdwan district (Collection No. 128),Sankarpur near Nabadwip, Nadia district (CollectionNo. 342).

Voucher *No. 77 (Collection No. 342).*

***Microcoleus paludosus*** (Kutz.)Gom.Monogr.Oscillariees, 358, pl.14, fig. 13, 1892; Desikachary, T.V. Cyanophyta.344, pl.56, fig. 2, 1959.

Thallus deep blue-green, expanded; filaments straight, sheath gelatinous, not stained violet by chlor-zinc-iodide; trichomes many closely aggregated and sometimes twisted into a cord-like structure, slightly tapering at the apices, constricted at the cross-walls; cells about 1.5-2 times as long as broad, 4.8-6.0μbroad, 7.2-9.6μ long, septa prominent, transverse walls not granulated, end cell not capitate, conical; cell contents finely granular, dark blue-green (Pl. 7, fig. 6).

Formerly recorded from Chilka lake, Orissa (Biswas, 1932); Calcutta, West Bengal (Biswas, 1925; Banerjee, 1938); Saharanpur district, Uttar Pradesh (Pal and Yadav, 1974); Meerut, Uttar Pradesh (Bendre and Kumar, 1975); Bankura district, West Bengal (Sinha and Mukherjee, 1975); Ludhiana, Punjab (Pandhol and Grover, 1976); Patiala, Punjab (Sarma and Kanta, 1978); Sagar, Madhya Pradesh (Mishra and Purohit, 1979); Karnataka (Bongale and Bharati, 1980); 24-Parganas and Howrah districts, West Bengal (Mukhopadhyay and Chatterjee, 1981);Garhbeta, Midnapore district, West Bengal (Pal and Santra, 1985); Santoshpur, Kolkata, West Bengal (Sen, 2006).Herobhanga, Kholnarkhal, South 24 Pgs. district, West Bengal (Barman *et al.*, 2015); Varanasi, U.P. (Singh *et al.*, 2015).

Collected from a ditch at Raj Chandrapur near Dankuni, Hooghly district (Collection No. 1); paddy fields at Osmanpur, near Kalna, Burdwan district (Collection No. 43), Halusa near Mogra, Hooghly district (Collection No 121), Debagram, Nadia district (Collection No.191).

Voucher No. 78 (Collection No. 43).

***Microcoleus subtorulosus*** (Breb.)Gom.Monogr.Oscillariees, 369,pl. 14, figs. 14, 15, 1892; Desikachary, T.V. Cyanophyta.345, pl. 56, figs. 8, 9, 1959.

Thallus greenish blue; filaments nearly straight, 14.0-18.0μ broad; sheath gelatinizing, uneven, colourless, coloured violet with chlor-zinc-iodide, having many closely grouped trichomes; trichomes straight, not attenuated at the apices, constricted at the crosswalls; cells nearly as long as broad or slightly longer, 6.0-8.4μbroad, 6.0-10.8μ long, septa prominent, transverse walls not granulated, end cell not capitate, cylindrical conical; cell contents greenish blue, granular (Pl. 7, fig 7).

Formerly recorded from Ludhiana, Punjab (Pandhol and Grover,1976); Karnataka (Bongale and Bharati, 1980); Bankura district, West Bengal (Sinha and Mukherjee, 1984).

Growing on rice field at Usunpur near Ghola, North 24-Parganas district (Collection No. 353).

Voucher No. 79 (Collection No. 353).

## SCHIZOTHRIX Kutz.

Phyc.gene., 230, 1843; Gomont, Monogr. Oscillariees, 292, 1892.

### Key to the Species

1. Sheath hyaline; trichome attenuated at the apical part; end cells pointed ........................................ **1. *Sch. arenaria***

1. Sheath yellowish or reddish; trichome not attenuated at the apex; end cell not pointed........................................ 2

2. Thallus with little calcium incrustation; sheath not lamellated, yellowish; cells 1.2-1.5μ broad........................................ **2. *Sch. lacustris***

2.Thallus without calcium incrustation; sheath lamellated, reddish; cells 4.8-7.2µb road...................................................................................................... ***3. Sch. telephoroides***

***Schizothrix arenaria*** (Berk.)Gom.Monogr.Oscillariees, 312, pl. 8,figs. 11 and 12, 1892; Desikachary, T.V. Cyanophyta. 327.1959.

Thallus membranous, thin, faintly olive-green, without calcium incrustation; filaments entangled at the basal part and separating towards the apices assuming branching; sheath firm, moderately thick,narrow at the tip forming more or less pointed ends, lamellated at the lower part, hyaline, outer surface more or less even, stained violet by chlor-zinc-iodide; trichomes few in a common sheath, more in number in the mature part, slightly attenuated towards the apices involving 2-3 cells, constrictions at the cross-walls not very prominent; cells 1.8-3.3 µ broad, 5.4-6.0 µ long, septa distinct, not granulated, end cell conical, pointed, cell contents homogeneous,greenish blue (Pl. 7, fig. l) .

Formerly reported from Maharashtra (Bombay State) (Gonzalves and Gangla, 1949); Bankura district, West Bengal (Sinha and Mukherjee, 1975); Lucknow, Uttar Pradesh (Prasad and Mehrotra, 1980);Marathwada, Maharashtra (Ashtekar and Kamat, 1980); Salt Lake, Kolkata, West Bengal (Sen, 2006).

Growing on paddy field at Simurali, Nadia district (Collection No. 139).

Voucher No. 80 (Collection No. 139).

***Schizothrix lacustris*** A. Br. ex Gom. A. Braun in Kutzing, Species Algarum, 320, 1849; Gomont, Monogr. Oscillariees, 301, pl.6, figs. 9-12, 1892; Desikachary, T.V. Cyanophyta.325, pl. 56, figs. 6 and 10, 1959.

Thallus firm, crustaceous, with very little calcium incrustation, olive-green; filaments closely aggregated and branched more towards the ends, not attached; sheath moderately thick, with pointed ends, yellowish, not lamellated, stained violet by chlor-zinc-iodide; trichomes a few or single in a common sheath towards the apices of filaments, more in number and often twisted around one another at the mature parts, not attenuated at the apices, constricted at the cross-walls; cells longer than broad, 1.2-1.5µbroad, 3.0-4.2µlong, septa prominent, not granulated, end cells not attenuated (Pl. 7, fig. 2).

Formerly recorded from Bombay (Gonzalves and Joshi, 1946); Chingleput near Madras (Desikachary and Ramakrishnan, 1953); Bankura district, West Bengal (Sinha and Mukheijee, 1975); Marathawada, Maharashtra (Ashtekar and Kamat, 1980).

Collected from paddy fields at Mogra, Hooghly district (Collection No. 120), Bakulla near Guptipara, Hooghly district (Collection No. 298).

Voucher No. 81 (Collection No. 120).

***Schizothrix telephoroides*** (Mont.) Gom.Monogr.Oscillariees, 319, pl. 10, figs. 1-4, 1892; Desikachary, T.V. Cyanophyta.330, pl. 57, figs. 2, 7 and 8, 1959.

Thallus crustaceous, felty, brown-red, without calcium incrustation; filaments more or less parallel aggregated, very closely tufted at the basal part and separating towards the apices assuming branching; aggregating filaments having individual sheaths;sheath very thick with pointed ends, lamellated, reddish but faintly so at the periphery, outer surface not uniform, coloured violet with chlor-zinc-iodide; trichomes mostly single in the sheath, not attenuated at the apices, constricted at the cross-walls; cells about 1.5 times longer than broad, 4.8-7.2μ broad, 7.8-10.8μ long,septa distinct, not granulated, end cell rotund, cell contents granular, blue-green (Pl. 7, fig. 3).

Formerly recorded from Cherrapunji, Assam (Biswas, 1934); Bankura district, West Bengal (Sinha and Mukherjee, 1975); Assam (Reddy *et al.*, 1986).

Growing on moist soil at the sides of the drying small pods at Chat Katala near Fulia Boyra, Nadia district(Collection No. 57).

Voucher No. 82 (Collection No. 57).

## CYLINPROSPERMUM Kutz.

Phyc.gene., 211, 1843; Bornet and Flahault, Revision des Nostocacees heterocystees, 249, 1888.

### Key to the Species

1. Spores having thick, papillose epispore ............................................ ***1.C. majus***
1. Spores having thick, smooth epispore ........................................................ 2
2. Spores spherical ........................................................................ ***2.C. sphaerica***
2. Spores not spherical ...................................................................***3. C. musicola***

***Cylindrospermum majus*** Kutzing ex Born.et Flah. Kutzing, Phyc. Gene., 212, 1843; Bornet and Flahault, Revision des Nostocacees heterocystees, 252, 1888; Desikachary, T.V, Cyanophyta. 360, pl. 80, fig. 1, 1959.

Thallus soft, mucilaginous, expanded, dirty green; trichomes entangled with each other, constricted at the cross-walls, more or less uniformly broad, 4.2-4.8μ broad, blue-green; cells barrel-shaped, almost as long as broad or sometimes slightly longer, 4.2-5.4μ long; heterocysts slightly broader than the trichome, one at each end of the trichome, 4.8-5.4μ broad, 7.2-7.8μ long; spores ellipsoidal, thick walled, epispore thick, papillose, brownish, 9.6-10.8μ broad, 20.4-21.6μ long (Pl. 8, fig. l).

Formerly reported from Rohilkhand division, Uttar Pradesh (Singh and Chaturvedi, 1970); Mysore state (Kamat, 1974); Bankura district, West Bengal (Sinha and Mukherjee, 1975); Meerut, Uttar Pradesh (Bendre and Kumar, 1975); Ludhiana, Punjab (Grover and Pandhol, 1975); Sagar, Madhya Pradesh (Mishra and Purohit, 1979); Karnataka (Bongale and Bharati, 1980); 24-Parganas and Howrah districts, West Bengal (Mukhopadhyay and Chatterjee, 1981); Allahabad, Uttar Pradesh (Pandey and Pandey, 1982); Burdwan district, West Bengal (Chatterjee and Chatterjee, 1981); Jammu and Kashmir (Goyal *et al.*, 1984); Vidarbha region,

Maharashtra (Kolte and Goyal, 1985); Meghalaya and Nagaland (Reddy *et al.*, 1986); Pusa, Bihar (Jha *et al.*, 1986); Bijoygarh, Kolkata, West Bengal (Sen, 2006).

Growing on rice fields at Kaliaganj near Debagram, Nadia district(Collection No. 169), Manosapota near Simurali, Nadia district (Collection No. 196), Ranibandh near Kalna, Burdwan district (Collection No. 280).

Voucher No. 83 (Collection No. 196).

***Cylindrospermurn sphaerica*** Prasad.B.N., J. Indian bot. Soc., 31: 358, figs.1-4, 1952; Desikachary, T.V. Cyanophyta.363, pl. 64, figs. 7, 8, 1959.

Thallus soft, mucilaginous, expanded, brownish; trichomes entangled, constricted at the cross-walls, 4.8-5.4μ broad; cells barrel-shaped or sometimes longer than the broad, 4.2-7.2μ long; heterocysts ellipsoidal or slightly elongated, one at each end of the trichome, 4.8-5.4μ broad, 6.6-9.6μ long; spores single, next to the terminal heterocyst, almost spherical or slightly longer, 15.3μ diam., thick walled, epispore light brownish, smooth (Pl. 8, fig. 2).

Formerly reported from Benaras, Uttar Pradesh (Prasad, 1952);River Cauvery, Karnataka (Somashekar, 1983); Burdwan district, West Bengal (Chatterjee and Chatterjee, 1983); Varanasi, U.P. (Singh *et al.*, 2015).

Growing on crop field at Palta, North 24-Parganas district (Collection No. 343), moist road side soil at Kaji para near Dum Dum, Kolkata (Collection No.381).

Voucher No. 84 (Collection No. 381).

***Cylindrospermum musicola*** Kutzing ex Born.et Flah. Kutzing, Phyc.germ., 173, 1845; Tab. Phycologieae, 1: 53, pl. 98, fig. 1, 1849; Bornet et Flahault, Revision des Nostocacees heterocystees, 254, 1888; Desikachary, T.V. Cyanophyta. 366, pl. 65, fig. 3, 1959.

Thallus soft, mucilaginous, expanded, dirty green; trichomes entangled, constricted at the cross-walls, 3.6-4.2μ broad; faint bluish-green; cells mostly quadrate or rarely longer than the broad,4.2-5.0μ long; heterocysts round, at the end of trichome, 4.2-5.0μ broad, 5.4-6.0μ long; spores ellipsoidal, sometimes slightly elongated, single, next to the terminal heterocyst, 8.4-10.8μ broad, 14.4-18.6μ long, thick walled, epispore brownish, smooth (Pl. 8, fig. 3).

Formerly recorded from Northern India (Randhawa, 1936); Benaras, Uttar Pradesh (Rao, 1937); Andhra Pradesh, Puri, Orissa (Rao, 1938); Mysore State (Kamat, 1974); Saharanpur district, Uttar Pradesh (Pal and Yadav, 1974); Bankura district, West Bengal (Sinha and Mukherjee, 1975); Lucknow, Uttar Pradesh (Prasad and Mehrotra, 1980); Karnataka (Bongale and Bharati, 1980); 24-Parganas and Howrah districts, West Bengal (Mukhopadhyay and Chatterjee, 1981); Allahabad, Uttar Pradesh (Pandey and Pandey, 1982); River Cauvery, Karnataka (Somashekar, 1983); Jammu and Kashmir (Goyal *et al.*, 1984); Vidarbha region, Maharashtra (Kolte and Goyal, 1985); Meghalaya (Reddy *et al.*, 1986); Tamilnadu (Anand and Revathi, 1987); Burdwan and Hooghly districts, West Bengal (Gupta and Sen,1987); Varanasi, U.P. (Singh *et al.*, 2015).

Collected from rice field at Jagadanandapur near Dainhat, Burdwan district (Collection No. 16); from water-logged soil in a crop field at Pandua, Hooghly district (Collection No. 93); from rice field at Patuli, Burdwan distrct (Collection No. BD-20), Singhee near Dainhat, Burdwan district (Collection No. BD-48), Bishnupur near Dainhat, Burdwan district (Collection No. BD-57), Gopalpur near Kalna, Burdwan district (Collection No. 183), Belpukur, Nabadwip, Nadia district (Collection No. 284).

Voucher No. 85 (Collection No. 93).

**AULOSIRA Kirchner**

Kryptogamenflora von Schlesien, 238, 1878; Bornet and Flahault, Revision des Nostocacees heterocystees, 256, 1888.

**Key to the Species**

1. Sheath thick; cells barrel-shaped, 6.0-7.2μ broad, 7.2-8.4μ long; heterocysts 8.4-8.7μ broad, 10.8-12.0 μ long ........................................................ 1. ***A. fertilissima***

1. Sheath thin, diffluent; cells longer than broad, 3.6-4.8μ broad, 6.0-16.2(-18)μ long; heterocysts 4.8-7.2μ broad, 6.0-17.4μ long ........................ 2. ***A. prolifica***

***Aulosira fertilissima*** Ghose, A systematic and ecological account of a collection of blue-green algae from Lahore and Simla, J. Linn. Soc. Bot., 46: 342, pl. 31, fig. 9, 1924; Desikachary, T.V. Cyanophyta 431, pl. 80, fig. 6, 1959.

Thallus expanded, blue-green; trichomes more or less parallel, generally slightly flexuous, sometimes straight; sheath thick, brownish; cells 6.0-7.2μ broad and 7.2-8.4μ long, barrel-shaped, cell content granular; heterocysts intercalary, ellipsoidal, 8.4-8.7μ broad, 10.8-12.0μ long; spores in series, interrupted by one or two dead cells, ellipsoidal, 9.6-10.8μ broad, 18.0-19.2μ long (Pl, 2,fig. 1).

Formerly recorded from Saharanpur, Uttar Pradesh (Randhawa, 1936); Uttar Pradesh and Bihar (Singh, 1942); Bombay, Maharashtra (Gonzalves and Joshi, 1946); Rohilkhand division, Uttar Pradesh (Singh and Chaturvedi, 1970); Saharanpur district, Uttar Pradesh(Pal and Yadav, 1974); Ludhiana, Punjab (Grover and Pandhol, 1975); Cuttack, Orissa (Singh, 1975); Meerut, Uttar Pradesh (Bendre and Kumar, 1975); Gadigarh stream, Jammu (Anand, 1979); Lucknow, Uttar Pradesh (Prasad and Mehrotra, 1960); Marathwada, Maharashtra (Ashtekar and Kamat, 1980); 24-Parganas and Howrah districts, West Bengal (Mukhopadhyay and Chatterjee, 1981); Rohilkhand division, Uttar Pradesh (Pandey, 1982); Jammu and Kashmir (Goyal *et al.*, 1984); Diamond- Harbour, South 24-Parganas district, West Bengal (Maity and Santra, 1985); Assam (Reddy *et al.*, 1986); Tamilnadu (Anand and Revathi, 1987); Burdwan and Hooghly district, West Bengal (Gupta and Sen, 1987); Diamond-Harbour, South 24-Parganas district, West Bengal (Santra *et al.*, 1988); Hatiara near Dum Dum airport, Kolkata, West Bengal (Sen, 2006).

Growing on paddy fields at Jagadanandapur near Dainhat, Burdwan district (Collection No. BD-39), on moist ridge of earth around an agricultural land at Mogra, Hooghly district (Collection No. 152); Matiari near Debagram, Nadia district (Collection No. 154).

Voucher No. 86 (Collection No. 154).

***Aulosira prolific*** a Bharadwaja, Contributions to our knowledge of Myxophyceae of India, Ann. Bot., Lond., 47: 131, figs. 5, 6, 1933a. Desikachary, T.V. Cyanophyta. 426, pl. 81, figs. 1-6, 1959.

Thallus forming a scum, blue-green; filaments generally parallel, straight or sometimes slightly curved, pale green; sheath thin, more or less diffluent; trichomes constricted at the cross-walls, cells longer than broad, 3.6-4.8μ broad, 6.0-16.2 (-18)μ long, end cell round, tapering; heterocysts intercalary, single, broader than the trichomes, filaments swelled where heterocysts present, ellipsoidal, 4.8-7.2μ broad, 6.0-17.4μ long; spores not found (Pl. 12, fig. 2).

This species has been reported from Benaras, Uttar Pradesh (Bharadwaja, 1933); Andhra Pradesh (Rao, 1938); Saharanpur district, Uttar Pradesh (Pal and Yadav, 1974); Ludhiana, Punjab (Pandhol and Grover, 1976); Patiala, Punjab (Sarma and Kanta, 1978); Marathwada, Maharashtra (Ashtekar and Kamat, 1980); Rohilkhand division, Uttar Pradesh (Pandey, 1982); Allahabad, Uttar Pradesh (Pandey and Pandey, 1982); Marathwada, Maharashtra (Chaporkar and Gangawane, 1984); Sul-tanpur, Uttar Pradesh (Khan, 1985); Vidarbha region, Maharashtra (Kolte and Goyal, 1985)

Collected from rice field at Gopalpur near Kalna, Burdwan district(Collection No. 234).

Voucher No. 87 (Collection No. 234).

**ANABAENA Bory**

Diet, class, d'hist, nat., 1: 307, 1822; Bornet and Flahault, Revision des Nostocacees heterocystees, 224, 1888.

**Key to the species**

1. Trichomes spirally coiled .......... **1. *A. spiroides***

1. Trichomes not spirally coiled .......... 2

2. Akinetes on both sides of the heterocysts .......... 3

2. Akinetes mostly remote from theheterocysts .......... 4

3. End cells of trichomes round .......... 5

3. End cells of trichomes conical .......... 6

4. Trichomes variously coiled; heterocysts 8.4-9.0 μ broad; akinetes single .......... **2. *A. circinalis***

4. Trichomes straight or variously curved; heterocysts 5.4-7.2μ broad; akinetes in chain .......... 7

5. Trichomes single, not held in common sheath; heterocysts barrel shaped .......... 8

5. Trichomes held in a common sheath; heterocysts spherical .......... 9

6. Cells barrel shaped, slightly shorter than broad; heterocysts spherical; akinetes cylindrical........................................................................................................***3.A. torulosa***

6. Cells cylindrical, about twice as long as broad; heterocysts not spherical; akinetes ellipsoidal.............................................................................................. **4. *A. orientalis***

7. Cells round, end cells acute or accuminate; akinetes almost twice as long as broad ......... .........................................................................................................**5. *A. naviculoldes***

7. Cells cylindrical or barrel-shaped, end cells rounded; akinetes much less longer than broad or slightly shorter ........................................................................... 10

8. Spores in long series........................................................................... **6. *A. iyengarii***

8. Spores on either side of heterocyst ................................................ **7. *A. sphaerica***

9. Heterocysts more or less at regular intervals; akinetes cylindrical, single on either sides of heterocysts ............................................................................ **8. *A. ambigua***

9. Heterocysts not at regular interval;akinetes ellipsoidal, single or 2-3 together on both sides of heterocysts........................................................... **9. *A. oscillarioides***

10. Trichomes not held in a common sheath, slightly narrower towards the apices; cells barrel-shaped, 5.4-6.0μ broad; heterocysts spherical, 6.0μ long ........................ .......................................................................................................... **10. *A. fertilissima***

10. Trichomes in a common sheath, same width throughout; cells cylindrical,3.6μ broad; heterocysts barrel-shaped, 9.6 μ long .......... **11.*A. variabilis var. ellipsospora***

***Anabaena spiroides*** Klebahn, Flora, 80: 125, pl.4, figs. 11-13, 1895; Desikachary, T.V. Cyanophyta.395, pl. 71, fig. 9,1959.

Thallus mucilaginous, thin, free-floating, pale blue-green; trichome spirally coiled, single, spirals 48.0-52.0μ broad, 39.0-48.0μ distant; cells more or less spherical, frequently shorter than the broad, 6.2-7.8μ broad, cell contents granular, blue-green; heterocysts spherical, slightly broader than the vegetative cell, intercalary, 7.2-8.4μ diam; spores spherical, broader than the heterocyst, position not related to heterocysts, generally l4.4μ diam., sometimes slightly smaller (Pl. 10, fig. 3).

Formerly reported from Hyderabad, Andhra Pradesh (Ghousuddin,1936); River Hooghly, Calcutta (Biswas, 1942); Saharanpur district, Uttar Pradesh (Pal and Yadav, 1974); Bhagalpur, Bihar (Nasar and Datta Munshi, 1976); Karnataka (Bongale and Bharati, 1980); Bhubaneswar, Orissa (Mohanty, 1982); Marathwada, Maharashtra (Chaporkar and Gangawane, 1984); Jammu and Kashmir(Goyal *et al.*, 1984); Bhagalpur, Bihar (Saha and Pandit, 1987).

Collected from water-logged soil at Kalyanpur near Baruipur, South 24-Parganas district (Collection No. 368).

Voucher No. 88 (Collection No. 368).

***Anabaena circinalis*** Rabenh.ex Born. et Flah. Algen Eur. Exs., no.209, 1852; Fl. Eur. Alg., 2: 183, 1865; Bornet and Flahault, Revision des Nostocacees heteroocystees, 230, 1888; Desikachary, T.V. Cyanophyta. 414, pl. 77, fig. 2, 1959.

Thallus free-floating, floccose; trichome variously coiled,without any sheath, 8.4-10.8μ broad; cells spherical, sometimes shorter than broad, cell contents vacuolated; heterocysts more or less globose, 8.4-9.0μ in diameter; spores cylindrical, 18.0-19.2μ broad, 33.0-36.0μ long, mostly remote from the heterocyst, epispore hyaline, smooth (Pl. 10, fig. 4).

Formerly reported from West Bengal (Biswas, 1927); Cuttack,Orissa (Singh, 1975); Rohilkhand division, Uttar Pradesh (Chaturvedi and Pandey, 1976); Karnataka (Bongale and Bharati, 1980); Bhubaneswar, Orissa (Mohanty, 1982, 1984); Nagpur, Maharashtra (Tarar and Shewale, 1984); Assam (Reddy *et al.*, 1986); Jabalpur, Madhya Pradesh (Srivastava and Sen, 1987).

Growing in a earthenware pot at Hirapur, Howrah district(Collection No. 86).

Voucher No. 89 (Collection No. 86).

***Anabaena torulosa*** (Carm.)Lagerh.ex Born. et Flah. Bidrag till Sveriges Algflora, K. Vet.Akad.Forh., 40(2): 47, 1883;Bornet and Flahault, Revision des Nostocacees heterocystees,236, 1888; Desikachary, T.V. Cyanophyta. 415, pl. 71, fig. 6,1959.

Thallus mucilaginous, attached to the submerged aquatic plants, some are free floating, blue-green; trichome 5.0-6.0μ in diameter, apical part gradually tapering, prominently constricted at the crosswalls; sheath hyaline, thin, diffluent; cells barrel-shaped, slightly shorter than the broad, 4.8-5.4μ long, end cell subacutely conical; cell contents granular, blue-green; heterocysts spherical, inter-calary, 6.0μ diam.; spores on both sides of the heterocysts, more or less cylindrical, slightly constricted in the centre, about twice as long as broad, 6.0-7.2μ broad, 12.0-13.8μ long, epispore, smooth, brownish (Pl. 10, fig. 6).

Formerly reported from Orissa (Biswas, 1932); Panihati near Kalcutta (Banerji, 1938); Saharanpur district, Uttar Pradesh (Paland Yadav, 1974); Ludhiana, Punjab (Grover and Pandhol, 1975, 1976);Bankura district, West Bengal (Sinha and Mukherjee, 1975); Patiala,Punjab (Sarma and Kanta, 1978); Karnataka (Bongale and Bharati,1980); Marathwada, Maharashtra (Ashtekar and Kamat, 1980); Contai,Midnapore district, West Bengal (Pal and Santra, 1985); Vidarbha region, Maharashtra (Kolte and Goyal, 1985),; Pusa, Bihar (Jha *et al.*,1986); Maharashtra (Patil and Satav, 1986).

Growing on paddy field at Guptipara, Hooghly district (Collection No. 258).

Voucher No. 90 (Collection No. 258).

***Anabaena orientalis*** Dixit, The Myxophyceae of the Bombay Presidency, I, Proc. Indian Acad. Sci., B, 3: 101, fig. 3D, E, 1936; Desikachary, T.V. Cyanophyta. 405, pl. 77, fig. 6, 1959.

Trichomes almost straight, not held in any common sheath, free-floating, 3.6-4.2μ broad, breadth uniform throughout the trichome; cells about twice as long as broad, 4.8-6.0μ long, cylindrical, end cell conical; heterocysts intercalary, solitary, generally barrel shaped, 5.4-6.0μ broad, 7.2-9.0μ long; spores on both sides of the heterocysts, ellipsoidal, not in chain, 7.2-8.4μ broad and 12.0-14.8μ long, epispore thick, smooth, yellowish brown (Pl. 11, fig. 2).

Formerly recorded from Bombay (Dixit, 1936; Gonzalves, 1947); Mysore State (Kamat, 1974); Bankura district, West Begal (Sinha and Mukherjee, 1975); River Cauvery, Karnataka (Somashekar, 1983); Vidarbha region, Maharashtra (Kolte and Goyal, 1985).

Growing as free floating masses in a small pond at Sashan Road near Baruipur, South 24-Parganas district (Collection No. 369).

Voucher No. 91 (Collection No. 369).

***Anabaena naviculoides*** Fritsch, The genus ***Anabaena*** etc., J. Indian bot. Soc., 28: 138, figs. 17-39, 1949; Desikachary, T.V. Cyanophyta. 410, pl. 72, fig. 2, 1959.

Thallus gelatinous; trichomes curved, sometimes more or less straight, breadth more or less uniform throughout the trichome, constriction at the cross-walls prominent; cells 4.2-5.4μ broad, as long as broad, almost round, apical cell accuminate, sometimes acute, cell content granular, blue-green; heterocysts scanty, intercalary, solitary, barrel-shaped, 5.4-6.0 μ broad, slightly longer than broad; spores in chain, not related to heterocyst, ellipsoidal, ends acute, 10.8-12μ long, 6.0-6.6μ broad, spores almost twice as long as broad, epispores hyaline, smooth (Pl. 11, fig. 3).

Formerly reported from Cuttack, Orissa (Singh, 1975); 24-Parganas and Howrah districts, West Bengal (Mukhopadhyay and Chatterjee, 1981); Burdwan district, West Bengal (Chatterjee and Chatterjee, 1983); Bankura district, West Bengal (Sinha and Mukherjee, 1984); Jammu and Kashmir (Goyal *et al.*, 1984); Arunachal Pradesh (Reddy, 1986).

Collected from paddy field at Kalitala near Simurali, Nadia district (Collection No. 250).

Voucher No. 92 (Collection No. 250).

***Anabaena iyengarii*** Bharadwaja, The Myxophyceae of the United Provinces, I, Proc. Indian acad. Sci., B, 2: 105, fig. 6H-K, 1935; Desikachary, T.V. Cyanophyta. 406, pl. 78, fig. 2, 1959.

Trichome variously curved, several not held in a common sheath, loosely arranged, constricted at the cross-walls, breadth more or less uniform throughout the trichome; cells barrel-shaped, 5.4-6.0μ broad, about as long as broad, 5.4-6.6μ long, end cell conical; heterocysts barrel-shaped, intercalary, solitary, 7.2-7.8 μ broad, 7.2-9.6μ long; spore in long series, ellipsoidal, found on both sides of the heterocysts, 8.4-9.6μ broad, 12.6-18.0μ long, epispore smooth, thick and faintly brown (Pl. 10, fig. 5).

Formerly reported from Benaras, Uttar Pradesh (Bharadwaja, 1935), Allahabad, Uttar Pradesh (Gupta, 1956); Meerut, Uttar Pradesh(Bendre and Kumar, 1975);

Rohilkhand division, Uttar Pradesh (Chaturvedi and Pandey, 1976); Lucknow, Uttar Pradesh (Prasad and Mehrotra, 1980); Balmikinagar, Bihar (Sinha and Srivastava, 1980);Konkan area, Maharashtra (Sardeshpande and Goyal, 1981); Bhubaneswar, Orissa (Mohanty, 1982); Rohilkhand division, Uttar Pradesh (Pandey, 1982); Allahabad, Uttar Pradesh (Pandey and Pandey, 1982); River Cauvery, Karnataka (Somashekar, 1983); Jammu and Kashmir (Goyal *et al.*, 1984); Tamilnadu (Anand and Revathi, 1987); Diamond Harbour,South 24-Parganas district, West Bengal (Santra *et al.*, 1988); Thakurpukur, Kolkata, West Bengal (Sen, 2006).

Collected from rice fields at Kusumgram near Purbasthali, Burdwan district (Collection No. 62); Chanditala, Dankuni, Hooghly district (Collection No. 98), Paglachandi near Debagram, Nadia district (Collection No. 146), Bishnupur near Dainhat, Burdwan district (Collection No. 208), Balagarh near Guptipara, Hooghly district (Collection No. 302); Osmanpur near Kalna, Burdwan district (Collection No. 322).

Voucher No. 93 (Collection No. 322).

***Anabaena sphaerica*** Bornet et Flahault, Revision des Nostocacees heterocystees, 228, 1888; Desikachary, T.V. Cyanophyta. 393, 1959.

Thallus mucilaginous, blue-green, floating; trichome straight, without any sheath; cells barrel-shaped, 4.8-5.4μ broad, 4.2-4.8μ long; end cell rounded; heterocyst more or less globose, 6.3-6.6μ broad, 6.6μ long; spores subspherical, on either side of heterocysts, 10.8-12.0 μ broad, 12.0-15.6μ long, epispore light brown,smooth (Pl. 11, fig. 1).

This species has been reported from Bombay, Maharashtra (Schmidle, 1900); Calcutta, West Bengal (Banerji, 1938; Biswas,1942); Madras, Tamilnadu (Ganapati, 1940); Mysore State (Kamat, 1974); Saharanpur district, Uttar Pradesh (Pal and Yadav, 1974); Meerut, Uttar Pradesh (Bendre and Kumar, 1975); Sagar, Madhya Pradesh (Mishraand Purohit, 1979); Karnataka (Bongale and Bharati, 1980); Bhubaneswar, Orissa (Mohanty, 1982); Sultanpur, Uttar Pradesh (Khan, 1985); Kalcutta, West Bengal (Santra, 1987); Khanakul, Hooghly district,West Bengal (Gupte and Sen, 1987); Basanti, Jharkhali, South 24-Pgs., West Bengal (Barman *et al.*, 2015).

Growing in stagnant water at Khanakul, Hooghly district (Collection No. 107), as floating masses in a small pond at Nandigram, Midnapore district (Collection No. 376).

Voucher No. 94 (Collection No. 107).

***Anabaena ambigua*** Rao, C.B.A new species of ***Anabaena.*** Proc. Indian Acad. Sci., B, 5: 101, figs. 1, 2, 1937; Desikachary, T.V.Cyanophyta. 400, pl. 76, fig. 2, 1959.

Thallus upto 0.5 mm long, 15.0-24.0μ broad; trichomes dark blue-green, completely or incompletely curved, enclosed in a mucilaginous sheath, a few aggregated together, attached to some aquatic plant, sheath hyaline, having uneven outer surface; cells constricted at the cross-walls, barrel-shaped, 4.8-

6.0μ broad,3.6-5.4μ long; heterocysts almost globular, more or less at regular intervals,slightly broader than the cells,5.4-7.2μ diam.;spores single on either side of the heterocysts,ellipsoidal, sometimes with lattened ends,7.2-9.0μ broad and 10.8-13.2μ long,epispore thick, hyaline (Pl.11,fig.4).

Formerly reported from Benaras, Uttar Pradesh (Rao, 1937);Uttar Pradesh and Bihar (Singh, 1942); Allahabad, Uttar Pradesh(Mitra, 1951); Ludhiana, Punjab (Grover and Pandhol, 1975); Bankuradistrict, West Bengal (Sinha and Mukherjee, 1975); Meerut, Uttar Pradesh (Bendre and Kumar, 1975); Karnataka (Bongale and Bharati,1980); Lucknow, Uttar Pradesh (Prasad and Mehrotra, 1980); 24-Parganas and Howrah districts, West Bengal (Mukhopadhyay and Chatterjee, 1981), Allahabad, Uttar Pradesh (Pandey and Pandey, 1982);Burdwan district, West Bengal (Chatterjee and Chatterjee, 1983);Jammu and Kashmir (Goyal *et al.*, 1984); Pawagiri Oon, Madhya Pradesh (Mahajan, 1987); Tamilnadu (Anand and Revathi, 1987).

Growing on paddy fields at Singhee near Dainhat, Burdwan district (Collection No. 15), Halusa near Mogra, Hooghly district (Collection No. 89); Chagram near Kalna, Burdwan district (Collection No. 213), Matiari near Debagram, Nadia district(Collection No. 264), Bakulla near Guptipara, Hooghly district (Collection No. 261), Siakhala near Dankuni, Hooghly district (Collection No. 309),Swarupganj near Nabadwip, Nadia district (Collection No. 328)

Voucher No. 95 (Collection N0. 309).

***Anabaena oscillarioides*** Bory ex Born.et Flah. Bory, Dict., Class, d'hist., nat., 1 : 308, 1822; Bornet and Flahault, Revision des Nostocacees hctcrocystees, 233, 1888; Desikachary, T.V. Cyanophyta. 417, pl. 71, fig. 7, 1959.

Thallus soft gelatinous, green, trichome uniformly broad, generally curved, sometimes more or less straight, 4.8-5.4μ broad; cells as long as or shorter than broad, 4.5-5.0μ long, barrel-shaped , constricted at the cross-walls, end cells rounded; heterocysts more or less spherical, intercalary, 5.4-7.2μ broad, 5.4-7.8μ long; spores single or 2-3 in a series, developed centripe -tally, on both sides of the heterocysts, up to 2-3 times as long as broad, cylindrical, 9.0-10.8μ broad, 24.6-30μ long, epispore light brown, smooth (Pl. 11, fig. 5).

Formerly reported from Assam (Parukutty, 1939); Bankura district, West Bengal (Sinha and Mukherjee, 1975); Meerut, Uttar Pradesh (Bendre and Kumar, 1975); 24-Parganas and Howrah districts,West Bengal (Mukhopadhyay and Chatterjee, 1981); Marathwada, Maharashtra (Chaporkar and Gangawane, 1984); Pusa, Bihar (Jha *et al.*, 1986).

Collected from rice fields at Purbasthali, Burdwan district(Collection No. 12), Hirapur, Howrah district (Collection No. 80), Singhee near Dainhat, Burdwan district(Collection No. 270), Guptipara, Hooghly district (Collection No. 289).

Voucher No. 96 (Collection No. 289).

***Anabaena fertilissima*** Rao, C.B. The Myxophyceae of the United Provinces, III, Proc. Indian Acad. Sci., B, 6: 363, fig. 6A-C,1937; Desikachary, T.V. Cyanophyta. 398, pl. 74, fig. 1, 1959.

Trichomes not aggregated, bent, sometimes straight, with an indistinct mucilaginous sheath, 5.4-6.0µ broad, slightly narrower towards the apices; cells barrel-shaped, 5.0-6.6µ long, end cell rounded, cell content granular; heterocysts intercalary, spherical,6.0-7.2µ broad, 6.0µ long; spores in long chain, position not related to heterocyst, more or less ellipsoidal, slightly broader than long, 5.4-7.8µ broad, 5.4-7.2µ long (Pl. 11, fig. 6).

Formerly recorded from Benaras, Uttar Pradesh (Rao, 1937); Assam (Parukutty, 1939); Uttar Pradesh and Bihar (Singh, 1942);Bombay, Maharashtra (Gonzalves and Joshi, 1946); Rohilkhand division, Uttar Pradesh (Singh and Chaturvedi, 1970); Meerut, Uttar Pradesh (Bendre and Kumar, 1975); Sagar, Madhya Pradesh (Mishra and Purohit, 1979); Karnataka (Bongale and Bharati, 1980); Lucknow, Uttar Pradesh (Prasad and Mehrotra, 1980); Marathwada, Maharashtra (Ashtekar and Kamat, 1980); 24-Parganas and Howrah districts, West Bengal (Mukhopadhyay and Chatterjee, 1981); Allahabad, Uttar Pradesh (Pandey and Pandey, 1982); Rohilkhand division, Uttar Pradesh (Pandey,1982); River Cauvery, Kamataka (Somashekar, 1983); Burdwan district,West Bengal (Chatterjee and Chatterjee, 1983); Jammu and Kashmir(Goyal *et al.*, 1984); Marathwada, Maharashtra (Chaporkar and Gangawane, 1984); Vidarbha region, Maharashtra (Kolte and Goyal, 1985); Pusa, Bihar (Jha *et al.*, 1986); Nagaland (Reddy *et al.*, 1986); Jalgaon region, Maharashtra (Bhoge and Ragothaman, 1986); Burdwan district and Hooghly district, West Bengal (Gupta and Sen, 1987); Kakurgachi, Kolkata, West Bengal (Sen, 2006).

Growing on paddy fields at Gopkhanji near Dainhat, Burdwan district (Collection No. 18), Rautgram near Purbasthali, Burdwan district (Collection No. 35), Haripal near Tarakeswar, Hooghly district (Collection No. 102), Santipalli near Simurali, Nadia district (Collection No. 144), Muragachi near Debagram, Nadia District (Collection No. 157), Charmaizdia near Nabadwip, Nadia district (Collection No. 220).

Voucher No. 97 (Collection No. 18).

***Anabaena variabilis*** Kutz.ex Born. et Flah. var. *ellipsospora* Fritsch. The genus *Anabaena* with special reference to the species etc.,J. Indian Bot. Soc., 28, 142, figs. 40-50, 1949; Desikachary, T.V. Cyanophyta.411, pl. 72, fig. 1, 1959.

Thallus gelatinous, faintly bluish, trichome curved, sometimes more or less straight, loosely arranged, breadth more or less uniform throughout the trichome; cells 3.6µ up to 4.0µ broad, length 6.0-8.4µ cylindrical, end cell rounded, cell content granular, more or less hyaline; heterocysts intercalary, solitary, barrel-shaped,5.4-6.0µ broad, 9.6µ long, slightly yellowish; spores in chain, remote from the heterocysts, 6.0µ broad, 7.2-8.4µ long, ellipsoidal (Pl. 11, fig. 7).

Formerly reported from Calcutta, West Bengal (Banerji, 1938); Hooghly river, West Bengal (Biswas, 1942); Calcutta, West Bengal (Gupta, 1975); Lucknow, Uttar Pradesh (Prasad and Mehrotra, 1980); Rohilkhand division, Uttar Pradesh (Pandey, 1982); Bankura district, West Bengal (Sinha and Mukherjee, 1984); Marathwada, Maharashtra (Chaporkar and Gangawane, 1984): Meghalaya (Reddy *et al.*,

1986);Pusa, Bihar (Jha *et al.*, 1986); Tamilnadu (Anand and Revathi, 1987);Burdwan district, West Bengal (Gupta and Sen, 1987), Purba Sinthee, Dum Dum, West Bengal (Sen, 2006).

Growing on paddy fields at Dhatrigram near Kalna, Burdwan district (Collection No. 11), Gobra near Dankuni, Hooghly district (Collection No. 100), Mogra, Hooghly district (Collection No. 106), Singhee, Burdwan district (Collection No. BD-46), Madanpur near Simurali, Nadia district (Collection No. 247), Jagadanandapur near Dainhat, Burdwan district (Collection No. 266), Amghata near Nabadwip, Nadia district (Collection No. 324).

Voucher No. 98 (Collection No. 106).

**NOSTOC Vaucher**

Histoire des Confervees d'eau douce, 203, 1803; Bornet and Flahault,Revision des Nostocacees heterocystees, 181, 1888.

**Key to the Species**

1. Trichomes loosely arranged ............ 2

1. Trichomes closely arranged............ 3

2. Trichomes more than 4.2μ broad, heterocysts subspherical............ **1. *N. piscinale***

2. Trichomes upto 4.2μ broad,heterocysts oblong............4

3. Sheath of trichomes not distinct, heterocysts spherical ............5

3.Sheath of trichomes distinct, heterocysts otherwise in shape ............6

4. Cells cylindrical............ 7

4. Cells barrel shaped or subspherical ............8

5. Trichomes upto 3.0μ broad, cells subelliptic, 1.5 times longer than bro ad............ **2. *N. humifusum***

5. Trichomes more than 3.0μ broad, cells barrel-shaped, slightly shorter than br oad............**3. *N. sphaericum***

6. Trichomes more than 4.8μ broad,sheath lamellated, heterocysts slightly broader than long............ **4. *N. commune***

6.Trichomes less than 4.8μ broad, sheath not lamellated, heterocysts slightly longer than broad............ 9

7. Heterocysts 6.6μ broad, spores 6.0μ broad, epispore hyaline............ **5. *N. carneum***

7. Heterocysts 8.4-9.6μ broad, spores 6.0-7.2 μ broad, epispore with opaque sheath ............**6. *N. ellipsosporum***

8. Filament sheath distinct, trichome more than 2.4μ broad, spores ovoid ............ ............**7. *N. paludosum***

8. Filament sheath not distinct, trichome upto 2.4μ broad, spores spherical ............ ............**8. *N. calcicola***

9. Trichomes less than 3.6μ broad, cells ovoid, spores more than 4.8μ broad ...... ...................................................................................................**9. N. muscorum**

9. Trichomes more than 3.6μ broad, cells ellipsoidal, spores not more than 4.8μ ......................................................................................................... **10. N. punctiforme**

***Nostoc piscinale*** Kutz ex Born, et Flah. Kutzing, Phyc. gene. 208, 1843; Bornet and Flahault, Revision des Nostccacees heterocystees, Ann. Sci. nat. Bot. Ser. 7, 7, 194, 1888; Desikachary, T.V. Cyanophyta. 377, pl. 69, fig. 3, 1959.

Thallus macroscopic, free-floating, gelatinous, yellowish brown, with irregular shape, 1-3 cm or more in breadth, with widely spersed tubercles; sheath light brownish to almost hyaline; filaments loosely arranged, flexuous, 4.2 to 4.8μ broad; cells as long as or shorter than broad; heterocysts intercalary, subspherical with flattened polar nodules, 6.0μ broad; spores globose, 6.0-6.6μ broad, in chains, epispore smooth, with hyaline sheath (Pl. 8, fig. 4).

Former records are from Bombay, Maharashtra (Beck and Zahlbruchner, 1897); Allahabad, Uttar Pradesh (Gupta, 1956); Marathwada, Maharashtra (Kamat, 1974); Mysore State (Kamat, 1974); Bankura district, West Bengal (Sinha and Mukherjee, 1975); Meerut, Uttar Pradesh (Bendre and Kumar, 1975); Kalcutta, West Bengal (Gupta, 1975); Patiala, Punjab (Sarma and Kanta, 1978; Sarma *et al.*, 1979); Lucknow, Uttar Pradesh (Prasad and Mehrotra, 1980); Karnataka (Bongale and Bharati, 1980); Jammu and Kashmir (Goyal *et al.*, 1984); Diamond Harbour, 24-Parganas district, West Bengal (Maity and Santra, 1985); Jalgaon region, Maharashtra (Bhoge and Ragothaman, 1986); Katwa, Burdwan district and Pandua, Hooghly district, West Bengal (Gupta and Sen, 1987); Diamond Harbour, South 24-Parganas district, West Bengal (Santra *et al.*, 1988).

Growing as mucilaginous scum in a shallow water filled pit at Pandua, Hooghly district (Collection No. 95); on paddy fields at Khajuti near Katwa, Burdwan district (Collection No. BD-14); Chanduria near Simurali, Nadia district (Collection No. 178); Amghata near Nabadwip, Nadia district (Collection No. 223); Bhandartikuri near Purbasthali, Burdwan district (Collection No. 240); Behula near Guptipara, Hooghly district (Collection No. 256); Jangalpara near Dankuni, Hooghly district (Collection No. 341).

Voucher No. 99 (Collection No. 256).

***Nostoc humifusum*** Carm.ex Born. et Flah. Carmichael ex Harvey, Hooker's British Flora, 2: 399, 1833; Born, et Flah. Revision des Nostocacees heterocystees, 201, 1888; Desikachary, T.V. Cyanophyta.384, 1959.

Thallus gelatinous, attached to the moist soil, irregular in shape and size, sometimes punctiform, occasionally several talli confluent, olive-green in colour; filaments closely entangled in thallus, flexuous; sheath generally indistinct, brownish; trichomes2.4-3.0μ broad, greenish-blue; cells more or less subelliptic, about 1.5 times longer than broad; heterocysts subspherical, 3.0μ broad; spores oblong 4.8μ broad, 5.4μ long, epispore smooth, light brown (Pl. 8, fig. 5).

Formerly recorded from neighbourhood of Bombay, Maharashtra (Gonzalves and Gangla, 1949); Mysore State (Kamat, 1974); Karnataka (Bongale and Bharati, 1980); Lucknow, Uttar Pradesh (Prasad and Mehrotra, 1980); 24-Parganas and Howrah district, West Bengal (Mukhopadhyay and Chatterjee, 1961); Rohilkhand division, Uttar Pradesh (Pandey, 1982); Marathwada, Maharashtra (Chaporkar and Gangawane, 1984).

Collected from paddy fields at Osmanpur near Kalna, Burdwan district (Collection No. 13), Singhee near Dainhat,Burdwan district (Collection No. 32), Gobra near Dankuni, Hooghly district (Collection No. 94),Putsuri near Purbasthali, Burdwan district (Collection No. 212), Bakulla near Guptipara, Hooghly district (Collection No. 291).

Voucher No. 100 (Collection No. 32).

***Nostoc sphaericum*** Vaucher ex Born.et Flah. Vaucher, Histoire desConferves d'eau douce, 223, pl. 16, fig. 2, 1803; Bornet and Flahault, Revision des Nostocacees, heterocystees, 208, 1888;Desikachary, T.V. Cyanopbyta. 390, 1959.

Thallus gelatinous, globose to oblong, smooth, free-floating in stagnant water in rice-fields, 5.5-6.5 cm in diameter, thick, greenish blue, peripheral part dense; sheath not very distinct; filaments closely entangled, flexuous; trichomes 4.2μ broad, cells slightly shorter than long, barrel-shaped; heterocyst nearly spherical, about 4.8μ broad; spores ellipsoidal, 5.4μ broad, 7.2-7.5μ long, epispore light brown, thick (Pl. 8, fig. 6).

Formerly recorded from Bombay, Maharashtra (Schmidle, 1900);Calcutta, West Bengal (Biswas, 1925; Banerji, 1938); Allahabad, Uttar Pradesh (Mitra, 1951); Rohilkhand division, Uttar Pradesh (Singh and Chaturvedi, 1970); Sagar, Madhya Pradesh (Mishra and Purohit, 1979); Lucknow, Uttar Pradesh (Prasad and Mehrotra, 1980); 24-Parganas and Howrah districts, West Bengal (Mukhopadhyay and Chatterjee, 1981); Konkan region, Maharashtra (Sardeshpande and Goyal, 1981); Bhubaneswar, Orissa (Mohanty, 1982, 1984); Rohilkhand division, Uttar Pradesh (Pandey, 1982); Allahabad, Uttar Pradesh (Pandey and Pandey, 1982); Burdwan district, West Bengal (Chatterjee and Chatterjee, 1981); Meghalaya (Reddy *et al*., 1986); Baghbazar, Kolkata, West Bengal (Sen, 2006).

Growing on rice fields at Begpur, Purbasthali, Burdwan district (Collection No. 71), Tribeni near Mogra, Hooghly district (Collection No. 111),Simlon near Kalna, Burdwan district (Collection No. 186), Belpukur near Nabadwip, Nadia district (Collection No. 187), Kumirmara near Dankuni,Hooghly district (Collection No. 272).

Voucher No. 101 (Collection No. 272).

***Nostoc commune*** Vaucher ex Born. et Flah. Vaucher, Histoire des Conferves d'eau douce, 222, pl. 16, fig. 1, 1803; Bornet etFlahault, Revision des Nostocacees heterocystees, 203, 1888;Desikachary, T.V. Cyanophyta. 387, pl. 68, fig. 3, 1959.

Thallus macroscopic, globose, fairly large, about 5-6 cm diameter but flattened and membranous when still larger, free floating or attached to aquatic plants at water level, firm, gelatinous, irregularly perforated, olive-green at periphery; filaments closely entangled, flexuous; sheath distinct at the periphery but less distinct inside the thallus, thick, yellowish brown, faintly so inside the thallus, lamellated; trichomes 4.8-5.4μ broad; cells spherical or nearly so, sometimes barrel-shaped, 5.0μ long; heterocysts slightly broader than long, 6.6μ broad, 6.0μ long (Pl. 9, fig. 1).

Formerly reported from Bombay, Maharashtra (Schmidle, 1900); Calcutta, West Bengal (Biswas, 1925; Banerji, 1938); Tamilnadu (Fremy, 1942); Mysore State (Kamat, 1974); Saharanpur district,Uttar Pradesh (Pal and Yadav, 1974); Meerut, Uttar Pradesh (Bendre and Kumar, 1975); Sagar, Madhya Pradesh (Mishra and Purohit, 1979); Lucknow, Uttar Pradesh (Prasad and Mehrotra, 1980); Karnataka (Bongale and Bharati, 1980); 24-Parganas and Howrah districts, West Bengal (Mukhopadhyay and Chatterjee, 1981); Bhubaneswar, Orissa (Mohanty, 1982); Burdwan district, West Bengal (Chatterjee and Chatterjee, 1983); Jammu and Kashmir (Goyal et al., 1984); Marathwada, Maharashtra (Chaporkar and Gangawane, 1984); Vidarbha region, Maharashtra (Kolte and Goyal, 1985); Pusa, Bihar (Jha *et al*., 1986); Nagaland (Reddy *et al*., 1986); Maharashtra (Patil and Satav, 1986); Bishnupur, Burdwan district and Khanakul, Hooghly district, West Bengal (Guota and Sen, 1987); Tamilnadu (Anand and Revathi, 1987); Basanti, Sarberia, South 24 Pgs. district, West Bengal (Barman et al., 2015); Varanasi, Uttar Pradesh (Singh *et al.,* 2015).

Growing at Gopkhanji near Dainhat, Burdwan district (Collection No. 17); attached with partly submerged aquatic plant in a pond at Khanakul near Arambagh, Hooghly district (Collection No. 73); Bethuadahari near Debagram, Nadia district (Collection No. 151); Tatigachi near Simurali, Nadia district (Collectiona No. 202); Dhatrigram near Kalna, Burdwan district (Collection No. 278); Putsuri near Purbasthali, Burdwan district (Collection No. 319); Belpukur near Nabadwip, Nadia district (Collection No. 323); Janai Road near Dankuni, Hooghly district (Collection No. 335).

Voucher No. 102 (Collection No. 202).

***Nostoc carneum*** Ag.ex Born.et Flah. Agardh, Syst. Alg., 22, 1824;Bornet and Flahault, Revision des Nostocacees heterocystees, 196, 1888; Desikachary, T.V. Cyanophyta.381, pl. 69, fig. 6, 1959.

Thallus globose, occasionally irregularly expanded, tuberculated, gelatinous, attached to the aquatic plants, reddish brown, filaments very loosely arranged in thallus, contorted, flexuous; sheath not distinct, hyaline; trichome 3.6-4.2μ broad; cells cylindrical, 1.5-2 times longer than broad; heterocysts oblong, 5.4μ broad, 6.6μ long; spores ellipsoidal, 6.0μ broad, 8.0μ long, epispore smooth, hyaline (Pl. 9, fig. 2).

Formerly recorded from Rohilkhand division, Uttar Pradesh(Singh and Chaturvedi, 1970); Mysore State (Kamat, 1974); Cuttack,Orissa (Singh, 1975);

Karnataka (Bongale and Bharati, 1980); Lucknow, Uttar Pradesh (Prasad and Mehrotra, 1980); Pusa, Bihar (Jha*et al.*,l986); Basanti, Sarberia, South 24 Pgs. district, West Bengal (Barman, 2015).

Collected from paddy field at Hirapur, Howrah district (Collection No. 80). Voucher No. 103 (Collection No. 80).

***Nostoc ellipsosporum*** (Desm.)Rabenh.ex Born. et Flah. Rabenhorst,Fl. Eur. Alg., 2: 169, 1865; Bornet and Thuret, Notes algologique, 2: 94, pl. 27, figs. 7-11, 1880; Desikachary, T.V.Cyanophyta. 383, pl. 69, fig. 5, 1959.

Thallus irregularly expanded, macroscopic, gelatinous, attached to the substratum, pale-brown; filaments loosely arranged in a thallus; trichomes faint olive-green, upto 3.6μ broad; cells 5.4-12.6μ long, cylindrical; heterocysts longer than broad, oblong or nearly so, 7.5μ broad, 8.4-9.6μlong; spores ellipsoidal in shape, in chain of 2-3, 6.0-7.2 μ broad, 12.0-15.0μ long, epispore smooth, with opaque sheath. The spores are sometimes shorter upto 10.0μ long (Pl. 9, fig. 3).

Formerly reported from Mogal thur, Andhra Pradesh (Rao, 1938); Benaras, Gorakhpur, Mirzapur and Bosti, Uttar Pradesh (Singh, 1939); Mysore State (Kamat, 1974); Saharanpur district, Uttar Pradesh (Pal and Yadav, 1974); Bankura district, West Bengal (Sinha and Mukherjee, 1975); Karnataka (Bongale and Bharati, 1980); Lucknow, Uttar Pradesh (Prasad and Mehrotra, 1980); Marathwada, Maharashtra (Ashtekar and Kamat, 1980); River Cauvery, Karnataka (Somashekar, 1983); Jalgaon region, Maharashtra (Bhoge and Ragothaman, 1986); Basanti, Sarberia, South 24 Pgs. district, West Bengal (Barman *et al.*, 2015).

Growing as mucilaginous masses attached to aquatic plants in small pond at Rajchandrapur near Dankuni, Hooghly district (Collection No. 4).

Voucher No. 104 (Collection No. 4).

***Nostoc paludosum*** Kutz.ex Born. et Flah. Kutzing, Tab. Phycologicae,2: 1, pl. 1, fig. 2, 1850; Bornet et Flahault, Revision des Nostocacees heterocystees, 191, 1888; Desikachary, T.V. Cyanophyta. 375, pl. 69, fig. 2, 1959.

Thallus small, not conspicuous even under the microscope, yellowish blue-green, adhering to the submerged aquatic plants; filaments loosely arranged in a thallus; sheath distinct, broad, hyaline, sometimes faintly brownish, 12.0- 15.0μ broad; trichomes upto 4.2μ broad, cell barrel-shaped, light blue-green; heterocysts ellipsoidal, broader than the vegetative cells, upto 5.4μ broad; spores ovoid 4.8-5.2μ broad, 7.2-7.8μ long, epispore smooth, hyaline (Pl. 9 fig. 4).

Formerly reported from Benaras, Uttar Pradesh (Rao, 1937); Allahabad, Uttar Pradesh (Mitra, 1951); Mysore State (Kamat, 1974); Bankura district, West Bengal (Sinha and Mukherjee, 1975); Nagpur, Maharashtra (Kamat and Harankhedkar, 1976); Patiala, Punjab (Sarmaand Kanta, 1978); Marathwada, Maharashtra (Ashtekar and Kamat,1980);Karnataka (Bongale and Bharati, 1980); Lucknow, Uttar Pradesh (Prasadand Mehrotra, 1980); Allahabad, Uttar Pradesh (Pandey and Pandey,1982);Burdwan district, West Bengal (Chatterjee and Chatterjee, 1983);

Digha, Midnapore district, West Bengal (Pal and Santra, 1985); Paddy field near Dum Dum airport, Kolkata, West Bengal (Sen, 2006).

Growing on rice fields at Dampalbanah near Purbasthali, Burdwan district (Collection No. 20), Gopalpur near Kalna,Burdwan district (Collection No. 46); Jagadanandapur near Dainhat, Burdwan district (Collection No. 129); Sonadanga near Debagram, Nadia district (Collection No. 171).

Voucher No. 105 (Collection No. 20).

***Nostoc calcicola*** Brebisson ex Born, et Flah. Brebisson in Meneghini,Monographia Nostochinearum italicarum, 121, 1843; Bornet and Flahault, Revision des Nostocacees heterocystees, 202, 1888; Desikachary, T.V. Cyanophyta. 384, pl. 68, fig. 1, 1959.

Thallus fairly large, mucilaginous, diffluent, blue-green; filaments loosely arranged in a thallus; sheath not distinct, hyaline or sometimes faintly brownish; trichomes 2.0-2.4μ broad, light blue-green; cells generally subspherical, 2.4μ long; heterocysts broader than long, subspherical, 4.8-5.4μ broad; spores more or less spherical, 4.2-5.0μ broad, with yellowish smooth membrane (Pl. 9, fig. 5).

Cultures of paddy field soil of Dainhat (Burdwan district) shows spores upto 6.5μ broad, with smooth hyaline membrane.

Formerly reported from Bombay, Maharashtra (Gonzalves and Gangla, 1949); Karnataka (Bongale and Bharati, 1980); Lucknow, UttarPradesh (Prasad and Mehrotra, 1980); 24-Parganas and Howrah districts, West Bengal (Mukhopadhyay and Chatterjee, 1981); Burdwan district,West Bengal (Chatterjee and Chatterjee, 1983); Jammu and Kashmir (Goyal *et al.*, 1984); Marathwada, Maharashtra (Chaporkar and Gangarwane, 1984); Vidarbha region, Maharashtra (Kolte and Goyal, 1985); Meghalaya (Reddy *et al.*, 1986); Tamilnadu (Anand and Revathi, 1987);Bhagalpur, Bihar (Saha and Pandit, 1987); Varanasi, Uttar Pradesh (Singh *et al.*, 2015).

Collected from rice-fields at Kaligunj near Simurali, Nadia district (Collection No. 195), Agradwip near Dainhat,Burdwan district (Collection No. 210), Chagram near Kalna, Burdwan district (Collection No. 218).

Voucher No. 106 (Collection No. 210).

***Nostoc muscorum*** Ag.ex Born, et Flah. Agardh, Dispositio Algar.Sueciae, 44, 1812; Bornet et Flahault, Revision des Nostocacees heterocystees, 200, 1888; Desikachary, T.V. Cyanophyta. 385, pl. 70, fig. 2, 4, 1959.

Thallus gelatinous, expanding, membranous, tuberculate, attached to the soil, light olive-brown in colour; filaments closely entangled; sheath hyaline, distinct only at the periphery, moderately thick, yellowish brown; trichomes irregular in arrangement, 3.6-4.8μ broad; cells ovoid, slightly longer than broad, 4.2-5.4μ long; heterocysts slightly longer than broad, 6.0μ broad, 6.6-7.0μ long; spores ellipsoidal, arranged in a long chain, 5.4-5.7μ broad, 7.2-7.8μ long, epispore yellowish, smooth (Pl. 10, fig. l).

Formerly recorded from Bombay, Maharashtra (Gonzalves and Gangla, 1949); Rohilkhand division, Uttar Pradesh (Singh and Chaturvedi, 1970); Mysore State (Kamat, 1974); Saharanpur district, UttarPradesh (Pal and Yadav, 1974); Meerut, Uttar Pradesh (Bendre and Kumar, 1975); Cuttack, Orissa (Singh, 1975); Karnataka (Bongale and Bharati, 1980); 24-Parganas and Howrah districts, West Bengal (Mukhopadhyay and Chatterjee, 1981); Burdwan district, West Bengal (Chatterjee and Chatterjee, 1983); Jammu and Kashmir (Goyal *et al.*,1984);Vidarbha region, Maharashtra (Kolte and Goyal, 1985); Maharashtra (Patil and Satav, 1986); Jalgaon region, Maharashtra (Bhoge and Ragothaman, 1986); Nagaland and Meghalaya (Reddy *et al.*, 1986); Kalcutta, West Bengal, (Santra, 1987); Hooghly district, West Bengal (Gupta and Sen, 1987); Varanasi, Uttar Pradesh (Singh *et al.*, 2015).

Growing on paddy field at Adisaptagram near Mogra, Hooghly district (Collection No. 84); Muragachi near Debagram, Nadia district (Collection No. 160); Bhandartikuri near Purbasthali, Burdwan district (Collection No. 214); Bakulla near Guptipara,Hooghlydistrict (Collection No. 303); Simlon near Kalna, Burdwan district (Collection No. 311), Swarupganj near Nabadwip, Nadia district (Collection No. 325), Kumirmara near Dankuni, Hooghly district (Collection No. 339).

Voucher No. 107 (Collection No. 214).

***Nostoc punctiforme*** (Kutz.)Hariot, J. de Bot., 5: 31, 1891; Desikachary, T.V. Cyanophyta.374, pl. 69, fig. 1, 1959.

Thallus having indefinite form, more or less globose, attached to the submerged aquatic plant, peripheral part not distinguishable from interior, associated with other green algae, blue-green; filaments densely entangled in thallus, flexuous; sheath generally distinct, hyaline, thin; trichomes 3.0-3.6μ broad, cells ellipsoidal, light blue-green; heterocysts 3.6μ broad, 4.8μ long;spores subglobose, 4.8μ diam., epispore thick, smooth (Pl. 10, fig.2)

Formerly reported from Bombay, Maharashtra (Schmidle, 1900); Meerut, Uttar Pradesh (Bendre and Kumar, 1975); Bankura district,West Bengal (Sinha and Mukherjee, 1975); Patiala, Punjab (Sarma and Kanta, 1978); Karnataka (Bongale and Bharati, 1980); 24-Parganas and Howrah districts, West Bengal (Mukhopadhyay and Chatterjee, 1981); Konkan area, Maharashtra (Sardeshpande and Goyal, 1981); River Cauvery, Karnataka (Somashekar, 1983); Marathwada, Maharashtra (Chaporkar and Gangawane, 1984); Jammu and Kashmir (Goyal *et al.*, 1984); Vidarbha region, Maharashtra (Kolte and Goyal, 1985); Sultanpur, Uttar Pradesh (Khan, 1985); 24-Parganas district, West Bengal (Maity and Santra, 1985); Assam and Arunachal Pradesh (Reddy *et al.*, 1986);Jalgaon region, Maharashtra (Bhoge and Ragothaman, 1986); Pusa, Bihar (Jha *et al.*, 1986); Diamond Harbour, South 24-Parganas district,West Bengal (Santra *et al.*, 1988).

Collected from rice fields at Dhatrigram near Kalna, Burdwan district (Collection No. 22), Agradwip near Dainhat, Burdwan district (Collection No. 209); Begpur near Purbasthali, Burdwan district (Collection No. 321).

Voucher No. 108 (Collection No. 22).

## SCYTONEMA Ag.

Systema Algarum, 26, 1824; Bornet and Flahault, Revision des Nostocacees heterocystees, 85, 1887.

### Key to the Species

1. Sheath lamellated; heterocysts subquadrate .......... 2

1. Sheath not lamellated; heterocysts round,ellipsoidal or cylindrical .......... 3

2. Trichomes not constricted at cross-walls; cells broader than long .......... **1. *S. ocellatum***

2. Trichomes constricted at cross-walls; cells longer than broad .......... **2. *S. mirabile***

3. Trichomes 15.6-21.6µ broad; cells broader than long .......... **3. *S. cincinnatum***

3. Trichomes 9.6-12.0µ broad; cells longer than broad .......... **4. *S. simplex***

***Scytonema ocellatum*** Lyngbye ex Born.et Flah. Lyngbye, Hydrophytologia Danica, 97, pl. 28a, 1819; Bornet and Flahault, Revision des Nostocacees heterocystees, 95, 1887; Desikachary, T.V. Cyanophyta. 467, pl. 92, fig. 3, 1959.

Thallus expanded, floccose, olivaceous; filaments long, 9.6-14.4 µ broad, entangled; false branches generally short; sheath stratified, brownish, thick, apical cells not dilated; trichome 6.0-10.2 µ broad, cross-walls not constricted; cells generally shorter than broad; heterocysts subquadrate, as broad as or slightly broader than the trichome, yellowish (Pl. 12, fig. 3).

Formerly reported from Rajmahal, Bihar (Martens, 1971); Mahabaleshwar and Poona (Schmidle, 1900); Kalcutta, West Bengal (Bruhl and Biswas, 1923; Biswas, 1932; Banerji, 1938; Biswas, 1942); Fyzabadand Lucknow, Uttar Pradesh (Bruhl and Biswas, 1923); Tamilnadu(Fremy, 1927); Bombay, Maharashtra (Dixit, 1936); Benaras, UttarPradesh (Rao, 1937); Puri, Orissa (Rao, 1938); Trivandrum, Kerala(Parukutty, 1940); Rohilkhand division, Uttar Pradesh (Singh and Chaturvedi, 1970); Lucknow, Uttar Pradesh (Prasad and Mehrotra,1980); Allahabad, Uttar Pradesh (Chadha and Pandey, 1982); Jagadanandapur, Burdwan district and Khanakul, Hooghly district, WestBengal (Gupta and Sen, 1987); Gariahat and Kamarhati, Kolkata, West Bengal (Sen, 2006).

Growing on moist soil near a road side drain at Khanakul, Hooghly district (Collection No. 104); rice fields at Jagadanandapur near Dainhat, Burdwan district (Collection No. BD-33), Bishnupur near Dainhat, Burdwan district (Collection No. 133), Dhatrigram near Kalna,Burdwan district (Collection No. 185), Dampalbandhnear Purbasthali, Burdwan district (Collection No.312).

Voucher No. 109 (Collection No. 185).

***Scytonema mirabile*** (Dillw.) Born.Nosto.heteroc. du Syst. Alg. de C.A. Agargh, Bull. Soc. bot. Fr., 36: 155, 1889; Desikachary, T.V. Cyanophyta.483, pl. 91, fig. 3 and pl. 98, fig. 2, 1959.

Thallus expanded, velvetty, deep greenish brown; filaments entangled, 14.4-17.4μ broad, 3.6-10.8 mm long; very often false branched, generally geminate; sheath thick, lamellated, stratification slightly divergent, faint brownish, sometimes colourless; trichome 7.2-9.6μ broad; olivaceous green, slightly constricted at the cross-walls; cells longer than broad, apical cells rotund;heterocysts quadrate or slightly longer, brown, single, intercalary(Pl. 12, fig. 4)

Formerly collected from Bombay and other places of Maharashtra (Schmidle, 1900); Calcutta, West Bengal (Bruhl and Biswas,1923; Biswas, 1925; Banerji, 1938); Berkuda island, Orissa (Biswas, 1924);Benaras, Uttar Pradesh (Rao, 1937); Andhra Pradesh (Rao, 1938);Tamilnadu (Fremy, 1942); Marathwada, Maharashtra (Kamat, 1974); Bankura district, West Bengal (Sinha and Mukherjee, 1975).

Growing on paddy-fields at Simlon near Kalna, Burdwan district (Collection No. 45); Khajuti near Dainhat, Burdwan district (Collection No. 127); Kanrale near Debagram, Nadia district (Collection No. 165).

Voucher No. 110 (Collection No. 127).

***Scytonema cincinnatum*** Thuret ex Born, et Flah. Thuret, Essai de classification des Nostochinees, Ann. Sci. nat., 6 ser., Bot., 1: 380, 1875; Bornet and Flahault, Revision des Nostocacees heterocystees, 89, 1887; Desikachary, T.V. Cyanophyta. 453, pl. 93, fig. 1, 1959.

Thallus entangled, floccose, olive-green to brownish green; filaments 18.0-24.0μ broad, very long, upto about 2.5 cm, wavy; false branched, often geminate; sheath moderately thick, slightly brownish, not stratified; trichome 15.6-21.6μ broad, prominently constricted at the cross-walls, light violet green; cells compressed, 3-4 times broader than long, apical cell rounded; heterocysts single, almost round to elliptical, with thick outer wall, as broad as the trichome (Pl. 13, fig. 1).

Reported earlier from Calcutta, West Bengal (Martens, 1871; Prain, 1905; Biswas, 1942); Assam (Biswas, 1934); Travancore, Kerala (Parukutty, 1940); Shembaganur, Tamilnadu (Fremy, 1942); Sagar, Madhya Pradesh (Mishra and Purohit, 1979); Sultanpur, Uttar Pradesh (Khan, 1985); Arunachal Pradesh (Reddy *et al.*, 1986)

Collected from moist land at Chatkatala near Fulia Boyra,Nadia district (Collection No. 56).

Voucher No. 111 (Collection No. 56).

***Scytonema simplex*** Bharadwaja, Rev. algol., Paris, 7: 157, fig. 1A,B, 1934; Desikachary, T.V. Cyanophyta.455, pl. 89, fig. 1,1959.

Thallus greenish blue, expanded, firm; filaments 12.6-14.4μ broad,loosely entangled; false branches geminate, long; sheath 1.8-2.4μ thick, colourless, not stratified; trichome 9.6-12.0μ broad,slightly constricted at the cross-walls; cells usually cylindrical,3-4 times as long as broad; heterocysts single, elongated, cylindrical, with thick convex end walls, as broad as the vegetative cells,28.8-36.0μ long (Pl. 12, fig. 5).

Reported earlier from Balmikinagar, Bihar (Sinha and Srivastava, 1980); 24-Parganas and Howrah districts, West Bengal (Mukhopadhyay and Chatterjee, 1981); River Cauvery, Karnataka (Somashekar, 1983); Vidarbha region, Maharashtra (Kolte and Goyal, 1985); Sultanpur, Uttar Pradesh (Khan, 1985); Maharashtra (Patil and Satav, 1986).

Collected from paddy fields at Ramchandrapur, Howrah district(Collection No. 77), Girat near Guptipara,Hooghly district (Collection No. 275).

Voucher No. 112 (Collection No. 275).

**TOLYPOTHRIX Kutzing**

Phyc.gene., 227, 1843; Bornet and Flahault, Revision des Nostocacees heterocystees, 118, 1887.

**Key to the Species**

1. Trichomes 4.8-5.2μ broad; sheath hyaline, not stratified; heterocyst single........................................ **1. *T. nodosa***

1. Trichomes 6.0-8.4μ broad; sheath faintly yellow, almost parallely stratified; heterocyst also in short chain........................................ **2. *T. tenuis***

***Tolyppthrix nodosa*** Bharadw. The Taxonomy of ***Scytonema*** and ***Tolypothrix*** ,Rev. algol., Paris, 7: 175, fig. 7C, 1934; Desikachary, T.V. Cyanophyta. 494, pl. 98, fig. 5, 1959.

Thallus mucilaginous, blue-green to brownish; filaments variously bent, closely entangled, false branching scanty, heterocyst at the base of branch keeps a space from the other trichome, 6.0-7.2μ broad; sheath thin, hyaline, close to the trichome, uneven at surface, not stratified; trichome slightly constricted at the cross-walls, uniformly broad throughout; cells very irregular in length, always longer than broad, about 1.5 to 4 times longer than broad,4.8-5.2μ broad; heterocysts both basal and intercalary, single, ovoid, 6.0-6.6μ broad, 9.0-10.2μ long (Pl. 13, fig. 2).

This species has been reported from Benaras, Uttar Pradesh(Bharadwaja, 1934; Rao, 1937); Andhra Pradesh (Rao. 1938); Bankura district, West Bengal (Sinha and Mukherjee, 1975); Rohilkhand division, Uttar Pradesh (Pandey, 1982); Kolaghat, Midnapore district, West Bengal (Pal and Santra, 1985).

Collected from a paddy field at Dainhat, Burdwan district (Collection No. 38).

Voucher No. 113 (Collection No. 38).

***Tolypothrix tenuis*** (Kutz.)Johs. Schmidt ern.Bot. Tidsskr., 22: 383, 1899; Fremy, Myxo.d'Afr.equat. franc., 286, fig. 251, 1929; Desikachary, T.V. Cyanophyta. 494, pl. 102, fig. 4,1959.

Thallus brownish green, cushion like; filaments false branched upto about 1.5 cm long, 7.2-12.0μ broad; sheath thin, close to the trichome, faintly yellowish, almost parallely stratified; cells 6.0-8.4μ broad, 1-1½ as long as broad, in the apical region quadrate, slightly constricted at the cross-walls; heterocysts ovoid, rarely cylindrical, 5.4-7.2μ broad, in a series of 2 to 4 forming short chain, yellowish (Pl. 13, fig. 3).

Formerly reported from Bombay, Maharashtra (Dixit, 1936);Benares, Uttar Pradesh (Rao, 1937); Travancore, Kerala (Parukutty,1940); Delhi (Rao, 1940); Mysore State (Kamat, 1974); Bankura district, West Bengal (Sinha and Mukherjee, 1975); Nagpur, Maharashtra(Kamat and Hgrankhedkar, 1976); Lucknow, Uttar Pradesh (Prasad andMehrotra, 1980); Rohilkhand division, Uttar Pradesh (Pandey, 1982);Burdwan district, West Bengal (Chatterjee and Chatterjee, 1983);Marathwada, Maharashtra (Chaporkar and Gangawane, 1984); Vidarbharegion, Maharashtra (Kolte and Goyal, 1985); Meghalaya (Reddy *et al.*, 1986); Maharashtra (Patil and Satav, 1986).

Growing on rice-fields at Balagarh near Guptipara, Hooghly district (Collection No. 287), Bhandartikuri near Purbasthali, Burdwan district(Collection No. 314).

Voucher No. 114 (Collection No. 314).

**AMPHITHRIX Kutzing**

*Amphithrix* Kutzing, 1843; emend., Bornet and Flahault, 1886.

***Amphithrixjanthina*** (Mont.) Bornet . Flah. Smith, G.M. The Fresh water algae of the United States, 598, 1950.

Thallus forming a thin layer on moist brick built wall of water reservoir, associated with pinnate diatoms, dark olive-green; trichomes unbranched, gradually narrow at the distal part, parallely arranged on a stratum of interwoven trichomes assuming irregularly arranged cells, 60.6-72.0μ (-80μ) long, 3.3-3.6μ broad at the base, blue-green; cells quadrate or slightly longer than broad, 3.0-4.2μ long; hormogones mostly at the distal part of trichomes (Pl. 13, fig. 4).

This species has been reported for the first time from West Bengal as well as from India.

Collected from unfilter water tank at Masjidbari Street, Kolkata (Collection No. 356).

Voucher No. 115 (Collection No. 356).

**RIVULARIA (Roth) Ag.**

Systema Algarum, 19, 1824; Bornet and Flahault, Revision des Nostocacees heterocystees, 345, 1887.

***Rivularia aquatic*** a De Wilde. Alg. rapp. J. Massart, Ann. Buitenzorg, suppl. 1: 40, 1897; Desikachary, T.V. Cyanophyta. 552, pl.116, figs. 2 & 3. 1959.

Thallus soft, more or less spherical, attached to the submerged aquatic plants, small, about 2-3 mm diam., olivaceous; filaments closely arranged in the thallus, sheath hyaline, thin; trichomes broad at the base and gradually narrow towards the apices, ending in a long fine hair; cells about 4 times longer than broad at the base and sometimes more towards the apex, basal cells 7.2-7.8 μ broad;heterocyst spherical, single, basal, 10.8-12.0 μ diam, (Pl 14, fig.5).

This species has been reported from Rohilkhand division, Uttar Pradesh (Singh and Chaturvedi, 1970); Saharanpur district, UttarPradesh (Pal and Yadav, 1974); Meerut, Uttar Pradesh (Bendre and Kumar, 1975); Balmikinagar, Uttar Pradesh (Sinha and Srivastava,1980); Allahabad, Uttar Pradesh (Pandey and Pandey, 1982).

Free floating in a pool at Palpara, Howrah district (Collection No. 78); in a pond at Sashan Road, South 24-Parganas (Collection No. 370).

Voucher No. 116 (Collection No. 78).

## GLOEOTRICHIA Ag.

Alg. Maris et Adriati., 8, 1842; Bornet and Flahault, Revision des Nostocacees heterocystees, 365, 1886.

### Key to the Species

1. Trichome 10.2-10.8μ broad at base;cells up to 3 times broader than long at base; spores 39.6-49.2μ long...............................................**1. *Gl. raciborskii var. kashiense***

1.Trichome 7.2-8.4μ broad at base; cells slightly longer than broad; spores176.4-194.4μ long ........................................................................................................ ***2. Gl. pisum***

***Gloeotrichia raciborskii*** Woloszynska var. ***kashiense*** Rao, C.B.The Myxophyceae of the United Provinces, III, Proc. Indian Acad. Sci., B, 6: 351, fig. 3 A-E, 1937; Desikachary, T.V. Cyanophyta. 563, pl. 117, figs. 2-6, 1959.

Thallus soft, globose or almost ellipsoidal, 1.5-3.5 cm. diam., pale blue-green; filaments less densely packed, sheath yellowish brown, stratified, prominent at the basal spore and a part at the base of trichome; trichome prominently constricted at the cross-walls, gradually narrow towards apex and ending in a long hair, 10.2-10.8μ broad at the base; cells up to 3 times broader than long at the base, almost quadrate near the middle and 8-12 times longer than broad at the apical hairy part, basal cells 10.2-10.8μ broad, 3.6-4.2μ long, pale blue-green; heterocyst spherical or slightly ovoid, single, basal, 10.2-10.8μ broad, 10.2-11.4μ long; spores long, cylindrical, some-times spindle shaped, next to the basal heterocyst, outer wall faintly yellow, smooth, 11.4-18.0μ broad, 39.6-49.2μ long, covered by sheath (Pl. 14, fig. 6).

Formerly reported from Benaras, Uttar Pradesh (Rao, 1937);Puri and Cuttack, Orissa (Rao, 1938); Rohilkhand division, UttarPradesh (Singh and Chaturvedi, 1970); Meerut, Uttar Pradesh (Bendreand Kumar, 1975); Bhubaneswar, Orissa (Mohanty, 19&2); Rohilkhand division, Uttar Pradesh (Pandey, 1982); Diamond Harbour, 24-Parganas district, West Bengal (Maity and Santra, 1985); Gosaba, South 24-Parganas, West Bengal (Santra *et al.*, 1988).

Collected from a shallow pond with aquatic plants at Gocharan,South 24-Parganas district (Collection No. 367).

Voucher No. 117 (Collection No. 367).

***Gloeotrichia pisum*** Thuret ex Born.et Flah. Thuret, Essai de Class.des Nostocacees, Ann. Sci. nat. Sci., ser. 6, 1: 382, 1875;Bornet and Flahault, Revision des Nostocacees heterocystees,366, 1886; Desikachary, T.V. Cyanophyta. 555, pl. 116, figs.4,5, 1959.

Thallus solid, round or sometimes cylindrical, 1.5 mm or so diam., rarely larger, attached on some aquatic plants, brownish;filaments more or less densely packed, pressed together; sheath thin, delicate, hyaline; trichome ending in a long hair, 7.2-8.4 μ broad at the base; cells slightly longer than broad, 8.4-10.8μ long,olivaceous; heterocysts almost spherical, 10.8-12.0 μ broad, 10.8-11.4μ long; spores cylindrical, next to heterocyst, 10.8μ broad,176.4-194.4μ long, thick-walled, with smooth nearly brown outerwall, covered by sheath, about 2.0-2.5μ thick (Pl. 14, fig. 7).

Formerly reported from Calcutta, West Bengal (Biswas, 1932,1942; Banerji, 1938); Bankura district, West Bengal (Sinha andMukherjee, 1975); Rohilkhand division, Uttar Pradesh (Pandey, 1982);Pusa, Bihar (Jha *et al.*, 1986); Gondalpara, Hooghly district, WestBengal (Gupta and Sen, 1987).

Growing on paddy fields at Gondalpara, Hooghly district (Collection No. 173), Chanditala, Dankuni, Hooghly district (Collection No. 337).

Voucher No. 118 (Collection No. 173).

## CALOTHRIX Ag.

Syst. Alg., 24, 1824; Bornet and Flahault, Revision des Nostocacees heterocystees, 345, 1886.

### Key to the Species

1. Sheath stratified; trichomes with a terminal hair ......................... **1. *C. parietina***

1.Sheath not stratified; terminal hair absent ......................................................2

2. Filaments epiphyte on other alga,short, less than 100μ in length.................................................................................................................................**2. *C. brevissima***

2. Filaments not epiphytic, longer than 100μ .....................................................3

3. Thallus tufted; filaments intertwined by their bases; sheath projecting beyond the trichome; trichome swollen at base ....................................................... ***3. C. elenkinii***

3. Thallus membranous; filaments laterally, closely, parallely arranged; sheath generally over-arching the trichome; trichome not swollen at the base.......................................................................................................... **4. *C. membranacea***

***Calothrix parietina*** Thuret ex Born, et Flah.Thuret, Essai class.Nostochinees, Ann. Sci. nat. Bot., ser. 6, 1: 381, 1875;Bornet and Flahault, Revision des Nostocacees heterocystees, 366, 1886; Desikachary, T.V. Cyanophyta.538, pl. 108, figs.6-8 and pl. 115, fig. 1, 1959.

Thallus deep brown; filaments sometimes erect, single but clumped together by sheaths into an expanded thallus; trichomes having individual sheaths, sometimes lateral ones held together by one's base in a common sheath forming false branching, straight or undulated, broadest but not so swollen at the base and gradually narrow towards the apical part, ending into a terminal prominent hair, 6.6-7.8μ broad; sheath fairly thick, close to the trichome, stratified, extended almost throughout the length, 1.2-1.5μ thick; cells broader than the length, about 1/3 as long as broad, quadrate at the narrowed upper part of trichome, elongated at the terminal attenuated part; heterocyst single basal, rarely intercalary, basal heterocysts hemispherical, 7.8-8.4μ broad, intercalary heterocysts cylindrical (Pl. 14, fig. 1)

Formerly reported from Calcutta, West Bengal (Martens, 1870;Prain, 1905); Shembaganur, Tamilnadu (Fremy, 1942); Lucknow, UttarPradesh (Prasad and Mehrotra, 1980); Burdwan district, West Bengal(Chatterjee and Chatterjee, 1981).

Growing on uncultivated paddy field near Nabadwip Dham Railway Station, Nadia district (Collection No. 54).

Voucher No. 119 (Collection No. 54).

***Calothrix brevissima*** West, G.S. Freshwater Alg. Tanganyika, J. Linn.Soc. Bot., Lond., 38: 180, pl. 10, fig. 8, 1907; Desikachary,T.V. Cyanophyta. 533, pl. 110, fig. 8 and pl. 114, fig. 1,1959.

Filaments epiphytic on free floating ***Pithophora*** sp., single or sometimes aggregated, adjacent more than one filaments held together at base by a common sheath, short, 48.0-84.0μ long, 5.2-6.6μ broad; trichomes short, 36.0-54.0μ long; breadth more or less uniform, sometimes slightly narrow towards the apices, without a hair, slightly constricted at the cross-walls; sheath close to the trichome, thin, colourless, projecting beyond the trichomes, not stratified, almost cylindrical; cells shorter than broad, basal one sometimes longer and almost as long as broad; heterocysts basal sub spherical, about 6.0μ broad (Pl. 14, fig. 2).

Formerly reported from Benaras, Uttar Pradesh (Rao, 1937); Karnataka (Bongale and Bharati, 1980); Burdwan district, West Bengal (Chatterjee and Chatterjee, 1983); Jammu and Kashmir (Goyal *et al.*, 1984); Vidarbha region, Maharashtra (Kolte and Goyal, 1985).

Collected from rice fields at Begpur near Purbasthali, Burdwandistrict (Collection No. 37); Kaliganj near Debagram, Nadia district (Collection No. 132); Kalna, Burdwan district (Collection No. 277).

Voucher No. 120 (Collection No. 132).

***Calothrix elenkinii*** Kossinskaja, Not. Syst. Crypt. Inst. Horti.Bot. Petropol., 3: 11, 1924, Desikachary, T.V. Cyanophyta,531, pl. 114, figs. 5, 6, 11, 1959.

Filaments fairly long, about 100μ or more in length, intertwined with one another by their bases into a long tufted thallus, nearly straight, bent near the swollen base, base 7.2-8.4 μ broad, 4.8- 5.4 μ in the middle; sheath closely covering the trichome and projecting beyond, hyaline, thin, not stratified; trichomes at the base 5.4-6.6μ broad, in the middle 3.6-4.2μ broad, blue-green, gradually attenuated, apical hair absent; cells constricted at the cross-walls, quadrate, sometimes slightly shorter than broad; heterocysts single, basal, 4.2-6.0μ broad (Pl. 14, fig. 3).

Formerly recorded from Travancore, Kerala (Parukutty, 1940); Saharanpur district, Uttar Pradesh (Pal and Yadav, 1974); Lucknow, Uttar Pradesh (Prasad and Mehrotra, 1980); Marathwada, Maharashtra (Ashtekar and Kamat, 1980); Karnataka (Bongale and Bharati, 1980); Burdwan district, West Bengal (Chatterjee and Chatterjee, 1983); Vidarbha region, Maharashtra (Kolte and Goyal, 1985); Nagaland and Tripura (Reddy *et al.*, 1986).

Growing on rice field at Dey Para near Dainhat, Burdwan district(Collection No. 265).

Voucher No. 121 (Collection No. 265).

***Calothrix membranacea*** Schmidle, Beitr. Zur Algenfl. Afrikas, Engler's Bot. Jahrb., 30: 61, pl. 2, figs. 12-14, 1901; Desikachary, T.V. Cyanophyta. 542, pl. 106, fig. 10, 1959.

Thallus membranous, composed of laterally and closely arranged filaments, blue-green; filaments long, upto 180μ long, arranged more or less parallely in a thallus; sheath close to the trichome, colourless, not lamellated, generally over-arching the trichome ends; trichomes not swollen at the base, torulose, deeply constricted at the cross-walls, apical part sometimes slightly narrow, not ending into a hair, 4.2-6.0μ broad; cells shorter than broad, sometimes cells at the basal part almost quadrate, apical cells narrower; heterocysts single, basal, slightly longer than broad, 4.2-5.4μ broad (Pl. 14, fig. 4).

Formerly recorded from Allahabad, Uttar Pradesh and Tamilnadu (Mitra, 1951); Karnataka (Bongale and Bharati, 1980); Lucknow, Uttar Pradesh (Prasad and Mehrotra, 1980); Konkan area, Maharashtra (Sardeshpande and Goyal, 1981); 24-Parganas and Howrah districts, West Bengal (Mukhopadhyay and Chatterjee, 1981); Rohilkhand division,Uttar Pradesh (Pandey, 1982); Jammu and Kashmir (Goyal *et al.*, 1984); Vidarbha region, Maharashtra (Kolte and Goyal, 1985).

Collected from paddy fields at Jagadanandapur near Dainhat, Burdwan district (Collection No. 29), Matiari near Debagram, Nadia district (Collection No. 190); Simlon near Kalna, Burdwan district (Collection No. 282); Swarupganj near Nabadwip,Nadia district (Collection No. 343).

Voucher No. 122 (Collection No. 29).

## STIGONEMA Ag.

Systema Algarum, 20, 1824; Bornet and Flahault, Revision des Nostocacees heterocystees, 62, 1887 (excl.*Fischerella* Born and Flah.).

***Stigonema hormoides*** (Kutz.) Born, et Flah. Revision des Nostocacees heterocystees, 68, 1887; Desikachary, T.V. Cyanophyta.604, pl. 134, figs. 1-4, 1959.

Thallus slightly chocolate brown, forming a membranous layer, composed of entangled filaments; filaments long, 8.4-10.8μ broad, irregularly branched, lateral branches erect, as broad as the main filament, ends not attenuated; sheath thick in mature filaments and in the branches, homogenous, yellowish, sometimes yellowish brown, outer surface smooth; cells mostly in a single row, subspheri cal; heterocysts very few, intercalary, single, broader than long (Pl.14, fig. 8).

Formerly collected from neighbourhood of Bombay, Maharashtra (Schmidle, 1900); Kodai Kanal, Tamilnadu (Fremy, 1927, 1942); Bakkhali, 24-Parganas district, West Bengal (Maity and Santra, 1985); Andaman and Nicobar Islands (Prasad and Srivastava, 1986); Bakkhali, South 24-Parganas district, West Bengal (Santra *et al.*, 1988)

In stagnant water of a small pond at Falta, South 24-Parganas district (Collection No. 366).

Voucher No. 123 (Collection No. 366).

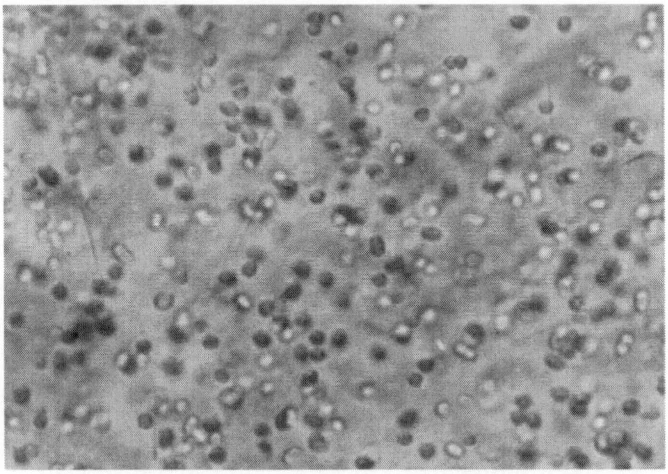

**Microcystis flos-aquae (Wittr.) Kirchner (X250)**

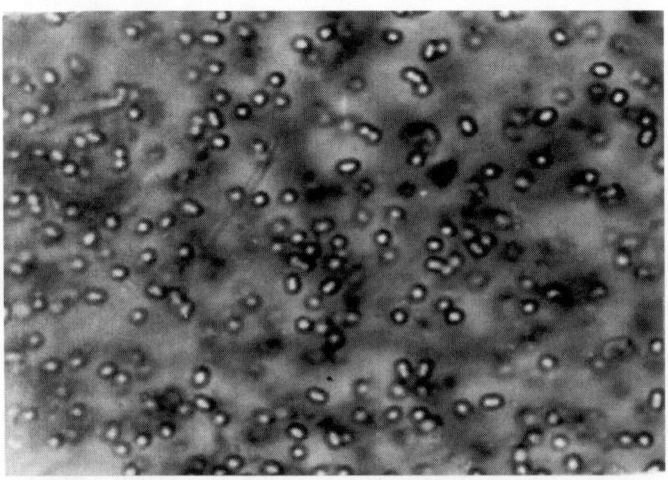

**Aphanocapsa grevillei (Hass.) Rabenh. (X200)**

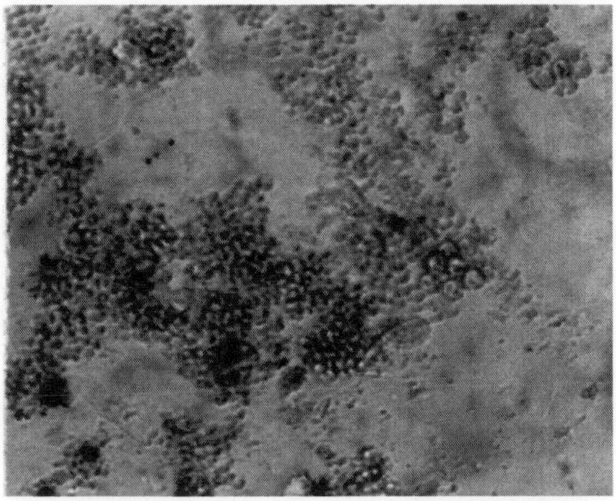

**Aphanocapsa montana Cramer (X250)**

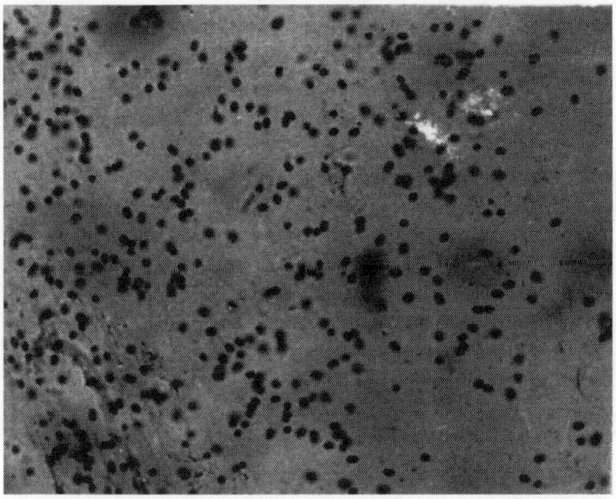

**Chlorogloea fritschii Mitra (X100)**

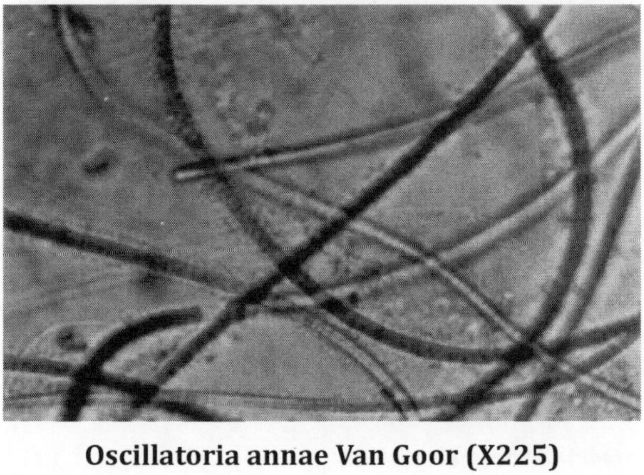

**Oscillatoria annae Van Goor (X225)**

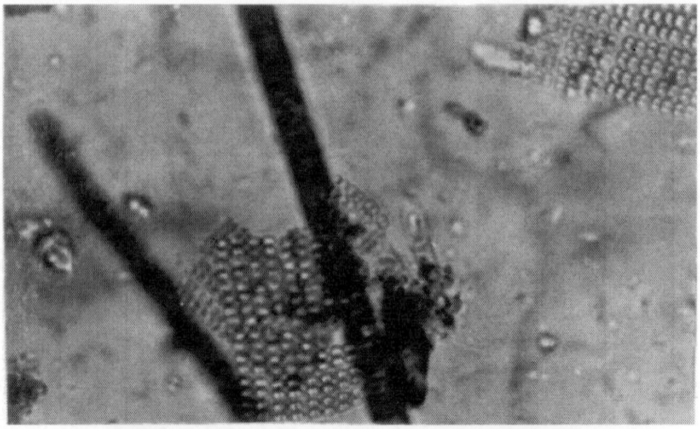

**Merismopedia tenuissima (X250)**

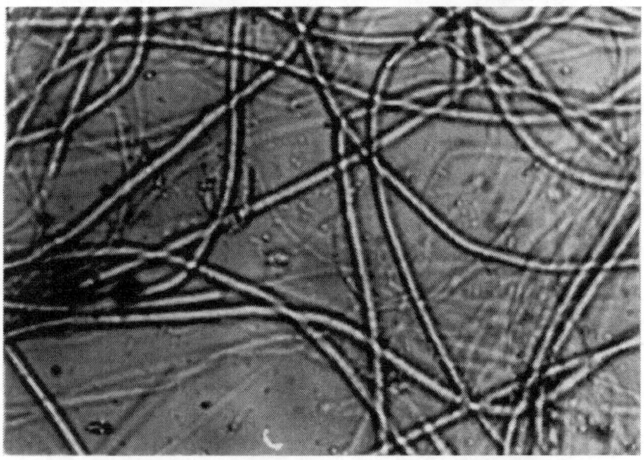

**Oscillatoria limosa Ag. ex Gom. (X110)**

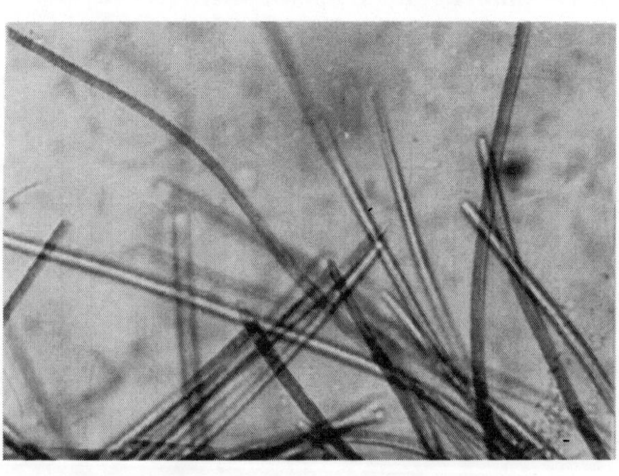

**Oscillatoria chalybea (Mertens) Gom. (X130)**

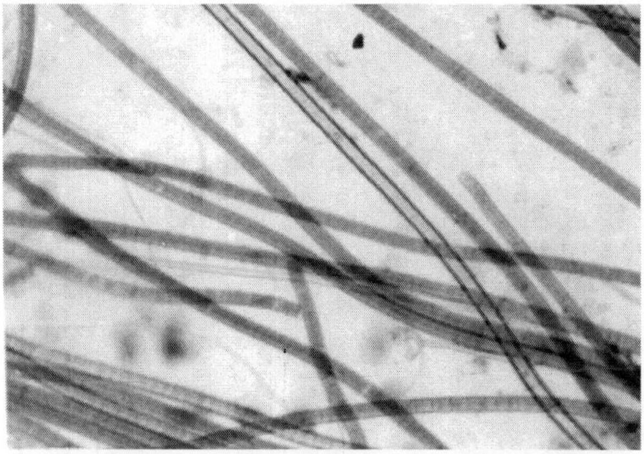

**Oscillatoria princeps Vouch. ex Gom. (X100)**

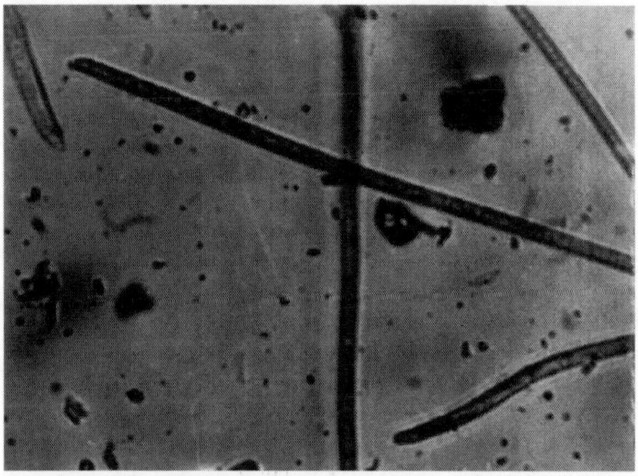

**Oscillatoria formosa Bory ex Gom. (X300)**

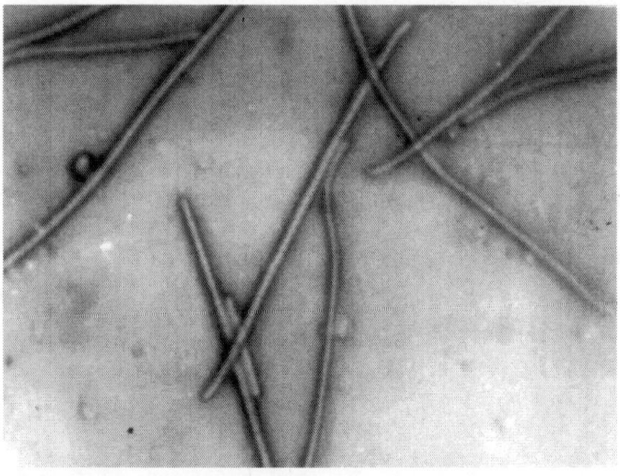

**Oscillatoria pseudogeminata G. Schmid. var. unigranulata Biswas (X500)**

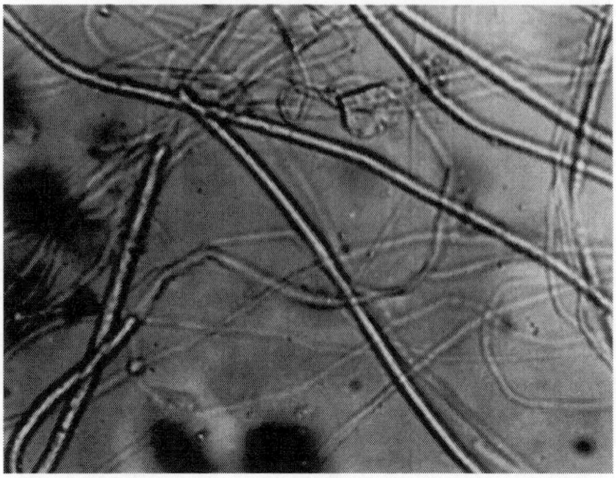

**Oscillatoria amphibia Ag. ex Gom. (X150)**

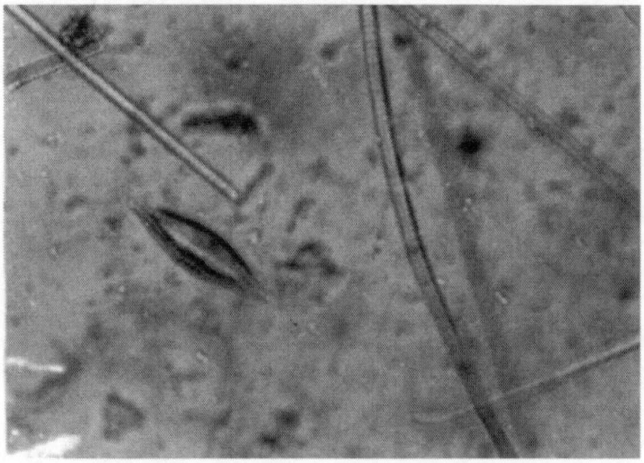

**Oscillatoria salina Biswas (X400)**

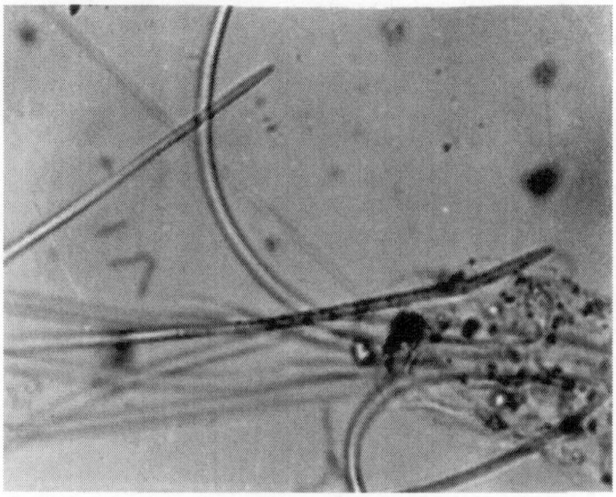

**Oscillatoria animalis Ag. ex Gom. (X300)**

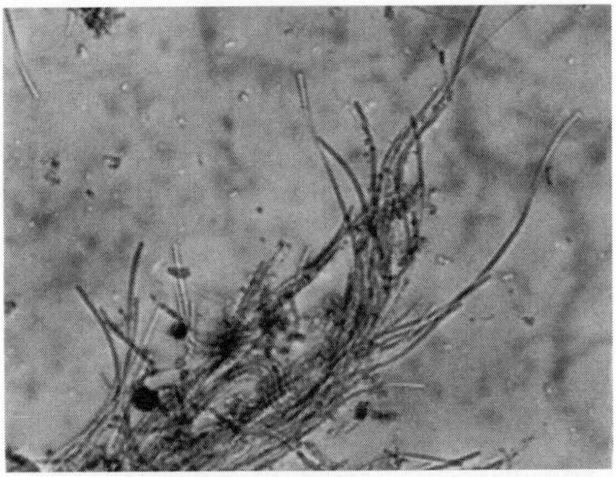

**Oscillatoria acuta Bruhl et Biswas, Orth. mut. Geitler (X100)**

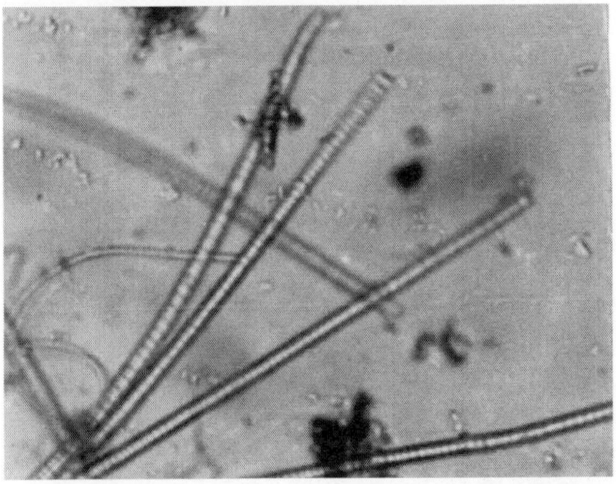

**Oscillatoria irrigua (Kutz.) Gom. (X250)**

**Oscillatoria rubescens DC. ex Gom. (X750)**

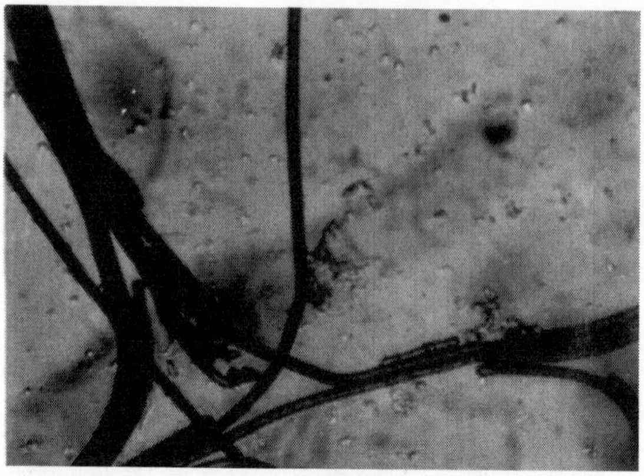

**Oscillatoria simplicissima Gom. (X125)**

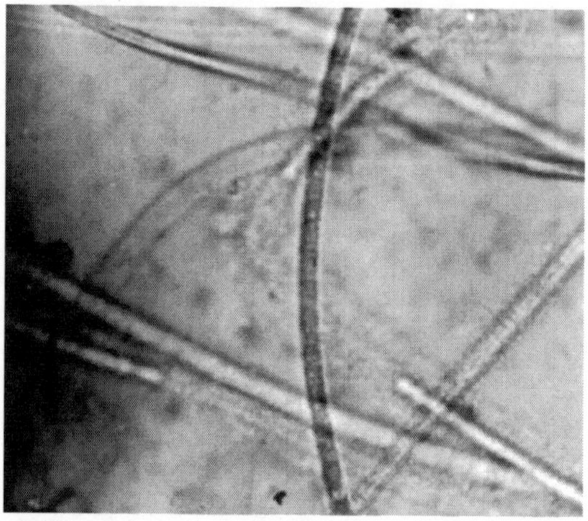

**Oscillatoria obscura Bruhl et Biswas (X500)**

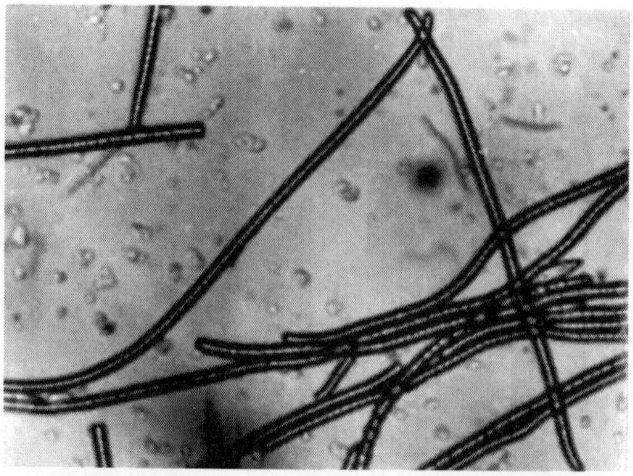

**Oscillatoria brevis (Kutz.) Gom. (X200)**

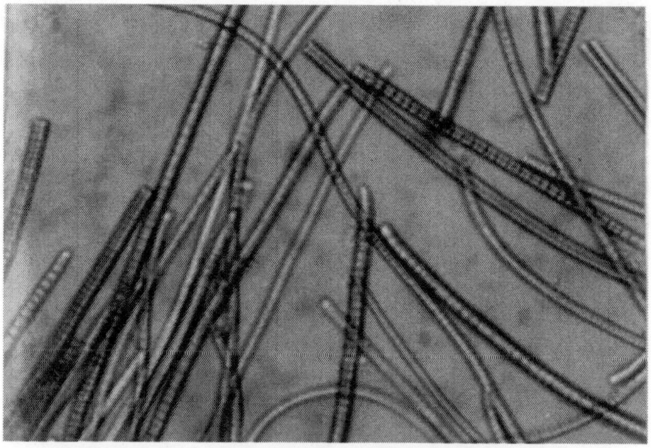

**Oscillatoria subbrevis Schmidle (X300)**

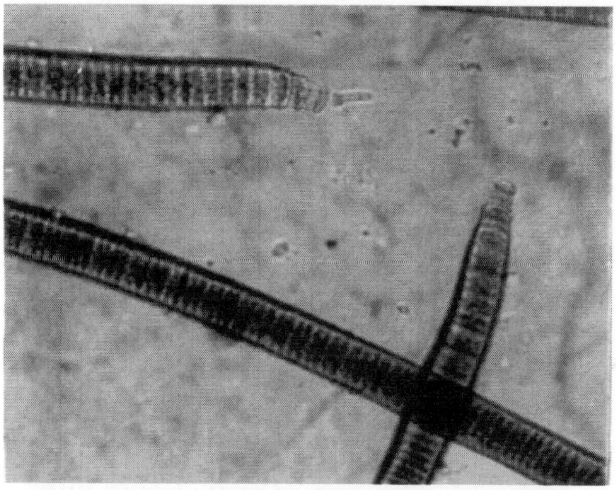

**Oscillatoria prolifica (Grev.) Gom. (X1250)**

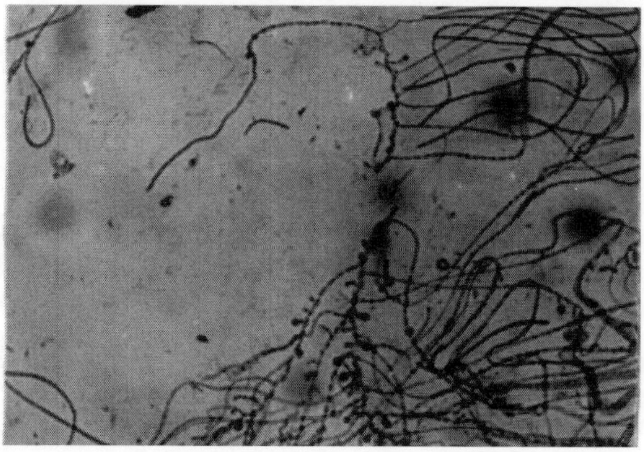

**Anabaena torulosa (Carm.) Lagerh. ex Born et Flah. (X100)**

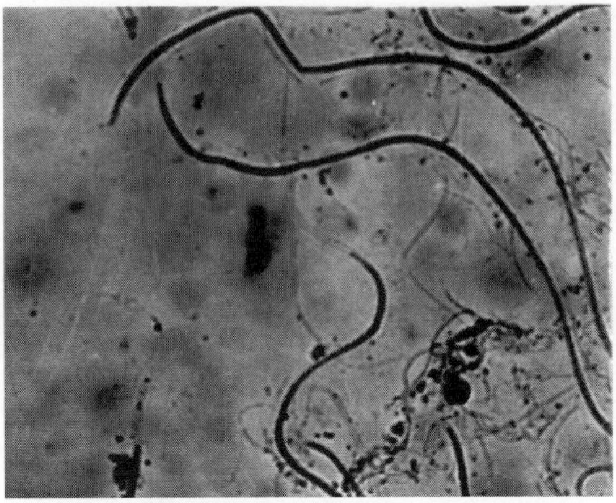

Anabaena iyengarii Bharadwaja (X150)

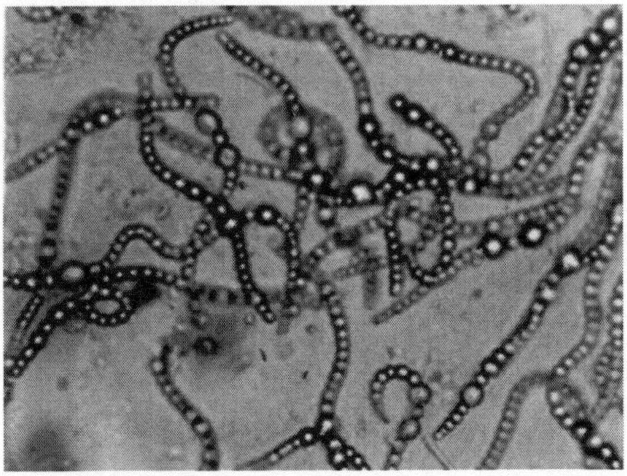

Anabaena sphaerica Born et Flah. (X200)

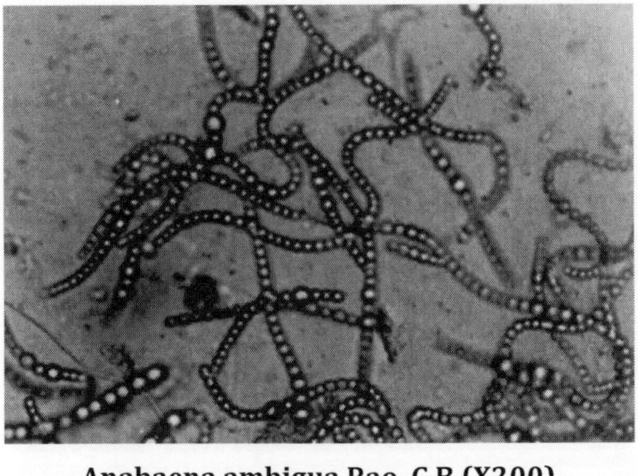

Anabaena ambigua Rao, C.B.(X200)

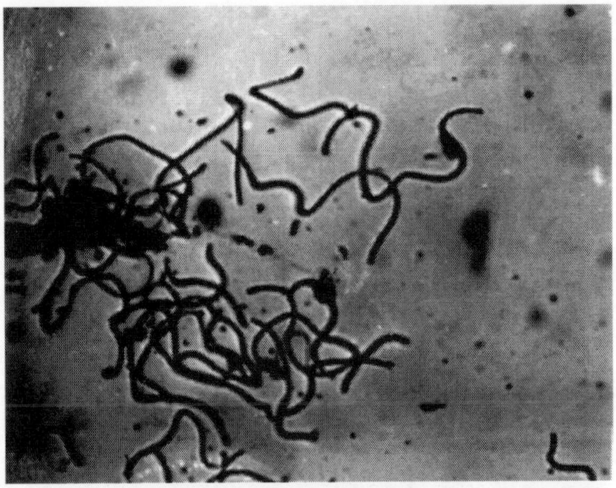

**Nostoc piscinale Kutz. ex Born et Flah. (X100)**

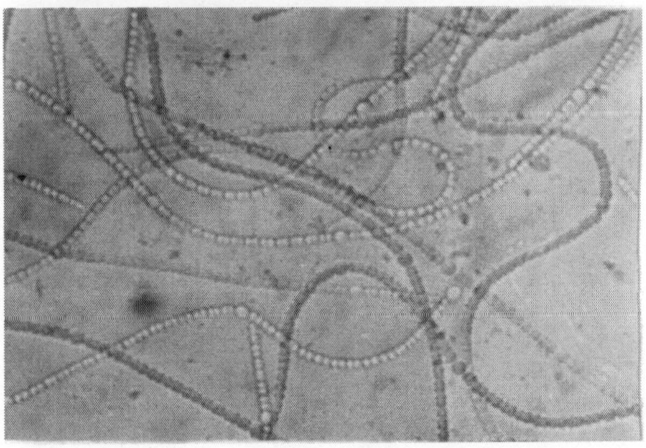

**Nostoc sphaericum Vouch. ex Born et Flah. (X250)**

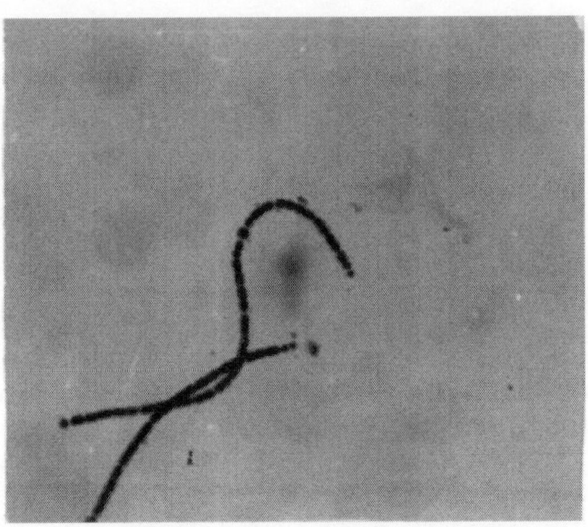

**Nostoc commune Vouch. ex Born et Flah. (X150)**

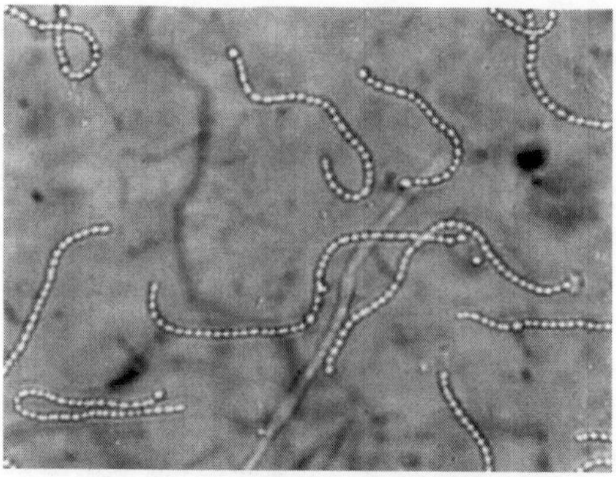

**Nostoc paludosum Kutz. ex Born et Flah. (X200)**

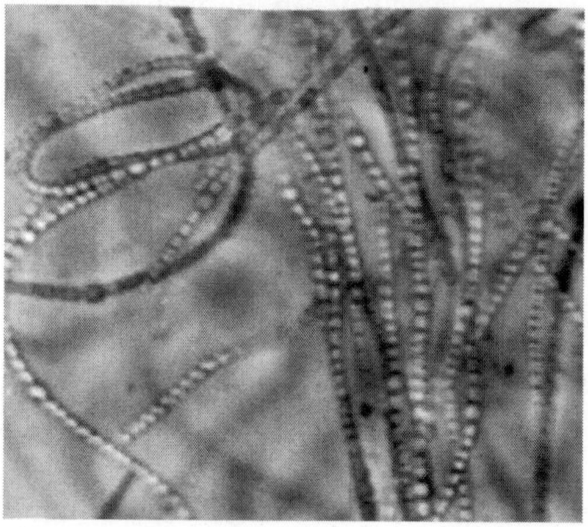

**Nostoc calicola Breb. ex Born et Flah. (X250)**

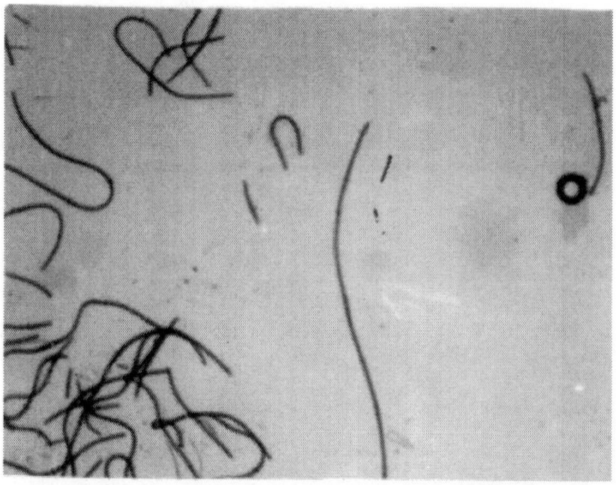

**Nostoc muscorum Ag. ex Born et Flah. (X50)**

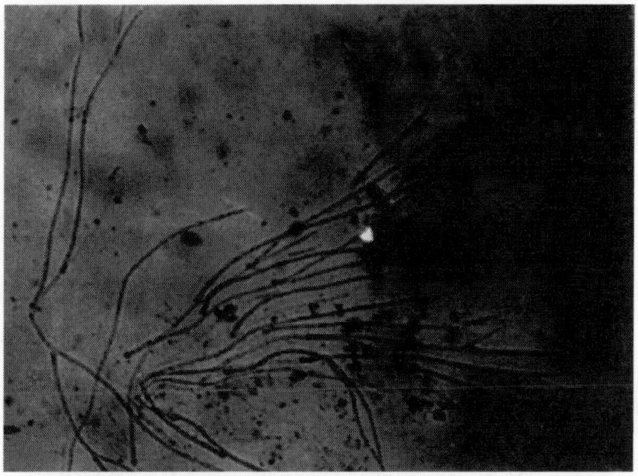

**Rivularia aquatica De Wilde (X50)**

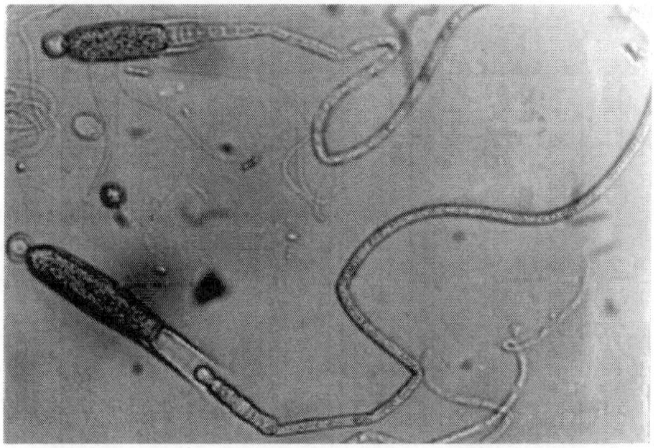

**Gloeotrichia raciborskii Woloszynska var. kashiense Rao, C.B. (X250)**

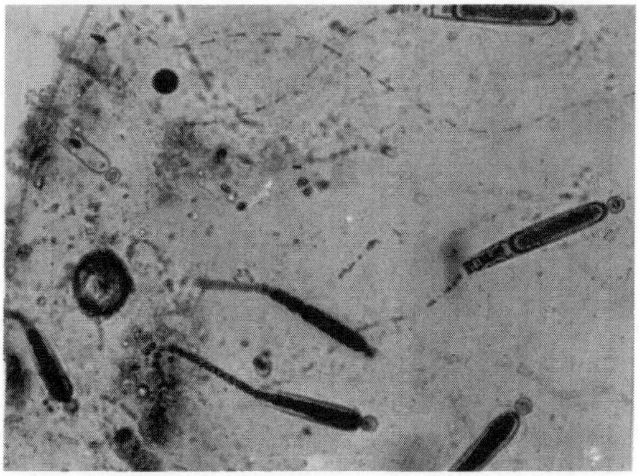

**Gloeotrichia pisum Thuret ex Born et Flah. (X200)**

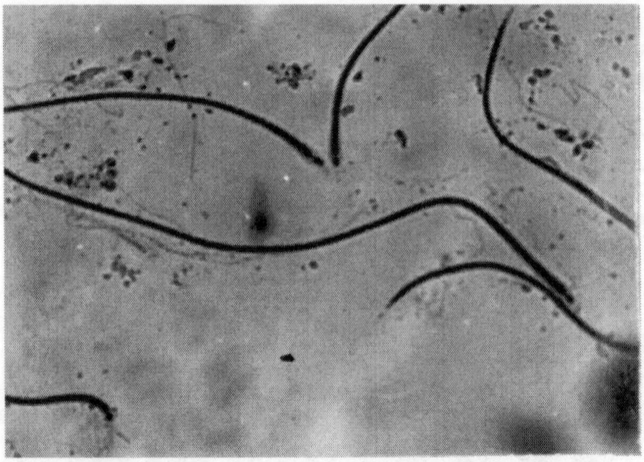

**Calothrix membranacea Schmidle (X100)**

## DISCUSSION

In the present study, documentation, enumeration and systematic accounts of the blue-green algae have been made following the monograph 'Cyanophyta' by Desikachary (1959), for the sake of convenience and following such taxonomic treatments made by different authors. However, how far the criteria used for such taxonomic classification are justified is still open to question. The situation has become more important since this prokaryotic division has been considered by the microbiologists as structu-rally diverse assemblage of Gram-negative photoautotroph eubacteria. The taxonomy of blue-green algae rests mainly on the description and identification of morphological characteristics which in general are far more distinct and recognizable than those of bacteria. Characterization were routinely made in the past on specimens collected in the field. This practice, however, introduced a measure of uncertainty as to the true value of morphological characters as taxonomic tools, since natural material is prone to considerable morphological variability depending on the prevailing environmental conditions. On the other hand, characterization based exclusively on axenic cultures may be equally inadequate, as the limited and constant physical-chemical conditions applied for the growth of laboratory cultures may restrict the full expression of the genome. It is also important to take into consideration characteristics other than morphology. Indeed, the use of as many well definable (structural, biochemical, physiological and ecological) characteristics as possible should aid to establish the whole spectrum of phenotypical expression of a particular genome.

Drouet (1968, '69, '73, '78, '81) classified the members of this group on the basis of the examination of type specimens and of thousands of living, preserved and herbarium specimens of blue-green algae from natural habitats. Geitler (1979) and Bourelly (1979) were also of the opinion that such procedures are the most fruitful in classifying the taxonomy of blue-green algae. Drouet's studies showed that organisms in one population could alter the structure of the sheath such that they could resemble many different species in several different genera. Thus he proposed that we rely on different characteristics in order to deal effectively with these genera and to produce workable means of separating genera. Drouet through his own investigations reduced the number of earlier described species to relatively few, the variants of which he considered to reflect environmental factors. Drouet (1981) classified all the unicellular and colonial forms in one order Chroococcales and the filamentous one in a second Hormogoniales. Pringsheim (1967) argued for a combination of field and laboratory studies of algae and the authors are of the opinion that this would be especially rewarding in achieving realistic classification of the blue-green algae.

During the last 25 years, following the unequivocal demonstration of the prokaryotic nature of these organisms, a number of Bacteriologists (Kantz and Bold, 1969; Baker and Bold, 1970;Stanier *et al.*, 1971, 1978, 1981; Stanier and Cohen-Bazine, 1977; Rippka *et al*, 1979; Oren and Tindall, 2005; Ramos *et al.*, 2017) emphasizing microbial techniques have focussed their efforts on growing blue-green algae in unialgal and axenic laboratory cultures. On the basis of populations in culture, they have attempted to establish the range of morphological variation under rigidly controlled environmental conditions and to elucidate biochemical and physiological characteristics of the organisms as a basis of their classification. Microbiologists have considered various other features. Among them are the nature and plane of divisions, viz. budding, binary or multiple fission, G+C percentage, ability to fix elementary nitrogen, presence or absence of special reproductive cells, constrictions between cells of filamentous form, gas vacuole, hormogone, heterocysts with their locations and other special features. So, the taxonomic representation of the organisms needs review on the perspective of present day considerations. As a result of these studies there are currently several hundred isolates of cyanobacteria maintained in pure culture. These isolates include representatives of many, but certainly not all, of the major groups of cyanobacteria described in the botanical literature.

The classification of cultured cyanobacteria has been problematic. It was hoped that phenotypic and genotypic characters, made accessible by the availability of pure cultures, would facilitate the classification of the group. However, until recently, this, in large part, had not proven to be the case. As in the traditional botanical taxonomic treatments, morphological and developmental features from the bases for the description of taxa at the level of genera and above (Castenholz, 1989a,1989b,1989c; Rippka, 1988b). Exceptions occur among some unicellular cyanobacteria that lack adequate morphological and developmental complexity to permit genera to be defined by these criteria alone (Waterbury and Rippka, 1989). However,because of the importance of structural and developmental characters for the classification of both field and cultured material,it will be possible for the system of classification currently being developed for the cyanobacteria, under the aegis of theBacteriological Code (Lapage, 1975 *et al.,*) to represent to a large degree, a logical extension and refinement of the classical botanical system.

The Cyanophycean genera worked out under the present study have also been classified into five major sections following Waterbury and Stanier (1978) and Rippka *et al.* (1979).

1. Unicellular, cells single or in aggregate, held by additional outer cell wall layers ........................................ 2

1. Filaments with intercalary cell divisions, reproduction by trichome breakage or by hormogone or by akinete ........................................ 3

2. Reproduction by binary fission or by budding........................................ Section I

2. Reproduction by multiple fission producing baeocytes or by both multiple and binary fission........................................ Section II

3. Trichomes composed of vegetative cells only, non-heterocystous, division in only one plane ........................................ Section III

3. Trichomes contain heterocystous and sometimes akinetes ........................................ 4

4. Division in only one plane ........................................ Section IV

4. Division in more than one plane........................................ Section V

The major groups or sections have been classified into subgroups. The genera obtained in present study have been assigned to the subgroups as follows:

## Section I: Unicellular cyanobacteria with budding or binary fission

1. Reproduction by binary fission, division in one plane ........................................ ***1. Aphanothece***

1. Reproduction by binary fission, division in two or three planes ........................................ ***2. Chroococcus***

***3.Microcystis***

***4. Aphanocapsa***

***5. Gloeocapsa***

***6. Chloroglea***

## Section II: Unicellular cyanobacteria having multiple fission (Pleurocapsa group)

1. Unicellular vegetative cell by multiple fission forms baeocyst.........***7. Dermocarpa***

1. Unicellular vegetative cell by binary and multiple fission forms cell aggregates, some of the cells form baeocytes by multiple fission........................................ ***8. Myxosarcina***

## Section III: Filamentous, non-heterocystous forms that divide in only one plane (Oscillatoria group)

1. Trichome helical, cells of trichome isodiametric, cylindrical or disc-like, little or no constriction between cells ........................................ ***9. Spirulina***

**10. *Arthrospira***

1. Trichome straight, cells of trichome disc-shaped, cells not separated by deep constriction. ........................................2

2.Trichome single ........................................ 3

2.Trichome in bundle ........................................ 4

3. Sheath absent........................................**11. *Schizothrix***

3.Sheath present ........................................**12. *Oscillatoria***

**13.** *Lyngbya*

**14.*Phormidium***

4. Trichomes parallel on basal stratum of cells ........................................ ***15. Amphithrix***

4.Trichomes not on any stratum of cells........................................ ***16. Microcoleus***

## Section IV: Filamentous, heterocystous form that divide in only one plane

1. Hormogone absent ........................................ 2

1. Hormogone present ........................................3

2. Heterocyst terminal ........................................ ***17. Cylindrospermum***

2. Heterocyst terminal or intercalary ........................................***18.Anabaena***

3. Heterocyst terminal or intercalary ........................................4

3. Heterocyst either terminal or intercalary ........................................ 5

4. Mature trichomes taper from base; bear terminal heterocyst ........... **19. *Calothrix***

4. Trichomes with even width ........................................ 6

5. Heterocyst intercalary ........................................ **20. *Aulosira***

5. Heterocyst exclusively terminal ........................................7

6. Trichomes not branched, vegetative cells spherical, ovoid or cylindrical; akinetes often in chain ........................................ **21. *Nostoc***

6. Trichome with pseudobranches ........................................ 8

7. Akinete adjacent to heterocyst ........................................ **22. *Gleotrichia***

7. Akinete absent ........................................ **23. *Rivularia***

8. Pseudobranches always bear terminal heterocyst ........................................ **24. *Tolypothrix***

8. Heterocysts predominently intercalary ........................................ **25. *Scytonema***

## Section V: Filamentous, heterocystous form that divide in more than one plane

Trichomes irregularly branched due to division in two planes .............**26. *Stigonema***

The unicellular forms are primitive and ancestral is evident from their simplicity and considered so by both the phycologists and the microbiologists. The other variations of forms have evolved from this basic stock. In Chroococcaceae, the unicells remain adhered by gelatinous matrix and the cells do not show any definite order of arrangement. In some members the arrangement of cells within gelatinous envelope is somewhat regular e.g., in ***Gloeocapsa.*** However, the members of Entophysalidaceae e.g. ***Chloroglea,*** present a colonial form in which the cell remain in radial rows of arrangement but do not form any typical filament. From such organisation the order Pleurocapsales might have been derived, where the parenchyma like thallus organisation is nothing but the aggregation of cells. Following Waterbury and Stanier (1978), we find a relationship between ***Dermocarpa*** and ***Myxosarcina*** where multiple fission is exhibited i.e., a series of successive binary fissions take place without intervening cell growth and the cell is cleaved into a number of baeocytes or daughter cells. In ***Chloroglea*** nannocytes are known which links the other unicellular forms with the Pleurocapsales. Origin of filamentous form of Oscillatoria group, without having heterocyst can be visualised from unicellular forms. That Pleurocapsa group might have evolved to the heterocystous form may be surmised from the fact that many of the members of the Pleurocapsa group produce nitrogenase under anaerobic con-dition (Stanier *et al.*, 1990). The existence of nif genes in non-heterocystous forms of Oscillatoria group also place their claim to derive heterocystous forms. But Pleurocapsales being a colonial form, each and every cells are capable to show nitrogenase activity and it is more probable that heterocyst might have developed from such cell. Desikachary assembled all the genera having unbranched trichomes together under the order Nostocales. He did not take the presence or absence of heterocyst into account. But Rippka *et al.* (1979) considered heterocyst as a criterion of primary classification and separated the filamentous forms into Oscillatoria group and heterocystous cyanobacteria. The schematic representation of the phylogenetical concept as proposed in the present study is placed at the next page.

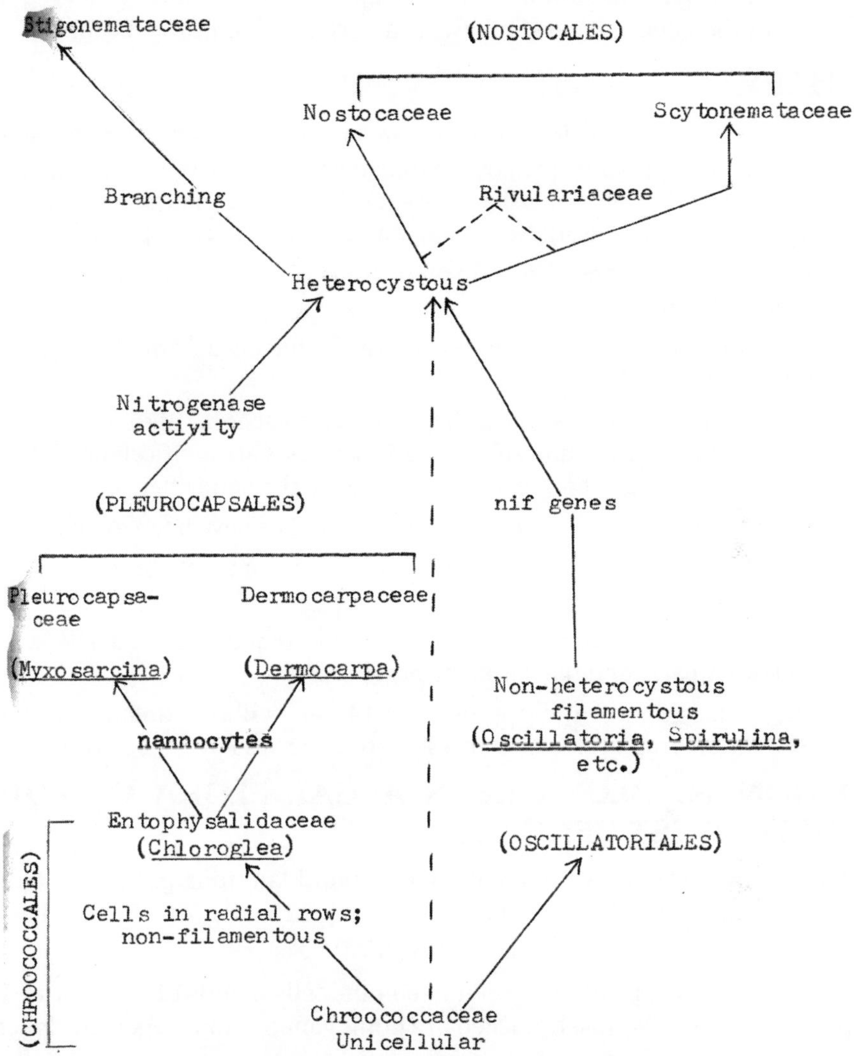


Schematic idea of suggested evolution

The classification of cyanobacteria following Rippka *et al.* (1979) and assignment of different genera shows close similarity to different sections with that made by Desikachary. The taxonomic treatment with ***Dermocarpa*** and ***Myxosarcina*** is different in classifications by Desikachary (1959) and Rippka *etal.* (1979). The former author gave emphasis on the capacity of binary fission in ***Dermocarpa*** to place it under separate order Chaemosiphonales instead of Chroococcales characterised with unicellular forms. He separated ***Myxosarcina*** from ***Dermocarpa*** on the basis of parenchymatous nature of thallus in ***Myxosarcina***. But Rippka *etal.* (1979) placed these two genera together under Pleurocapsa group as he considered ***Myxosarcina*** thallus as an aggregation of unicellular vegetative cells and both the genera have similar characteristic of multiple fission.

## SUMMARY

- The blue-green algal flora in lower Gangetic plain of West Bengal includes the districts of Midnapore (beside the Hooghly river), Howrah, Hooghly, Burdwan (Kalna and Katwa subdivisions); greater Kalcutta, 24-Parganas (North); part of 24-Parganas (South); Nadia and Murshidabad (Berhampur subdivision) .
- The main geographical features and climate of the area under study have been included.
- In taxonomical studies attention is given to the algal flora of rice fields and other cultivated lands.
- 123 species of blue green algae belonging to 26 genera have been recorded of which 19 species are unicellular and 104 species are multicellular, 70 species of blue-green algae have been obtained from the paddy fields.
- Desikachary's classification and order of families have been followed.
- Descriptions of the species are given, key to the families, genera and species has been provided.
- An account of collection of the species by different authors from West Bengal and other parts of India has been incorporated.
- The dissertation contains altogether 14 plates illustrating all the species. Microphotographs of some representative species have been given.

## ADDITION TO BLUE GREEN ALGAL FLORA OF LOWER GANGETIC WEST BENGAL

***Merismopedia tenuissima*** Lemm.Beitr.Kennt.Planktonalg., II, Bot. Centralbl. 76: 154, 1898; Geitler, Kryptogamenflora, 263, fig.129a,b, 1932; Desikachary, T.V. Cyanophyta. 154, pl. 29, fig. 7 & pl. 30, figs. 8, 9, 1959.

Thallus free floating, bluish green in colour; cells ovoid, 1.8μbroad, 16-64 or multiple number of cells closely packed in a homogeneous mucilage; multiplication by divisions in two planes.

This species has been reported earlier from Benaras, Uttar Pradesh (Rao, 1937); Andhra Pradesh (Rao, 1938b); Berhampur, Orissa (Rao, 1938a); Delhi (Rao, 1940); Meerut, Uttar Pradesh (Bendre and Kumar, 1975); Jaipur, Rajasthan (Trivedy, 1982); Nagpur, Maharashtra (Nandkar *et al.*, 1983); Khandesh, Maharashtra (Barbate and Tarar, 1983); Bankura district, West Bengal (Sinha and Mukherjee, 1984); Gosaba, South 24-Parganas district, West Bengal (Santra *et al.*,1988).

Planktonic in a pond at Rajchandrapur near Dankuni, Hooghly district (Collection No. 386). Voucher No. 124 (Collection No. 386).

# PART III
# ECOLOGICAL STUDIES OF BLUE-GREEN ALGAL FLORA OF PADDY FIELDS

# A. DISTRIBUTIONAL PATTERN IN THE RICE FIELDS AND THEIR ASSOCIATION AND SUCCESSION IN RELATION TO THE GROWTH STAGES OF RICE

## INTRODUCTION

Understanding of the rice field ecosystem has gained a considerable importance and the algal flora, specially the blue-green algae plays an important role of biological significance, beside the characters of soil and environment. The importance of blue-green algae in relation to the fertility of the rice field due to their ability to fix atmospheric nitrogen (De, 1936, 1939; Fritsch and De, 1938; Fogg and Wolfe, 1954) needs no emphasis. Besides the nitrogen fixing ones, other blue-green algae in the rice field are also important because of their effects through the addition of different organic compounds to the soil by their death and decay, by extracellular secretions, by adding certain substances which are stimulatory or inhibitory to the co-existing algae, fungi and other microorganisms. Moreover they may compete with other nitrogen fixing blue-green algae for space, nutrition and other factors needed for their growth. The antimicrobial activity of certain blue-green algae against rival bacteria (Harder, 1917; Davidson, 1961; Moikeha and Chu, 1971; Stetsenko and Stetsenko, 1977) has also been recorded. Larvicidal activities of blue-green algae has also been detected (Allen *et al*., 1957). So, for understanding of the rice field ecosystem, a study of the blue-green algae and their distributional pattern in the rice fields is imperative.

Survey of the blue-green algal flora of paddy field has been made in certain States of India (Singh, 1942, 1961; Aiyer, 1965; Pandey, 1965; Kamat and Patel, 1973; Grover and Pandhol, 1975; Sinha and Mukherjee, 1975; Prasad and Mehrotra, 1980; Sardeshpande and Goyal 1981; Chatterjee and Chatterjee, 1983; Goyal *et al*., 1984; Nayak and Prasana, 2007; Singh *et al.,* 2015; Dhanya and Ray, 2015; Basavaraja and Naik, 2018; Ghadage and Karande, 2020).In West Bengal, the survey of paddy field in respect of blue-green alga is meagre and so it demands attention.

Study of the formation, association and succession of blue-green algae in rice field has not been emphasised in West Bengal. Report on seasonal fluctuation of blue-green algae and the relative abundance of different species in rice fields with respect to the growth and maturity of the rice crop is little. It is necessary for proper understanding of the rice field ecosystem.

For the purpose of enumeration of the blue-green algae from the paddy field soil of lower Gangetic West Bengal, collections have been made from different areas of the districts - Murshidabad (excluding the Kandi subdivision), Nadia, Howrah, Hooghly, North 24-Parganas, part of South 24-Parganas and the eastern parts of Burdwan and Midnapore. Detailed distributional pattern of the blue- green algae in rice field soil and their study with respect to the growth and development of rice crop have been made in selected lowland area of three districts viz., Nadia, Burdwan and Hooghly.

Blue-green algae were collected from paddy fields in different seasons of the year. For the study of distributional pattern and the relationship between the occurrence and seasonal succession of blue-green algae and the different stages of growth of the rice crop, collections of soil for isolating the algae in culture, as well as the algal colonies on the soil have been made at regular intervals from the time of preparation of the field for cultivation to the time of harvesting of paddy. The ecological parameters e.g., pH, temperature, relative humidity and light/dark cycle were also considered to get an idea of the abiotic factors influencing the algal colonization.

# METHODOLOGY

## Collection and preservation of rice field blue-green algae

### Places of collection

Paddy cultivated fields located in and around Debagram, Nabadwip and Simurali of Nadia district; Dainhat, Purbasthali and Kalna at eastern part of Burdwan district; Guptipara, Mogra and Dankuni of Hooghly district were selected for studying distributional pattern of rice field blue-green algae. The places were selected which were not too far from river Ganga and more or less equidistant from one another along the course of the river. In those fields rice was mainly cultivated as a 'Kharif' crop (June-October). In some fields specially in Hooghly district 'Rabi' crop of rice (February-June) is also cultivated where irrigation facilities are available.

### Habit and habitat of the algae collected

Collections were made from both high-land and low-land fields, The paddy fields depended for cultivation both on rain water during the rice growing seasons as well as on irrigation. Cultivation of rice during 'Rabi' season chiefly depended on irrigation. The fields were irrigated by water available from deep and shallow tube-wells, canal and ponds. Some of the fields were regularly irrigated by water. The fields remained water-logged during and after transplantation of seedlings and also at the early stage of growth. The watering of the fields continued till the grains matured. Frequently the water level of the field was kept at 15 to 25 cm, sometimes even after ripening of the paddy where 'Ratna' variety was cultivated.

A number of blue-green algae remained at the bottom of the inundated fields forming scum covering wide area. Few of them formed planktonic flora.

### Preservation of collected materials

The collected materials were thoroughly washed with tube-well water to remove the clay particles and preserved in 2% formalin.

For retaining the natural colour, another set of the materials were treated with colouring solution (50% ethanol 90 ml, 40% formalin 4 ml, glycerol 3 ml,glacial acetic acid 3 ml and copper acetate 10g) for 7 days and then transferred to formalin-acetic acid-alcohol for preservation.

### Morphological studies and identification of collected algae

Diagnostic characters of the collected algae were studied under the microscope for the purpose of identification. In some cases prior staining of the specimens with appropriate stains were necessary.

### Stains and staining procedures

The blue-green algae were stained with 0.01% aqueous solution of ruthenium red or methylene blue for 30 minutes to study the presence or absence of mucilaginous sheath. For understanding of the nature of cell wall and the gelatinous sheath the materials were treated with chloro-zinc-iodide solution (soln. A: $I_2$– 1g, $H_2O$ - 100 ml; soln. B: $ZnCl_2$ - 2g, $H_2O$ - 1 ml) for 15 to 20 minutes. To stain the Cyanophycean granules Janus green (1:10,00,000) treatment for four hours was made. To stain heterocysts, alcian blue was used following Tuffery (1969).

The preparations were studied under light microscope.

Identification of the algal forms was based mainly on the identification key given by Desikachary (1959) and other relevant literatures when necessary.

## Study of association and succession of blue-green algae in the rice fields with respect to the environmental factors

Fortnightly visits to the rice fields under consideration in the present study were made from the period of preparation of fields for rice cultivation to the time of harvesting. Three farmer's fields were considered, one in each of the three districts. In Nadia district, the field was located at Simurali, in Burdwan district at Dainhat and in Hooghly district at Guptipara. Random sampling of the algal materials, soil and water samples were made to study the succession of algal forms. Data on the prevailing ecological condition viz., light, soil and water pH, temperature of soil and water, waterlogging, nature of irrigation, relative humidity etc. were also recorded. The succession of algal components in the fields was studied by identifying the algae in laboratory from the collections. Comparative data of the different growth phases of the rice were also taken.

### Collection of soil and water samples

Random collections of surface soil and water samples were made. With the help of sterile cork borer (diameter 2.5cm) soil cylinders were taken out and the samples kept in sterile culture tubes. Water from the field was collected in sterile culture tubes.

### Isolation of algal components

For isolation of blue-green algae in culture, about 1g of soil sample was diluted to $10^1$, $10^2$ and $10^3$ times by sterile glass distilled water and aliquots from these dilutions were used as inoculum.1 ml of each dilution was seeded on solidified Chu-10 (modified) agar medium (with or without nitrogen).

Aseptic measures and special care was taken to avoid contamination. Duplicate plates were made for each dilution. Same procedure was taken for isolation of alga from field water samples. Fogg's (modified) medium was also used for soil samples. Inoculated petri plates were kept at 20°C ± 2°C under illumination of 8,000 lux. Dark and light cycles of 14 hours by 10 hours were operated. Several small colonies of algae were found to appear on the surface of medium after 7 days. These were examined under the microscope afterwards.

### Composition of growth media

Chu-10 (modified) medium consists of: Calcium nitrate 40 mg/l, $K_2HPO_4$ 10 mg/l, $MgSO_4$,$7H_2O$ 25 mg/l, $Na_2CO_3$ 20 mg/l, $Na_2SiO_3$ 25 mg/l, Fe-citrate 3 mg/l, Citric acid 3 mg/l and 1 ml/l of A5 micronutrient solution (Boric acid 2.86 g/l, $MnCl_2$ 1.81 g/l, $ZnSO_4$ 0.222 g/l, Sodium molybdate 0.0177 g/l, $CuSO_4$ 0.079 g/l); pH of the medium was 7.0.

Fogg's (modified) medium consists of: Soil extract 15 ml/l, $KH_2PO_4$ 0.2 g/l, $MgSO_4$ 0.2 g/l, $CaCl_2$ 0.1 g/l, $A_5$ micronutrient solution 1 ml/l, Fe-EDTA 1 ml/l; pH of the medium was kept at 7.5. Soil extract was prepared by boiling 1 kg of garden soil in 1 litre of distilled water for 30 minutes followed by filtration. To prepare Fe-EDTA the following procedure was followed: 26.1 g of EDTA and 24.9g of $FeSO_4$ were dissolved in 268 ml of 1(N) KOH. The volume was made up to 1000 ml. The solution was aerated overnight to prepare a stable complex marked by the change of colour to dark brown. The solution was filtered and the volume was again made up to 1000 ml by adding glass distilled water.

## Study of ecological factors:

### Measurement of pH of soil and water samples

The pH of the soil samples was measured from 1:2 suspensions of soil in 0.01M $CaCl_2$ (Schofield and Taylor, 1955) by Beckman pH meter. In case of water samples, the pH readings were taken directly by a pH meter.

### Measurement of water temperature

With the help of an ordinary thermometer, the temperature of the water samples were recorded in the field. The atmospheric temperature was obtained from a maximum-minimum thermometer.

### Measurement of relative humidity

Relative humidity was recorded by hygrometer directly.

### Recording of light/dark cycle

From the duration of light period of the sampling dates, the light/dark cycle was recorded.

### Study of the distributional pattern of blue-green algae in the rice field soils

Soil samples were collected from Nadia district, eastern part of Burdwan district and Hooghly district of West Bengal. The regions under the present investigation were along the course of the river Ganga.

Seventy-five soil samples from nine places in the three districts were collected from farmers field during April-May, 1984. Of the 75 samples, 30 were from Nadia district, 30 from Burdwan district and 15 from Hooghly district. Collections of soil samples were made from 10 spots in a field, sun dried and mixed together thoroughly and 500 g from this was taken as representative soil sample of the field. Enrichment cultures were raised in culture medium to identify and isolate different algal forms present in the soil samples. The algal forms appearing in the culture flasks were categorised as very common, common, rare and very rare depending upon the visual observations on growth and microscopical observations of 5 to 7 slides from each flask.

The occurrence of different species of blue-green algae were numerically scored to get a quantitative picture of the distributional pattern. Scoring was done by using the numerical values attributed to the relative proportion of occurrence of different species as very common (4), common (3), rare (2) and very rare (1). Numerical total was calculated from the number of soil samples where a particular algal species was present without considering its proportionate occurrence. The number of soil sample herbouring a particular genus was multiplied by the corresponding numerical value of proportionate occurrence to obtain the rated total. Rated scores were calculated by dividing the rates total by the total number of soil samples in a particular district and the values were expressed in terms of percentage to obtain a more clear picture of the proportionate distribution of a species.

## OBSERVATIONS AND RESULTS

### Blue-green algae of paddy fields of lower Gangetic plain of West Bengal

During the study of blue-green algae of lower Gangetic plain of West Bengal, special emphasis was given to those forms growing on the paddy field soil. The following is the enumeration of the species obtained:

### Systematic enumeration

#### Non-heterocystous forms

Genus ***Microcystis*** Kutz.

- ***M.aeruginosa.*** Kutz.
- ***M. flos-aquae*** (Wittr.) Kirchn.

Genus ***Chroococcus*** Nag.

- ***C. minutus*** (Kutz.) Nag.
- ***C. Pallidus*** Nag.
- ***C. turgidus*** (Kutz.) Nag.

Genus ***Aphanocapsa*** Nag.

- ***A. biformis*** A. Br.
- ***A. montana*** Cramer

Genus ***Aphanothece*** Nag.

- *A.pallida (Kutz.) Rabenh.*
- *A. stagnina (Spreng.) A. Br.*

Genus ***Arthrospira*** Stizenb.

- ***A. Spirulinoides*** Ghose

Genus ***Spirulina*** Turpinem. Gardner

- ***S. meneghiniana*** Zanard. exGom.
- ***S. subsalsa*** Oerstd.exGom.

Genus ***Oscillatoria***Vaucher

- ***O. amoena*** (Kutz.) Gom.
- ***O.amphibia*** Ag. ex Gom.
- ***O. brevis*** (Kutz.) Gom.
- ***O. chalybea*** (Mertens) Gom.
- ***O. formosa*** Bory ex Gom.
- ***O.limosa*** Ag. exGom.
- ***O. obscura*** Bruhl et Biswas
- ***O. princeps*** Vauch. ex Gom.
- ***O. proteus*** Skuja
- ***O.pseudogeminata*** *var.* ***unigranulata*** Biswas
- ***O. splendida*** Grev. exGom.

Genus ***Phormidium*** Kutz.

- ***P. anomala*** Rao, C.B.
- ***P.corium*** (Ag.) Gom.
- ***P. molle*** (Kutz.) Gom.
- ***P. retzii*** (Ag.) Gom.
- ***P. stagnina*** Rao, C.B.
- ***P. subfuscum*** Kutz. exGom.
- ***P. tenue*** (Menegh.) Gom.

Genus ***Lyngbya*** Ag

- ***L.aerugineo-coerulea*** (Kutz.) Gom

- ***L. ceylonica*** Wille
- ***L. hieronymusii*** Lemm.
- ***L. lagerheimii*** (Mob.) Gom.
- ***L. major*** Menegh. exGom.
- ***L. spiralis*** Geitler
- ***L. stagnina*** Kutz.

Genus ***Schizothrix*** Kutz.

- ***S. arenaria*** (Berk.) Gom.
- ***S. lacustris*** A. Br. ex Gom.

Genus ***Microcoleus*** Desm.

- ***M. chthonoplastes*** Thuret ex Gom.
- ***M. paludosus*** (Kutz.) Gom.
- ***M.subtorulosus*** (Breb.) Gom.

**Heterocystous forms**

Genus ***Cylindrospermum*** Kutz.

- ***C. majus*** Kutz. ex Born.et Flah.
- ***C. musicola*** Kutz. ex Born.et Flah.

Genus ***Nostoc***Vaucher

- ***N. Calcicola*** Breb. ex Born.et Flah.
- ***N. commune*** Vouch. ex Born. et Flah.
- ***N. humifusum*** Carm. ex Born.et Flah.
- ***N. muscorum*** Ag.ex Born.et Flah.
- ***N. paludosum*** Kutz. ex Born.et Flah.
- ***N. piscinale*** Kutz.ex Born.et Flah.
- ***N.punctiforme*** (Kutz.) Hariot
- ***N. sphaericum*** Vouch.ex Born. et Flah.

Genus ***Anabaena*** Bory

- ***A. ambigua*** Rao, C.B.
- ***A. fertilissima*** Rao, C.B.
- ***A.iyengarii*** Bharad.
- ***A. naviculoides*** Fritsch
- ***A. oscillarioides*** Bory ex Born. etFlah.
- ***A. torulosa*** (Carm.) Lagerh. ex Born. etFlah.
- ***A. variabilis*** *var.* ***ellipsospora*** Fritsch

Genus ***Aulosira*** Kirch.

- ***A. fertilissima*** Ghose
- ***A. prolifica*** Bharad.

Genus ***Scytonema*** Ag.

- **S. mirabile** (Dillw.) Born.
- ***S. ocellatum*** Lyngb. ex Born. etFlah.
- ***S. simplex*** Bharad.

Genus ***Tolypothrix*** Kutz.

- ***T. nodosa*** Bharad.
- ***T. tenuis*** (Kutz.) Johs, Schmidt em.

Genus ***Calothrix*** Ag.

- ***C. brevissima*** West, G.S.
- **C. elenkinii** Kossinskaja
- ***C. membranacea*** Schmidle

Genus ***Gloeotrichia*** Ag.

- ***G. pisum*** Thur. ex Born. et Flah.

## Other algal association of the paddy fields

Following is the enumeration of algal association other than blue-green algae of the paddy fields of lower Gangetic plain of West Bengal.

## Systematic enumeration

a. Chlorophyta

- ***Chlamydomonas*** sp.
- ***Microspora stagnorus*** (Kutz.) Lagerheim
- ***Microspora*** sp.
- ***Stigeoclonium*** *sp.*
- ***Rhizoclonium reparium***
- ***Cladophora*** sp.
- ***Pithophora*** sp.
- ***Oedogonium*** sp.
- ***Spirogyra submaxima*** Transsean
- ***Spirogyra varians*** (Hassal) Kutz.
- ***Mougeotia*** sp.
- ***Zygnema*** sp.
- ***Cosmerium granulatum*** Breb.
- ***Closterium dianae*** Ehrb.
- ***Hydrodictyon reticulatum*** Roth.

- ***Chlorella vulgaris*** Biejerinck
- ***Scenedesmus bijugatus*** (Turpin) Kutz.
- ***Chara braunii*** Gmelin
- ***Chara setosa*** Klein ex Willd.
- ***Chara vulgaris*** *var.* ***nitelloides*** (A. Br.) R.D.W.
- ***Chara zeylanica*** Klein ex Willd.

b. Euglenophyta

- ***Euglend polymorpha*** Dangeard

c. Xanthophyta

- ***Vaucheria geminata*** (Vauch.) De Candolle

d. Bacillariophyta

- ***Navicula*** sp.
- ***Surirella*** sp.
- **Diatoms** (unidentified)

## Distributional pattern of blue-green algae in rice field soil in three districts of Lower Gangetic West Bengal

Proportionate distribution of different blue-green algae in three districts viz., Nadia, Burdwan and Hooghly of Lower Gangetic West Bengal is presented in Tables 2.1, 2.2, 2.3. Table 3 presents the comparative distribution of different genera in the three districts and the different genera have been arranged in descending order on the basis of their rated score.

## Association and succession of blue-green algae in the rice fields of three districts of Lower Gangetic West Bengal with respect to teh environmental factors

Association and succession of different blue-green algae in the rice fields of the three districts have been studied in respect to the abiotic ecological factors as well as in relation to the different growth phases of the concurrently growing rice plants in those fields. The observations on association and succession of blue-green algae at the three places in the three districts are presented in Tables 4.1, 4.2 and 4.3.

The vertical columns of the tables represent qualitative display of algal association as well as relative abundance between the species in the rice field. The horizontal columns indicate quantitative measure of a species. The sign + and its numerical represent relative magnitude of algal growth and the absence of the species is denoted by - sign.

Enumeration of different species of paddy field blue-green algae in the three districts are presented in Table 5.1. Table 5.2 presents species distribution of paddy field blue-green algae in three representative districts.

**Table 2.1: Proportionate distribution of different algal species in three districts of Gangetic West Bengal**

**District: Nadia**

| Species | Places | | | | | | | | | | | | | | | | | | | | | | | | | | | | | |
|---|---|---|---|---|---|---|---|---|---|---|---|---|---|---|---|---|---|---|---|---|---|---|---|---|---|---|---|---|---|---|
| | Debagram | | | | | | | | | | Nabadwip | | | | | | | | | | Simurali | | | | | | | | | |
| | 1 | 2 | 3 | 4 | 5 | 6 | 7 | 8 | 9 | 10 | 1 | 2 | 3 | 4 | 5 | 6 | 7 | 8 | 9 | 10 | 1 | 2 | 3 | 4 | 5 | 6 | 7 | 8 | 9 | 10 |
| Anabaena ambigua | +3 | - | - | - | - | - | +2 | +4 | - | - | - | - | +3 | - | - | - | +3 | - | - | - | - | - | +3 | - | - | +4 | - | - | - | - |
| Anabaena fertilissima | - | - | - | - | - | - | - | +3 | - | - | - | - | - | - | +3 | - | - | +3 | - | - | - | - | - | - | - | - | - | +3 | +2 | +4 |
| Anabaena iyengarii | - | - | - | +1 | - | - | - | - | - | - | - | - | - | - | +3 | - | - | - | - | - | - | - | - | - | - | - | - | +3 | - | - |
| Anabaena naviculoides | - | - | - | - | - | - | - | - | - | - | - | - | - | - | - | - | - | - | - | - | +1 | - | - | - | - | - | - | - | - | - |
| Anabaena oscillaroides | - | - | - | - | - | - | - | - | - | - | - | - | - | - | - | - | - | - | - | - | - | - | - | - | - | - | - | - | - | - |
| Anabaena torulosa | - | - | - | - | - | - | - | - | - | - | - | - | - | - | - | - | - | - | - | - | - | - | - | - | - | - | - | - | - | - |
| Anabaena variabilis var. ellipsoporsa | - | - | +4 | - | - | +1 | +3 | - | - | - | - | - | - | - | - | +2 | - | - | +3 | - | - | - | - | - | +4 | - | - | - | - | - |
| Aphanocpsa biformis | - | - | - | - | - | - | - | - | - | - | - | - | -- | - | - | - | - | - | - | - | - | +2 | - | - | - | - | - | - | - | - |
| Aphanocpsa montana | - | - | - | - | - | - | - | - | - | - | - | - | - | - | - | - | - | - | - | - | - | - | - | - | - | - | - | - | - | - |
| Aphanothece pallida | - | +4 | - | - | - | - | - | - | +2 | +1 | - | +5 | - | +4 | - | +1 | - | - | - | +2 | - | - | +3 | +3 | - | - | - | - | - | - |
| Aphanothece stagnina | - | - | +3 | +2 | - | - | +3 | - | - | - | +2 | - | - | - | - | - | - | - | - | - | - | - | - | - | - | - | +2 | +1 | - | - |
| Arathrospira spirulinoides | - | - | - | - | - | - | - | - | - | - | - | - | - | - | - | - | - | - | - | - | - | - | - | - | - | - | - | - | - | - |
| Aulosira fertilissima | - | - | - | - | - | - | - | - | +2 | - | - | - | - | - | - | - | - | - | - | +1 | - | - | - | - | - | - | - | +2 | - | - |
| Aulosira prolifica | - | - | - | - | - | - | - | - | - | - | - | - | - | - | - | - | - | - | - | - | - | - | - | - | - | - | - | - | - | - |
| Calothrix brevissima | - | - | - | - | - | +3 | - | - | - | - | - | - | - | - | - | - | - | - | - | - | - | - | - | - | - | - | - | - | - | - |
| Calothrix elenkinii | - | - | - | - | - | - | - | - | - | - | - | - | - | - | - | - | - | - | - | - | - | - | - | - | - | - | - | - | - | - |
| Calothrix membranacea | - | - | - | +2 | +3 | - | - | - | - | - | - | +3 | - | - | +3 | - | - | - | +2 | - | - | - | - | - | - | - | - | - | - | - |
| Chroococcus minutus | - | - | - | - | - | - | - | - | +1 | - | - | - | - | - | - | - | - | - | - | - | - | - | - | - | - | - | +3 | - | +3 | - |
| Chroococcus pallidus | +2 | - | - | +3 | +2 | - | - | - | - | - | - | - | - | - | - | - | - | - | - | - | - | - | - | - | - | - | - | - | - | - |
| Cylindrospermum majus | - | +3 | - | +2 | - | - | +3 | +3 | - | - | - | - | - | - | - | - | - | - | - | - | - | - | - | - | - | +3 | - | - | - | - |
| Cylindrospermum musicola | - | - | - | - | - | - | - | - | - | - | - | - | - | - | - | - | +2 | - | +1 | - | - | - | - | - | - | - | - | - | - | - |

| Species | Places | | | | | | | | | | | | | | | | | | | | | | | | | | | | | |
|---|---|---|---|---|---|---|---|---|---|---|---|---|---|---|---|---|---|---|---|---|---|---|---|---|---|---|---|---|---|---|
| | Debagram | | | | | | | | | | Nabadwip | | | | | | | | | | Simurali | | | | | | | | | |
| | 1 | 2 | 3 | 4 | 5 | 6 | 7 | 8 | 9 | 10 | 1 | 2 | 3 | 4 | 5 | 6 | 7 | 8 | 9 | 10 | 1 | 2 | 3 | 4 | 5 | 6 | 7 | 8 | 9 | 10 |
| *Gloeotrichia pisum* | - | - | - | - | - | - | - | - | - | - | - | - | - | - | - | - | - | - | - | - | - | - | - | - | - | - | - | - | - | - |
| *Lyngbya aerugineo-coerulea* | - | - | - | - | - | - | - | - | - | - | - | - | - | - | - | - | - | - | - | - | - | - | - | - | - | - | - | - | - | - |
| *Lyngbya ceylanica* | - | - | - | - | - | - | - | - | - | - | - | - | - | - | - | - | - | - | - | - | - | - | - | - | - | - | +3 | - | - | - |
| *Lyngbya hieronymusii* | - | - | - | - | - | - | - | - | - | - | - | - | - | - | - | - | - | - | - | - | - | | - | - | - | - | +1 | - | - | - |
| *Lyngbya lagerheimii* | - | - | - | - | - | - | - | - | - | - | - | - | - | - | - | - | - | - | - | - | - | - | - | - | - | - | - | - | - | - |
| *Lyngbya major* | - | - | - | - | - | - | - | - | - | +2 | - | - | - | - | - | - | - | - | - | - | - | - | - | - | - | - | - | - | - | - |
| *Lyngbya stagnina* | - | - | - | - | - | - | - | - | - | - | - | - | - | - | - | - | - | - | - | +1 | - | - | - | - | - | - | - | - | - | - |
| *Microcoleus chthonoplastes* | - | - | - | - | - | - | - | - | - | - | - | - | - | - | - | - | - | +2 | - | - | - | - | - | - | - | - | - | - | - | - |
| *Microcoleus paludosus* | +2 | - | +3 | - | - | - | - | - | - | +1 | - | - | - | - | - | - | - | - | - | - | - | - | - | - | - | - | - | - | - | - |
| *Microcoleus subtorulosus* | - | - | - | - | - | - | - | - | - | - | - | - | - | - | - | - | - | - | - | - | - | - | - | - | - | - | - | - | - | - |
| *Microcystis aeruginosa* | - | - | - | - | - | - | - | - | - | - | - | - | - | +2 | - | - | - | - | - | - | - | - | - | - | - | - | - | - | - | - |
| *Microcystis flosaquae* | - | - | - | - | - | - | - | - | - | - | - | - | - | +1 | - | - | - | - | - | - | - | - | - | - | - | - | - | - | - | - |
| *Nostoc calicola* | - | - | - | - | - | - | - | - | - | - | - | - | - | - | - | - | - | - | - | - | - | - | - | - | - | - | - | - | - | - |
| *Nostoc commune* | - | - | +1 | - | +3 | - | - | - | - | - | - | +2 | - | - | - | +2 | +2 | - | - | - | - | - | - | - | - | - | - | - | - | - |
| *Nostoc humifusum* | - | - | - | - | - | - | - | - | - | - | - | - | - | - | - | - | - | - | - | - | - | - | - | - | - | - | - | - | - | - |
| *Nostoc muscorum* | +4 | - | - | - | - | - | - | +3 | - | - | - | +2 | - | +4 | - | - | - | - | - | - | - | - | - | - | - | - | - | - | - | - |
| *Nostoc paludosum* | - | +2 | - | - | - | +1 | - | - | - | - | - | - | - | - | - | - | +[illegible] | - | - | - | - | - | - | - | - | - | - | - | - | - |
| *Nostoc piscinale* | - | - | - | - | - | +2 | - | - | +2 | +1 | - | - | - | - | - | +3 | - | - | - | - | - | - | - | - | - | - | - | - | - | - |
| *Nostoc punctiforme* | - | - | - | - | - | - | - | - | - | - | - | - | - | - | - | - | - | - | - | - | - | - | - | - | - | - | - | - | - | - |
| *Nostoc sphaericum* | - | - | - | - | - | - | - | - | - | - | - | - | - | +3 | - | +2 | - | - | - | +3 | - | - | - | - | - | - | - | - | - | - |
| *Oscillatoria amoena* | - | - | - | - | - | - | +3 | - | - | - | +1 | - | - | - | - | - | - | - | - | - | - | - | - | - | - | - | - | - | - | - |
| *Oscillatoria amphibia* | - | - | - | - | - | - | - | - | - | - | - | +2 | - | - | - | - | - | +2 | - | - | - | - | - | - | - | - | - | - | - | - |
| *Oscillatoria brevis* | - | - | - | - | - | - | - | - | - | - | - | - | - | - | - | - | - | - | - | - | - | - | - | - | - | - | - | - | - | - |
| *Oscillatoria chalybea* | - | - | - | - | - | - | - | - | - | +2 | - | - | - | - | - | - | - | - | - | - | - | - | - | - | - | - | - | - | - | - |

| Species | Places | | | | | | | | | | | | | | | | | | | | | | | | | | | | | |
|---|---|---|---|---|---|---|---|---|---|---|---|---|---|---|---|---|---|---|---|---|---|---|---|---|---|---|---|---|---|---|
| | Debagram | | | | | | | | | | Nabadwip | | | | | | | | | | Simuralí | | | | | | | | | |
| | 1 | 2 | 3 | 4 | 5 | 6 | 7 | 8 | 9 | 10 | 1 | 2 | 3 | 4 | 5 | 6 | 7 | 8 | 9 | 10 | 1 | 2 | 3 | 4 | 5 | 6 | 7 | 8 | 9 | 10 |
| *Oscillatoria formosa* | - | - | - | +1 | - | - | - | - | - | - | - | - | +1 | - | - | - | - | - | +1 | - | - | - | - | - | - | - | - | - | - | - |
| *Oscillatoria limosa* | - | - | - | - | - | - | - | - | - | +2 | - | - | +2 | - | - | - | - | - | - | +3 | - | - | - | - | - | - | - | - | - | - |
| *Oscillatoria obscura* | - | - | - | - | +2 | - | - | - | - | - | - | - | - | - | - | - | - | +2 | - | - | - | - | - | - | - | - | - | - | - | - |
| *Oscillatoria princeps* | +1 | - | - | - | - | - | - | +2 | - | - | - | - | - | - | +2 | - | - | - | - | - | - | - | - | - | - | - | - | - | - | - |
| *Oscillatoria proteus* | - | - | - | - | - | - | - | - | - | - | - | - | - | - | - | - | - | - | - | - | - | - | - | - | - | - | - | - | - | - |
| *Oscillatoria pseudogeminta var. unigranulata* | +3 | - | +2 | - | - | - | - | - | - | - | - | - | - | - | - | - | - | - | - | - | - | - | - | - | - | - | - | - | - | - |
| *Oscillatoria splendida* | - | - | - | +1 | - | - | +2 | - | - | - | - | - | - | - | - | - | +2 | - | - | - | - | - | - | +3 | - | - | - | - | - | - |
| *Phormidium anomala* | - | +1 | - | +2 | - | - | - | +1 | - | - | +1 | - | - | - | +1 | - | - | +1 | +3 | - | +2 | - | - | - | - | - | - | +1 | - | - |
| *Phormidium corium* | +1 | - | - | - | - | - | - | - | - | - | - | - | - | +2 | - | - | - | - | - | - | +1 | - | - | - | - | - | - | - | - | - |
| *Phormidium molle* | +2 | - | - | - | - | +1 | +1 | - | - | - | +2 | - | - | +1 | - | +2 | - | - | - | - | - | +1 | - | - | - | +1 | - | +2 | - | - |
| *Phormidium retzii* | +2 | +2 | - | - | - | - | - | - | - | - | - | - | - | - | +2 | - | - | - | - | - | - | - | - | - | - | - | - | - | - | - |
| *Phormidium stagnina* | - | - | - | +1 | - | - | +2 | - | - | - | - | - | - | - | - | - | - | - | +1 | - | - | +2 | - | - | - | - | - | - | - | - |
| *Phormidium subfuscum* | - | - | - | - | - | - | - | - | - | - | - | - | - | - | - | - | - | - | - | - | - | - | - | - | - | - | - | - | - | - |
| *Phormidium tenue* | - | +3 | - | - | - | +2 | - | - | +3 | - | +1 | - | - | - | - | - | - | - | - | - | +2 | - | - | - | - | - | - | +1 | - | - |
| *Schizothrix arenaria* | - | - | - | - | - | - | - | - | - | - | - | - | - | - | - | - | - | - | - | - | - | - | - | +1 | - | - | - | - | +2 | - |
| *Schizothrix lacustris* | - | - | - | - | - | - | - | - | - | - | - | - | - | - | - | - | - | - | - | - | - | - | - | - | - | - | - | - | - | - |
| *Scytonema mirabile* | - | - | - | - | - | - | - | +3 | - | - | - | - | - | - | - | - | - | - | - | - | - | - | - | - | - | - | +2 | - | - | - |
| *Scytonema ocellatum* | - | - | - | - | - | - | - | - | - | - | - | - | - | - | - | - | - | - | - | - | - | - | - | - | - | - | - | - | - | - |
| *Spirulina meneghiniana* | - | - | - | - | - | - | - | - | - | - | - | - | - | - | +1 | - | - | - | - | - | +1 | - | - | - | +2 | - | - | - | - | - |
| *Tolypothrix nodosa* | - | - | - | - | - | - | - | - | - | - | - | - | - | - | - | - | - | - | - | - | - | - | - | - | - | - | - | - | - | - |
| *Tolypothrix tenuis* | - | - | - | - | - | - | - | - | - | - | - | - | - | - | - | - | - | - | - | - | - | - | - | - | - | - | - | - | - | - |

# Table 2.2: Proportionate distribution of different algal species in three districts of Gangetic West Bengal

## District: Burdwan

| Species | Places | | | | | | | | | | | | | | | | | | | | | | | | | | | | | |
|---|---|---|---|---|---|---|---|---|---|---|---|---|---|---|---|---|---|---|---|---|---|---|---|---|---|---|---|---|---|---|
| | Dainhat | | | | | | | | | | Purbasthali | | | | | | | | | | Kalna | | | | | | | | | |
| | 1 | 2 | 3 | 4 | 5 | 6 | 7 | 8 | 9 | 10 | 1 | 2 | 3 | 4 | 5 | 6 | 7 | 8 | 9 | 10 | 1 | 2 | 3 | 4 | 5 | 6 | 7 | 8 | 9 | 10 |
| *Anabaena ambigua* | +4 | - | +2 | +1 | - | - | - | - | - | - | - | - | - | - | - | - | +3 | - | - | - | - | - | - | - | +3 | - | - | - | +2 | - |
| *Anabaena fertilissima* | +2 | - | - | - | - | - | +2 | +2 | - | +3 | - | - | - | +3 | - | - | - | - | - | - | +3 | - | - | +1 | - | - | - | - | - | - |
| *Anabaena iyengarii* | - | - | - | - | - | - | - | - | - | - | - | +3 | - | - | +3 | - | - | +4 | +3 | - | - | - | - | - | - | +3 | - | - | - | - |
| *Anabaena naviculoides* | - | - | - | - | - | - | - | - | - | - | - | - | - | - | - | - | - | - | - | - | - | - | - | - | - | - | - | - | - | - |
| *Anabaena oscillaroides* | - | +1 | +2 | - | +2 | - | - | - | - | - | +1 | +2 | - | - | - | - | - | - | - | - | - | - | - | - | - | +1 | - | +2 | - | - |
| *Anabaena torulosa* | - | - | - | - | - | - | - | - | - | - | - | - | - | - | - | - | - | - | - | - | - | - | - | - | - | - | - | - | - | - |
| *Anabaena variabilis var. ellipsoporsa* | - | - | - | - | - | - | +2 | - | - | - | - | - | - | - | - | - | +3 | - | - | +3 | +2 | - | - | - | - | - | - | - | - | - |
| *Aphanocpsa biformis* | - | - | - | - | - | - | - | - | - | - | - | - | - | - | - | - | - | - | - | - | - | - | - | - | - | - | - | - | - | - |
| *Aphanocpsa montana* | - | - | - | - | - | - | - | - | - | - | - | - | - | - | - | - | - | - | - | - | - | - | - | - | - | - | - | - | - | - |
| *Aphanothece pallida* | - | - | - | - | - | +5 | - | - | - | - | - | +4 | - | - | - | +2 | - | - | - | +4 | - | - | - | - | - | - | - | - | - | - |
| *Aphanothece stagnina* | - | - | - | - | - | +2 | - | - | - | - | - | - | - | - | - | - | - | - | - | - | - | - | - | - | - | - | +3 | - | - | - |
| *Arathrospira spirulinoides* | - | - | - | - | - | - | - | - | - | - | - | +2 | - | - | - | - | - | - | - | - | - | - | - | - | - | - | - | - | - | - |
| *Aulosira fertilissima* | - | - | - | +2 | - | - | - | - | - | - | - | - | - | - | - | - | - | - | - | - | - | - | - | - | - | - | - | - | - | - |
| *Aulosira prolifica* | - | - | - | - | - | - | - | - | - | - | - | - | - | - | - | - | - | - | - | - | - | - | - | - | +3 | - | - | - | - | - |
| *Calothrix brevissima* | - | - | - | - | - | +3 | - | - | - | - | - | +3 | - | - | - | - | - | - | +3 | +3 | - | - | - | - | - | - | - | - | +2 | +3 |
| *Calothrix elenkinii* | - | - | - | - | - | - | - | +3 | - | - | - | - | - | - | - | - | - | - | - | - | - | - | - | - | - | - | - | - | - | - |
| *Calothrix membranacea* | - | +4 | - | +2 | - | - | - | - | +4 | - | +3 | - | - | - | +3 | - | +3 | - | - | - | - | +2 | +4 | - | - | - | +3 | - | - | - |
| *Chroococcus minutus* | - | - | - | - | - | - | - | - | - | - | - | - | −2 | - | - | - | - | - | - | - | - | - | - | - | - | - | - | - | - | - |
| *Chroococcus pallidus* | - | - | - | - | - | - | - | +3 | - | - | - | - | - | - | - | - | - | +3 | - | - | - | - | - | - | - | - | - | +3 | - | - |
| *Cylindrospermum majus* | - | - | - | - | - | - | - | - | - | - | - | - | - | - | - | - | - | - | - | - | - | - | - | - | - | - | +3 | - | - | - |
| *Cylindrospermum musicola* | - | - | - | - | - | - | - | - | - | - | - | - | - | - | - | - | - | - | - | - | - | - | - | - | - | - | - | - | - | - |

| Species | Places | | | | | | | | | | | | | | | | | | | | | | | | | | | | | |
|---|---|---|---|---|---|---|---|---|---|---|---|---|---|---|---|---|---|---|---|---|---|---|---|---|---|---|---|---|---|---|
| | Dainhat | | | | | | | | | | Purbasthali | | | | | | | | | | Kalna | | | | | | | | | |
| | 1 | 2 | 3 | 4 | 5 | 6 | 7 | 8 | 9 | 10 | 1 | 2 | 3 | 4 | 5 | 6 | 7 | 8 | 9 | 10 | 1 | 2 | 3 | 4 | 5 | 6 | 7 | 8 | 9 | 10 |
| *Gloeotrichia pisum* | - | - | - | - | - | - | - | - | - | - | - | - | - | - | - | - | - | - | - | - | - | - | - | - | - | - | - | - | - | - |
| *Lyngbya aerugineo-coerulea* | - | - | - | - | - | - | - | - | - | - | - | - | - | - | - | - | - | - | - | - | - | - | +2 | - | - | - | - | - | - | - |
| *Lyngbya ceylanica* | - | - | - | - | - | - | - | - | - | - | - | - | - | - | - | - | - | - | - | - | - | - | - | - | - | - | - | - | - | - |
| *Lyngbya hieronymusii* | - | - | - | - | - | - | - | - | - | - | - | - | - | - | - | - | - | - | - | - | - | - | +2 | - | - | - | - | - | - | - |
| *Lyngbya lagerheimii* | - | - | - | - | - | - | - | - | - | - | - | - | - | - | - | - | - | - | - | - | - | - | - | - | - | - | - | - | - | - |
| *Lyngbya major* | - | - | - | - | - | - | - | - | - | - | - | - | - | - | - | - | - | - | - | - | - | - | - | - | - | - | - | - | - | - |
| *Lyngbya stagnina* | - | - | - | - | - | - | - | - | - | - | - | - | - | - | - | - | - | - | - | - | - | - | - | - | - | - | - | - | - | - |
| *Microcoleus chthonoplastes* | - | - | - | - | +2 | - | - | - | - | - | - | - | - | - | - | - | - | - | - | - | - | - | - | - | - | - | - | - | - | - |
| *Microcoleus paludosus* | - | - | - | - | - | - | - | - | - | - | - | - | - | - | - | - | - | - | - | - | - | - | - | - | - | - | - | +2 | - | - |
| *Microcoleus subtorulosus* | - | - | - | - | - | - | - | - | - | - | - | - | - | - | - | - | - | - | - | - | - | - | - | - | - | - | - | - | - | - |
| *Microcystis aeruginosa* | - | - | - | - | - | - | - | - | - | - | - | - | - | - | - | - | - | - | - | - | - | - | - | - | - | - | - | - | - | - |
| *Microcystis flosaquae* | - | - | - | - | - | - | - | - | - | - | - | - | - | - | - | - | - | - | - | - | - | - | - | - | - | - | - | - | - | - |
| *Nostoc calicola* | - | - | - | - | +3 | - | - | - | - | - | - | - | - | - | - | - | - | - | - | - | - | +2 | +3 | - | - | - | - | - | - | - |
| *Nostoc commune* | +2 | - | - | - | +2 | - | - | - | - | +3 | - | - | - | +3 | - | - | +1 | - | - | - | - | - | - | - | - | - | - | - | - | +1 |
| *Nostoc humifusum* | - | +3 | - | - | - | +2 | - | - | +2 | - | - | - | - | - | - | - | +2 | - | - | +2 | - | - | - | - | - | - | - | - | - | +2 |
| *Nostoc muscorum* | - | - | - | - | +2 | - | - | - | +1 | - | +3 | - | - | +2 | - | - | - | - | - | - | - | - | - | - | +2 | - | - | - | - | - |
| *Nostoc paludosum* | - | - | +2 | - | - | - | - | - | - | - | - | - | - | - | +1 | - | - | - | - | - | +2 | - | - | - | - | - | - | - | - | - |
| *Nostoc piscinale* | - | - | +1 | - | - | - | - | +1 | - | - | - | - | +2 | - | +1 | - | - | - | - | +2 | - | - | - | - | +1 | +1 | - | - | - | - |
| *Nostoc punctiforme* | - | - | - | +2 | - | - | - | - | - | - | - | - | - | - | - | +3 | - | - | - | - | - | - | - | - | - | - | - | +1 | - | - |
| *Nostoc sphaericum* | - | +1 | - | +3 | - | - | - | - | - | - | - | - | +2 | - | - | - | - | - | - | - | - | +1 | - | - | - | - | - | +2 | +3 | - |
| *Oscillatoria amoena* | +2 | - | - | - | - | - | +1 | - | - | - | - | +2 | - | - | - | +2 | - | +1 | - | - | - | - | - | +3 | - | - | +1 | - | - | - |
| *Oscillatoria amphibia* | - | - | - | - | - | - | - | - | - | - | - | - | - | - | - | - | - | - | - | - | - | - | - | - | - | - | - | - | - | - |
| *Oscillatoria brevis* | - | - | - | - | - | - | - | - | - | - | - | - | - | - | - | - | - | - | - | - | - | - | - | - | - | - | - | - | - | - |
| *Oscillatoria chalybea* | +1 | - | - | - | +3 | - | - | - | - | - | - | - | - | - | - | +1 | - | - | - | - | - | - | - | - | - | - | +2 | - | - | - |
| *Oscillatoria formosa* | - | - | - | - | - | - | - | - | - | - | - | - | - | - | - | - | - | - | - | - | - | - | - | - | - | - | - | - | - | - |

| Species | Places | | | | | | | | | | | | | | | | | | | | | | | | | | | | | |
|---|---|---|---|---|---|---|---|---|---|---|---|---|---|---|---|---|---|---|---|---|---|---|---|---|---|---|---|---|---|---|
| | Dainhat | | | | | | | | | | Purbasthali | | | | | | | | | | Kalna | | | | | | | | | |
| | 1 | 2 | 3 | 4 | 5 | 6 | 7 | 8 | 9 | 10 | 1 | 2 | 3 | 4 | 5 | 6 | 7 | 8 | 9 | 10 | 1 | 2 | 3 | 4 | 5 | 6 | 7 | 8 | 9 | 10 |
| *Oscillatoria limosa* | - | - | - | - | - | - | - | - | - | - | - | - | - | - | - | - | - | - | - | - | - | - | - | - | - | - | - | - | - | - |
| *Oscillatoria obscura* | - | - | - | +1 | - | - | - | - | - | - | - | - | - | - | - | - | - | +3 | - | - | - | - | - | - | - | - | - | - | - | - |
| *Oscillatoria princeps* | +2 | - | - | - | - | - | +2 | - | - | - | - | - | - | - | - | - | - | - | - | - | - | - | - | - | - | - | +2 | - | - | - |
| *Oscillatoria proteus* | - | - | - | - | - | - | - | - | - | - | - | +2 | - | - | - | - | - | - | - | - | - | - | - | - | - | - | - | - | - | - |
| *scillatoria pseudogeminta var. unigranulata* | - | - | - | +2 | - | - | - | - | - | - | - | - | - | - | - | - | - | - | - | - | - | - | - | - | - | - | - | - | - | - |
| *Oscillatoria splendida* | - | - | - | - | - | - | - | - | - | - | - | - | - | - | - | - | - | - | - | - | - | - | - | - | - | - | - | - | - | - |
| *Phormidium anomala* | - | - | - | +2 | - | - | - | - | - | - | +1 | - | - | - | - | - | - | - | - | - | - | - | - | - | - | - | - | - | - | - |
| *Phormidium corium* | - | - | - | - | - | - | - | - | - | - | - | - | - | - | - | - | - | - | - | - | - | - | - | - | - | - | - | - | - | - |
| *Phormidium molle* | +1 | +3 | - | - | - | - | - | +1 | - | - | +2 | +1 | - | - | - | - | - | - | - | - | +3 | - | +1 | - | - | - | - | - | - | +2 |
| *Phormidium retzii* | - | - | - | - | - | - | - | +3 | - | - | - | - | - | - | - | - | - | - | +1 | - | - | - | - | - | - | +2 | - | - | - | - |
| *Phormidium stagnina* | +2 | - | - | - | - | +1 | - | - | - | - | - | - | - | - | - | - | - | - | - | - | - | - | - | - | - | - | - | - | +3 | - |
| *Phormidium subfusum* | - | - | - | - | - | - | - | - | - | +2 | - | - | - | - | - | - | - | - | - | - | - | - | +1 | - | - | - | - | - | - | - |
| *Phormidium tenue* | - | - | +2 | - | - | +2 | - | - | - | +2 | +1 | - | - | - | - | - | - | - | +2 | - | - | - | - | - | - | - | +1 | - | - | - |
| *Schizothrix arenaria* | - | - | - | - | - | - | - | - | - | - | - | - | - | - | - | - | - | - | - | - | - | - | - | - | - | - | - | - | - | - |
| *Schizothrix lacustris* | - | - | - | - | - | - | - | - | - | - | - | - | - | - | - | - | - | - | - | - | - | - | - | - | - | - | - | - | - | - |
| *Scytonema mirabile* | - | - | - | - | - | - | +1 | - | - | - | - | - | - | - | - | - | - | - | - | - | +3 | - | - | - | - | - | - | - | - | - |
| *Scytonema ocellatum* | - | - | - | - | +3 | - | - | - | - | - | - | - | - | +2 | - | - | - | - | - | - | - | - | - | - | +2 | - | - | - | - | - |
| *Spirulina meneghiniana* | - | - | - | - | - | - | - | - | - | - | - | - | - | - | - | - | - | - | +2 | +3 | - | - | - | - | - | - | - | - | - | - |
| *Tolypothrix nodosa* | - | - | - | - | - | - | - | +3 | - | - | - | - | - | - | - | - | - | - | - | - | - | - | - | - | - | - | - | - | - | - |
| *Tolypothrix tenuis* | - | - | - | - | - | - | - | - | - | - | - | - | - | - | +3 | - | - | - | - | - | - | - | - | - | - | - | - | - | - | - |

**Table 2.3: Proportionate distribution of different algal species in three districts of Gangetic West Bengal District: Hooghly**

| Species | Places | | | | | | | | | | | | | | |
|---|---|---|---|---|---|---|---|---|---|---|---|---|---|---|---|
| | Guptipara | | | | | | | Dankuni | | | | | Mogra | | |
| | 1 | 2 | 3 | 4 | 5 | 6 | 7 | 1 | 2 | 3 | 4 | 5 | 1 | 2 | 3 |
| Anabaena ambigua | - | - | - | +4 | - | - | +2 | - | +3 | - | - | - | +2 | - | - |
| Anabaena fertilissima | - | - | - | - | - | +1 | - | - | - | - | - | - | - | - | - |
| Anabaena iyengarii | - | - | - | +2 | - | - | - | - | - | - | - | +3 | - | - | - |
| Anabaena naviculoides | - | - | - | - | - | - | - | - | - | - | - | - | - | - | - |
| Anabaena oscillaroides | - | - | +2 | - | - | - | - | - | - | - | - | - | - | - | - |
| Anabaena torulosa | +3 | - | - | - | - | - | - | - | - | - | - | - | - | - | - |
| *Anabaena variabilis* var. *ellipsoporsa* | - | - | - | - | - | - | - | - | +2 | - | - | - | +2 | - | - |
| Aphanocpsa biformis | - | - | - | - | - | - | - | - | - | - | - | - | - | - | - |
| Aphanocpsa montana | - | - | - | - | +3 | - | - | - | - | - | - | - | - | - | - |
| Aphanothece pallida | - | +3 | - | - | - | - | - | - | - | - | - | - | - | - | - |
| Aphanothece stagnina | - | - | - | - | - | - | - | - | - | - | - | +2 | - | - | - |
| Arathrospira spirulinoides | - | - | - | - | - | - | - | - | - | - | - | - | - | - | - |
| Aulosira fertilissima | - | - | +2 | - | - | - | - | - | - | - | - | - | - | - | - |
| Aulosira prolifica | - | - | - | - | - | - | - | - | - | - | - | - | - | - | - |
| Calothrix brevissima | - | - | - | - | - | - | - | - | - | - | - | - | - | - | - |
| Calothrix elenkinii | - | - | - | - | - | - | - | - | - | - | - | - | - | - | - |
| Calothrix membranacea | - | - | - | - | - | - | - | - | - | - | - | - | - | - | - |
| Chroococcus minutus | - | - | - | - | - | - | - | - | - | - | - | - | - | - | - |
| Chroococcus pallidus | - | +3 | - | - | - | - | +2 | - | - | - | - | - | - | - | - |
| Cylindrospermum majus | - | - | - | - | - | - | - | - | - | - | - | - | - | - | - |
| Cylindrospermum musicola | - | - | - | - | - | - | - | - | - | - | - | - | - | - | - |
| Gloeotrichia pisum | - | - | - | - | - | - | - | - | - | +2 | - | - | - | - | - |

| Species | Places | | | | | | | | | | | | | | |
|---|---|---|---|---|---|---|---|---|---|---|---|---|---|---|---|
| | Guptipara | | | | | | | Dankuni | | | | | Mogra | | |
| | 1 | 2 | 3 | 4 | 5 | 6 | 7 | 1 | 2 | 3 | 4 | 5 | 1 | 2 | 3 |
| Lyngbya aerugineo-coerulea | - | - | +3 | - | - | - | - | +2 | - | - | - | - | - | - | - |
| Lyngbya ceylanica | - | - | - | - | - | - | - | - | - | - | - | - | - | - | - |
| Lyngbya hieronymusii | - | - | - | - | - | - | - | +2 | - | - | - | - | - | - | - |
| Lyngbya lagerheimii | - | - | - | - | - | - | - | - | - | - | +2 | - | - | - | +3 |
| Lyngbya major | - | - | - | - | - | - | - | - | - | - | - | - | - | - | +1 |
| Lyngbya stagnina | - | - | - | - | - | - | - | - | - | - | - | - | - | - | - |
| Microcoleus chthonoplastes | - | - | - | - | - | - | - | - | - | - | +3 | - | - | - | - |
| Microcoleus paludosus | - | - | - | - | - | - | - | - | - | - | - | - | - | +2 | - |
| Microcoleus subtorulosus | - | - | - | - | - | - | - | - | - | - | - | - | - | - | - |
| Microcystis aeruginosa | - | - | - | +3 | - | - | - | - | - | - | - | - | - | - | - |
| Microcystis flosaquae | - | - | - | - | - | - | - | - | - | - | +3 | - | - | - | - |
| Nostoc calicola | - | - | - | - | - | - | - | - | - | - | - | - | - | - | - |
| Nostoc commune | - | +3 | +2 | - | - | +3 | - | - | +1 | - | - | - | - | - | - |
| Nostoc humifusum | - | - | - | - | - | - | +2 | - | - | - | - | +2 | - | - | - |
| *Nostoc muscorum* | - | - | - | - | +1 | - | - | - | - | +2 | - | - | - | - | +1 |
| *Nostoc paludosum* | - | - | - | - | - | - | - | - | - | - | - | - | - | - | - |
| *Nostoc piscinale* | - | - | - | - | - | +1 | - | - | - | - | +3 | - | - | - | - |
| *Nostoc punctiforme* | - | - | - | - | - | - | - | - | - | - | - | - | - | - | - |
| *Nostoc sphaericum* | - | - | - | +1 | - | - | - | - | - | +3 | - | - | - | - | +2 |
| *Oscillatoria amoena* | +2 | +4 | - | - | - | - | - | +2 | - | - | - | - | - | - | - |
| *Oscillatoria amphibia* | - | - | - | +3 | - | - | - | - | - | +3 | - | - | +1 | - | - |
| *Oscillatoria brevis* | - | +1 | - | +1 | - | - | +1 | - | - | - | +2 | +1 | - | - | - |
| *Oscillatoria chalybea* | +2 | - | +1 | - | - | - | +2 | - | - | +2 | +3 | - | - | +1 | - |
| *Oscillatoria formosa* | - | - | +2 | - | - | - | - | - | - | - | - | +3 | - | +2 | - |
| *Oscillatoria limosa* | +1 | - | - | - | - | - | - | +2 | - | - | - | - | - | - | - |

| Species | Places | | | | | | | | | | | | | | |
|---|---|---|---|---|---|---|---|---|---|---|---|---|---|---|---|
| | Guptipara | | | | | | | Dankuni | | | | | Mogra | | |
| | 1 | 2 | 3 | 4 | 5 | 6 | 7 | 1 | 2 | 3 | 4 | 5 | 1 | 2 | 3 |
| *Oscillatoria obscura* | - | +1 | - | - | - | - | - | +3 | +1 | - | - | - | - | - | - |
| *Oscillatoria princeps* | - | +2 | +1 | - | - | - | - | - | +2 | +2 | - | - | +4 | - | - |
| *Oscillatoria proteus* | - | - | - | - | - | - | - | - | - | - | - | - | - | - | - |
| *Oscillatoria pseudogeminta* var. *unigranulata* | - | - | +2 | - | - | - | - | - | - | - | - | - | +1 | - | - |
| *Oscillatoria splendida* | - | - | - | - | - | - | - | - | - | - | - | - | - | - | - |
| *Phormidium anomala* | - | - | - | - | - | - | - | - | - | - | - | - | - | - | - |
| *Phormidium corium* | - | - | - | - | - | - | - | - | - | - | - | - | - | - | - |
| *Phormidium molle* | - | - | - | - | - | - | - | - | +1 | - | +1 | - | - | - | +2 |
| *Phormidium retzii* | - | - | - | - | - | - | - | - | - | - | - | - | - | - | +2 |
| *Phormidium stagnina* | - | - | - | - | - | - | - | - | - | - | - | - | - | - | - |
| *Phormidium subfusum* | - | - | - | - | - | - | - | - | - | - | - | - | - | - | - |
| *Phormidium tenue* | - | - | - | - | - | - | - | - | - | - | +2 | - | - | - | - |
| *Schizothrix arenaria* | - | - | - | - | - | - | - | - | - | - | - | - | - | - | - |
| *Schizothrix lacustris* | - | - | +2 | - | - | - | - | - | - | - | - | - | - | +2 | - |
| *Scytonema mirabile* | - | - | - | - | - | - | - | - | - | - | - | - | - | - | - |
| *Scytonema ocellatum* | - | - | +2 | - | - | - | - | - | - | - | - | - | - | - | - |
| *Spirulina meneghiniana* | - | - | - | - | +3 | - | - | - | - | - | - | +3 | - | - | - |
| *Tolypothrix nodosa* | - | - | - | - | - | - | - | - | - | - | - | - | - | - | - |
| *Tolypothrix tenuis* | - | - | - | - | +2 | - | - | - | - | - | - | - | - | - | - |

## Table 3: Comparative distribution of different algal genera in the three districts

| Genus | Numerical Total Districts | | | Rated Total Districts | | | Rated Score Districts | | |
|---|---|---|---|---|---|---|---|---|---|
| | Nadia | Burdwan | Hooghly | Nadia | Burdwan | Hooghly | Nadia | Burdwan | Hooghly |
| Anabaena | 19 | 22 | 8 | 65 | 68 | 26 | 217 | 227 | 173 |
| Aphanocapsa | 1 | 0 | 1 | 2 | 0 | 3 | 7 | 0 | 20 |
| Aphanothece | 15 | 6 | 2 | 38 | 20 | 5 | 127 | 67 | 33 |
| Arthrospira | 0 | 1 | 0 | 0 | 2 | 0 | 0 | 7 | 0 |
| Aulosira | 3 | 2 | 1 | 5 | 5 | 2 | 17 | 17 | 13 |
| Calothrix | 6 | 16 | 0 | 16 | 47 | 0 | 53 | 157 | 0 |
| Chroococcus | 6 | 4 | 2 | 14 | 11 | 5 | 47 | 37 | 33 |
| Cylindrospermum | 7 | 3 | 0 | 17 | 8 | 0 | 57 | 27 | 0 |
| Gloeotrichia | 0 | 0 | 1 | 0 | 0 | 2 | 0 | 0 | 13 |
| Lyngbya | 3 | 1 | 4 | 7 | 4 | 13 | 23 | 13 | 87 |
| Microcoleus | 4 | 2 | 2 | 8 | 4 | 5 | 27 | 13 | 33 |
| Microcystis | 1 | 0 | 2 | 3 | 0 | 6 | 10 | 0 | 40 |
| Nostoc | 19 | 24 | 11 | 64 | 76 | 27 | 213 | 253 | 180 |
| Oscillatoria | 21 | 9 | 12 | 59 | 33 | 61 | 197 | 110 | 407 |
| Phormidium | 17 | 15 | 3 | 54 | 41 | 8 | 180 | 137 | 53 |
| Schizothrix | 2 | 0 | 2 | 3 | 0 | 4 | 10 | 0 | 27 |
| Scytonema | 2 | 5 | 1 | 5 | 11 | 2 | 17 | 37 | 13 |
| Spirulina | 3 | 2 | 2 | 4 | 5 | 6 | 13 | 17 | 40 |
| Tolypothrix | 0 | 2 | 1 | 0 | 6 | 2 | 0 | 20 | 13 |

**Table 4.1: Relative magnitude of growth of blue-green algae in relation to growth of 'Boro' rice (local variety) cultivated in a field at Dainhat (District Burdwan, West Bengal) during the year 1984, with reference to certain ecological factors.**

| GROWTH PHASES OF RICE PLANT | Late May | Early June | Mid June | Early July | Mid July | Early August | Late August | Early September | Late September |
|---|---|---|---|---|---|---|---|---|---|
| | Preparation of the field | Transplantation of Seedlings | Acclimatization and establishment of seedlings | Growing phase of seedlings | Pre-flowering stage | Flowering and graining | Grain maturing | Grain ripening | Harvesting |
| **ABIOTIC FACTORS** | | | | | | | | | |
| Water pH | 6.5 | 6.5 | 6.5 | 6.4 | 6.2 | 6.6 | 7.0 | 7.0 | 6.8 |
| Soil pH | 6.2 | 6.2 | 6.2 | 6.0 | 6.0 | 6.5 | 6.5 | 6.8 | 6.5 |
| Water temperature (°C) | 32 | 34 | 34 | 31 | 28 | 26 | 26 | 25 | 25 |
| Temperature range (°C) | 32-36 | 32-37 | 31-37 | 29-33 | 29-32 | 28-31 | 28-32 | 27-31 | 27-34 |
| Relative humidity (%) | 74 | 81 | 74 | 87 | 84 | 87 | 84 | 84 | 77 |
| Light/Dark cycle (hrs.) | 13-12/10-48 | 13-19/10-41 | 13-25/10-35 | 13-23/10-37 | 13-20/10-40 | 13-03/10-57 | 12-35/11-25 | 12-22/11-38 | 12-04/11-56 |
| **ABUNDANCE OF BLUE-GREEN ALGAE** | - | | | | | | | | |
| *Anabaena ambigua* | - | - | - | - | - | + | + | + | - |
| *Anabaena fertilissima* | - | - | - | - | + | - | + | ++ | + |
| *Cylindrospermum musicola* | - | - | - | - | + | + | ++ | +++ | + |
| *Nostoc commune* | - | + | ++ | ++ | ++ | + | - | - | - |
| *Oscillatoria amoena* | - | - | + | ++ | ++ | + | + | + | + |
| *Oscillatoria chalybea* | - | ++ | +++ | ++++ | +++ | +++ | ++ | + | - |
| *Oscillatoria princeps* | - | - | - | + | + | + | + | + | + |
| *Phormidium molle* | - | - | - | + | + | + | + | + | + |
| *Phormidium stagnina* | - | - | - | - | - | - | + | + | ++ |

**Table 4.2: Relative magnitude of growth of blue-green algae in relation to growth of 'Jaya' rice cultivated in a field at Simurali (District Nadia, West Bengal) during the year 1985, with reference to certain ecological factors.**

| GROWTH PHASES OF RICE PLANT | Mid July | Early August | Mid August | Late August | Early September | Late September | Early October | Late October | Early November | Late November |
|---|---|---|---|---|---|---|---|---|---|---|
| | Preparation of the field | Transplantation of Seedlings | Acclimatization and establishment of seedlings | Growing phase of seedlings | Pre-flowering stage | Flowering stage | Graining stage | Grain maturing | Grain ripening | Harvesting |
| ABIOTIC FACTORS | | | | | | | | | | |
| Water pH | 6.5 | 6.5 | 6.5 | 6.5 | 6.0 | 6.2 | 6.5 | 6.8 | 6.8 | 7.0 |
| Soil pH | 6.5 | 6.5 | 6.3 | 6.5 | 6.2 | 6.2 | 6.2 | 6.5 | 6.4 | 6.5 |
| Water temperature (°C) | 30 | 30 | 32 | 29 | 32 | 32 | 28 | 28 | 27 | 25 |
| Temperature range (°C) | 26-33 | 28-32 | 28-35 | 27-34 | 27-34 | 27-34 | 25-30 | 28-30 | 26-28 | 24-34 |
| Relative humidity (%) | 88 | 88 | 80 | 80 | 80 | 76 | 87 | 77 | 70 | 49 |
| Light/Dark cycle (hrs.) | 13-17/10-43 | 13-04/10-56 | 12-49/11-11 | 12-40/11-20 | 12-19/11-41 | 12-00/12-00 | 11-50/12-10 | 11-28/12-32 | 10-59/13-01 | 9-42/14-18 |
| **ABUNDANCE OF BLUE-GREEN ALGAE** | | | | | | | | | | |
| *Anabaena fertilissima* | - | - | + | + | + | + | + | + | + | + |
| *Anabaena iyengarii* | - | - | + | + | + | + | - | - | - | - |
| *Aphanothece stagnina* | - | + | + | + | + | + | + | + | + | - |
| *Aulosira fertilissima* | - | - | - | + | ++ | ++ | - | - | - | - |
| *Oscillatoria obscura* | - | - | ++ | +++ | +++ | +++ | + | + | + | - |
| *Oscillatoria princeps* | - | - | + | - | - | - | - | - | - | - |
| *Phormidium anomala* | - | - | - | - | + | ++ | + | + | + | - |
| *Phormidium molle* | + | + | + | + | ++ | ++ | + | + | - | - |
| *Phormidium tenue* | + | + | + | + | + | - | - | - | - | - |
| *Schizothrix arenaria* | - | - | - | + | + | + | - | - | - | - |
| *Scytonema mirabile* | - | - | - | + | + | + | + | ++ | ++ | +++ |

| | | | | | | | | | | |
|---|---|---|---|---|---|---|---|---|---|---|
| *Spirulina meneghiniana* | - | - | + | ++ | ++ | ++ | ++ | + | + | + |

**Table 4.3: Relative magnitude of growth of blue-green algae in relation to growth of 'Ratna' rice cultivated in a field at Guptipara (District Hooghly, West Bengal) during the year 1986, with reference to certain ecological factors.**

| GROWTH PHASES OF RICE PLANT | Early February | Late February | Early March | Late March | Early April | Late April | Early May | Late May | Mid June | Late June |
|---|---|---|---|---|---|---|---|---|---|---|
| | Preparation of the field | Transplantation of Seedlings | Acclimatization and establishment of seedlings | Growing up of young seedlings | Pre-flowering stage | Flowering stage | Grain initiation | Grain maturing | Grain ripening | Harvesting |
| ABIOTIC FACTORS | | | | | | | | | | |
| Water pH | 6.5 | 6.8 | 7.0 | 7.0 | 6.8 | 6.8 | 7.0 | 7.5 | 8.0 | 8.0 |
| Soil pH | 6.5 | 6.5 | 6.8 | 6.2 | 6.0 | 6.2 | 6.5 | 6.8 | 6.8 | 6.8 |
| Water temperature (°C) | 25 | 28 | 32 | 25 | 30 | 25 | 29 | 30 | 35 | 33 |
| Temperature range (°C) | 21-30 | 23-32 | 30-33 | 26-28 | 28-34 | 25-32 | 31-36 | 32-38 | 34-38 | 32-35 |
| Relative humidity (%) | 58 | 75 | 42 | 60 | 71 | 69 | 68 | 68 | 59 | 61 |
| Light/Dark cycle (hrs.) | 11-04/12-56 | 11-20/12-40 | 11-34/12-26 | 11-52/12-08 | 12-18/11-42 | 12-46/11-14 | 13-03/10-57 | 13-16/10-44 | 13-24/10-36 | 13-26/10-34 |
| **ABUNDANCE OF BLUE-GREEN ALGAE** | | | | | | | | | | |
| *Anabaena oscillarioides* | - | - | + | + | + | + | + | + | + | - |
| *Aulosira fertilissima* | - | - | - | - | - | - | - | + | + | + |
| *Lyngbya aerugino-coerulea* | - | - | - | + | ++ | ++ | ++ | + | - | - |
| *Nostoc commune* | + | + | + | ++ | ++ | ++ | ++ | ++ | ++ | + |
| *Oscillatoria chalybea* | +++ | +++ | ++ | + | ++ | + | - | - | - | - |
| *Oscillatoria formosa* | + | + | ++ | ++ | +++ | +++ | +++ | ++ | ++ | ++ |
| *Oscillatoria princeps* | - | - | - | + | + | + | - | - | - | - |
| *Oscillatoria pseudogeminata* | - | - | - | - | + | ++ | + | + | - | - |
| *Schizothrix lacustris* | - | - | - | + | ++ | ++ | + | + | + | - |
| *Scytonema simplex* | - | - | - | - | - | - | + | + | + | + |

**Table 5.1: Enumeration of different species of paddy-field blue-green algae according to predominance**

| NADIA | BURDWAN | HOOGHLY |
|---|---|---|
| Aphanoathece pallida | Calothrix membranacea | Oscillatoria chalybea |
| | | Oscillatoria princeps |
| | | Anabaena ambigua |
| Anabaena ambigua | Calothrix brevissima | Nostoc commune |
| Anabaena fertilissima | Anabaena fertilissima | Oscillatoria amoena |
| | Anabaena iyengarii | |
| Anabaena variabilis var. ellipsospora | Anabaena ambigua | Oscillatoria amphibia |
| Nostoc muscorum | Aphanothece pallida | Oscillatoria formosa |
| Cylindrospermum majus | Phormidium molle | Nostoc sphaericum |
| | | Oscillatoria brevis |
| | | Spirulina meneghiniana |
| Aphanaothece stagnina | Nostoc humifusum | Anabaena iyengarii |
| Calothrix membranacea | | Chroococcus pallidus |
| Nostoc piscinale | | Lyngbya aerugineo-coerulea |
| Phormidium anomala | | Lyngbya lagerheimii |
| Phormidium molle | | Oscillatoria obscura |
| Nostoc commune | Nostoc commune | Nostoc humifusum |
| Phormidium tenue | Nostoc sphaericum | Nostoc muscorum |
| | Oscillatoria amoena | Nostic piscinale |
| | | Phormidium molle |
| | | Shizthrix lacustris |
| Nostoc sphaericum | Anabaena oscillaroides | Anabaena torulosa |
| Oscillatoria amoena | | Aphanocapsa montana |
| | | Aphanothece pallida |
| | | Microcystis aeruginosa |
| | | Microcystis flosaque |
| | | Oscillatoria limosa |
| | | Oscillatoria pseudogeminata var. unigranulata |
| Oscillatoria chalybea | Anabaena variabilis var. ellipsospora | Anabaena oscillaroides |
| Oscillatoria splendida | Nostoc muscorum | Aphanothece stagnina |
| | Nostoc piscinale | Aulosira fertilissima |
| | Phormidium tenue | Gloeotrichia pisum |
| | | Lyngbya hieronymusii |
| | | Microcoleus paludosus |
| | | Phormidium retzii |
| | | Phormidium tenue |
| | | Scytonema ocellatum |
| | | Tolypothrix tenuis |
| Anabaena lyengarii | Chroococcus pallidus | Anabaena fertilissima |
| Chroococcus minutus | | Lyngbya major |

| NADIA | BURDWAN | HOOGHLY |
|---|---|---|
| Chroococcus pallidus | | |
| Oscillatoria limosa | | |
| Osillatoria princeps | | |
| Micrcoleus paludosus | Nostoc calicola | |
| Nostoc calicola | | |
| Oscillatoria obscura | | |
| Phormidium retzii | | |
| Phormidium stagnina | | |
| Aulosira fertilissima | Oscillatoria chalybea | |
| Oscillatoria formosa | Scytonem ocellatum | |
| Oscillatoria pseudogeminata var. unigranulata | | |
| Scytonema mirabile | | |
| Nostoc paludosum | Nostoc punctiforme | |
| Oscillatoria amphibia | Oscillatoria princeps | |
| Phormidium corium | Phormidium retzii | |
| Spirulina meneghiniana | Phormidium stagnina | |
| **Calothrix brevissima** | **Aphanothece stagnina** | |
| Cylindrospermum musicola | Nostoc paludosum | |
| Lyngbya ceylanica | Spirulina meneghiniana | |
| Schizothrix arenria | | |
| Aphanocapsa biformis | Oscillatoria obscura | |
| Lyngbya majaor | Scytonema mirabile | |
| Microcoleus chthonoplastes | | |
| Microcystis aeruginosa | | |
| Anabaena naviculoides | Aulosira prolifica | |
| Lyngbya hieronymusii | Calothrix elenkinii | |
| Lyngbya stagnina | Cylindrospermum majus | |
| Microcystis flosaquae | Phormidium anomala | |
| | Phormidium subfuscum | |
| | Tolypothrix nodosa | |
| | Tolypothrix tenuis | |
| | Arthrospira spirulinoides | |
| | Aulosira fertilissima | |
| | Chroococcus minutus | |
| | Lyngbya aerugineo-coerulea | The txa grouped together indicate same degree of |
| | Lyngbya hieronymusii | Predominace. |
| | Microcoleus chthonoplastes | |
| | Microcoleus paludosus | |
| | Oscilltoria proteus | |
| | Oscilltoria pseudogeminata var unigranulata | |

## Table 5.2: Species distribution of paddy-field blue-green algae in three representative districts

| NADIA | BURDWAN | HOOGHLY |
|---|---|---|
| Anabaena ambigua | Anabaena ambigua | Anabaena ambigua |
| Anabaea fertillissima | Anabaena fertillissima | Anabaena fertilissima |
| Anabaena iyengarii | Anabaena iyengarii | Anabaena iyengarii |
| Anabaena naviculoides | Anbaena oscillaroides | Anbaena oscillaroides |
| Anabaena variabilis var. ellipsospora | Anabaena variabilis var. ellipsospora | Anabaena torulosa |
| Aphanocapsa biformis | Aphanothece pallida | Anabaena variabilis var. ellipsospora |
| Aphanothece pallida | Aphanaothece stagnina | Anabaena montana |
| Aphanothece stagnina | Arthrospira spirulinoides | Aphanothece pallida |
| Aulosira fertilissima | Aulosira fertilissima | Aphanothece stagnina |
| Calothrix brevissima | Aulosira prolifica | Aulosira fertilissima |
| Calothrix membranacea | Calothrix brevissima | Chroococcus pallidus |
| Chroococcus minutus | Alothrix elenkinii | Gloeotrichia pisum |
| Chroococcus pallidus | Calothrix membranacea | Lyngbya aerugineo-coerulea |
| Cylindrospermum majus | Chroococcus minus | Lyngbya hieronymusii |
| Cylindrospermum musicola | Chroococcus pallidus | Lyngbya lagerheimii |
| Lyngbya ceylanica | Cylindrospermum majus | Lyngbya major |
| Lyngbya hieronymusii | Lyngbya aerugineo-coerulea | Microcoleus chthonoplastes |
| Lyngbya major | Lyngbya hieronymusii | Microcoleus paludosus |
| Lyngbya stagnina | Microcoleus chthonoplastes | Microcystis aruginosa |
| Microcoleus chthonoplastes | Microcoleus paludosus | Microcysis flosaquae |
| | | |
| Microcoleus paludosus | Nostoc calicola | Nostoc commune |
| Microcystis aeruginosa | Nostoc commune | Nostoc humifusum |
| Microcystis flosaquae | Nostoc humifusum | Nostoc muscorum |
| Nostoc calicola | Nostoc muscorum | Nostoc piscinale |
| Nostoc commune | Nostoc paludosum | Nostoc sphaericum |
| Nostoc muscorum | Nostoc piscinale | Oscillatoria amoena |
| Nostoc paludosum | Nostoc punctiforme | Oscillatoria amphibia |
| Nostoc piscinale | Nostoc sphaericum | Oscillatoria brevis |
| Nostoc sphaericum | Oscillatoria amoena | Oscillatoria chalybea |
| Oscillatoria amoena | Oscillatoria chalybea | Oscillatoria formosa |
| Oscillatoria amphibia | Oscillatoria obscura | Oscillatoria limosa |
| Oscillatoria chalybea | Oscillatoria princeps | Oscillatoria obscura |
| Oscillatoria formosa | Oscillatoria proteus | Oscillatoria princeps |
| Oscillatoria limosa | Oscillatoria pseudogeminata var. unigranulata | Oscillatoria pseudogeminata var. unigranulata |
| Oscillatoria obscura | Phormidium anomala | Phormidium molle |
| Oscillatoria princeps | Phormidium molle | Phormidium retzii |
| Oscillatoria pseudogeminata var. unigranulata | Phormidium retzii | Phormidium tenue |
| Oscillatoria splendida | Phormidium stagnina | Schizothrix lacustris |

| NADIA | BURDWAN | HOOGHLY |
|---|---|---|
| Phormidium anomala | Phormidium subfuscum | Scytonema ocellatum |
| Phormidium corium | Phormidium tenue | Spirulina meneghiniana |
| Phormidium molle | Scytonema mirabile | Tolypothrix tenuis |
| Phormidium retzii | Scytonema oellatum | |
| Phormidium atagnina | Spirulina meneghiniana | |
| Phormidium tenue | Tolypohrix nodosa | |
| Schizothrix arenaria | Tolypothrix tenuis | |
| Scytonema mirabile | | |
| Spirulina meneghiniana | | |

## DISCUSSION

From the rice fields in the three districts of Gangetic West Bengal, a total of 70 different species have been recorded which belongs to 19 genera of blue-green algae. The algal species other than blue-green algae, chiefly belong to Chlorophyta, Euglenophyta and Bacillariophyta. A total of 26 species have been obtained from the rice fields under study of which 21 species were green algae.

The distributional pattern of blue-green algae in rice field soil of three districts of West Bengal shows that the number of blue-green algal species recorded from the three districts is nearly same. Nadia district records 47 species, 46 species have been recorded from Burdwan district and Hooghly district presents a total of 41 species.

Of the 19 genera of blue-green algae, ***Oscillatoria*** presents maximum number of species (11), followed by ***Nostoc*** (8), ***Phormidium*** (7), ***Anabaena*** (7), ***Lyngbya*** (6). ***Calothrix*** and ***Scytonema*** have 3 species of each genus. ***Aphanocapsa, Aphanothece Aulosira, Chroococcus, Cyllndrospermum,Microcoleus. Microcystis, Schizothrix, Spirulina*** and ***Tolypothrix*** are representedby 2 species under each genus. ***Gloeotrichia*** and ***Arthrospira*** have single species present in the fields under study.

The numerical totals of different genera (Table 3) indicate that the maximum representation of blue-green algae is present in Nadia district, though those in other districts do not deviate much.

The rated total gives an idea of the density of growth of different algal genera. 11 genera are distributed in all the three districts under investigation. Those are ***Oscillatoria, Nostoc, Anabaena, Phormidium, Aphanothece, Microcoleus, Chroococcus, Aulosira, Lyngbya. Scytonema*** and ***Spirulina.*** The rated scores (Table 3) indicate that the genera ***Oscillatoria, Nostoc*** and ***Anabaena*** show wide distribution in the lower Gangetic plain and ***Phormidium, Aphanothece*** and ***Calothrix*** are also distributed considerably. The genera ***Aphanocapsa,Arthrospira,Cylindrospermum Gloeotrichia, Microcystis, Schizothrix*** and ***Tolypothrix*** have very restricted distribution.

When the rated scores of different algal forms are considered together, the variations in the three districts i.e., Nadia (1215), Burdwan (1139) and Hooghly (1178) are not much. It indicates that the environmental conditions for the growth and development of blue-green algae in the lower Gangetic plain including the three districts are almost similar.

Among the 19 genera, 10 possessed the capacity of fixing atmospheric nitrogen. They are ***Anabaena, Aulosira, Calothrix, Cylindrospermum, Lyngbya, Nostoc, Schizothrix, Scytonema, Spirulina*** and ***Tolypothrix.*** Of them the rated score indicates that ***Anabaena*** and ***Nostoc*** are widely distributed throughout the three districts and their density of growth is considerable. ***Calothrix*** comes next in this respect.

Considering the rated scores of the N2 fixing and non N2 fixing forms, it becomes evident that the distribution and growth of both these types of blue-green algae are almost similar. The total rated score of N2 fixer and non N2 fixer in Nadia district is 1620 and 595; that in Burdwan district is 768 and 371 and in Hooghly district is 546 and 632 respectively. It indicates that the effects of environmental factors on the growth, development and distribution of the N2 fixing and non N2 fixing blue-green algae are not very different, though some genera may not be represented in any district.

The seasons of cultivation in the three fields in the three districts are not similar because the varieties of rice cultivated in these three different fields are different and so the growing seasons are also different. In the field located at Dainhat of Burdwan district months of observations are between late May to late September and the cultivated rice plants are 'Boro' variety.

But in the field at Simurali of Nadia district, 'Jaya' variety is cultivated between mid-July and late November. In the field of Guptipara of Hooghly district, 'Ratna' variety is cultivated and the seasons of cultivation are from early February to late June.The data of the different abiotic factors are considered on the basis of the growth phases of the concurrently growing rice plants.

## In rice field at Dainhat of Burdwan district (Table 4.1)

The water-logged rice field at Dainhat of Burdwan district was under cultivation of 'Boro' rice. The field was prepared in the later half of May and transplantation of rice seedlings was done in early June, within a fortnight of field preparation. No growth of blue-green algae was found before transplantation of the rice seedlings. In the middle of June, when the rice seedlings became acclimatized and established themselves in the field, a detectable growth of blue-green algae was observed for the first time. In the early July, with the growth of the seedlings, the algal patches were found to occupy more areas of the field. The growth of the algal components was not substantially increased till middle of July, but the number of species became maximum in the field from early August when flowering and grain initiation of the paddy started.

The broadest spectrum (9 species) of blue-green algae associated with the rice plants in the field under study was observed during the season of the grain maturing of the crop plants. But the abundance of different species was not maximum. Abundant algal growth was obtained during the growing phase of the seedlings in early July till the flowering and graining stage in early August. During August, the number of species increased though some of the species became less abundant. The growth of different species was different during the entire period of rice cultivation Of the different species *Anabaena ambigua, Anabaena fertilissima, Cylindrospermum musicola and Phormidium stagnina* appeared in the field during the flowering and graining periods of the crop plants. Among them the growth of *Cylindrospermum musicola* gained momentum soon after its appearance. The other algae viz., *Nostoc commune* and *Oscillatoria chalybea* appeared in the field at the time of transplantation of the seedlings. *Oscillatoria amoena* appeared when the seedlings established in the field.

***Oscillatoria princeps*** and ***Phormidium molle*** appeared slightly later and at the time of growing phase of the seedlings. However, the abundance of these two species remained same throughout their period of occurrence. ***Phormidium stagnina*** appeared very late during the period of rice cultivation and could be obtained at the grain maturing phase of the crop.

It is remarkable that during this period, mainly early July and middle of July the pH of the soil and of water came down to the lowest values 6.0 and 6.2 respectively. Algae attained climax vegetation when the grains of the paddy started maturing and ripening. During that period the pH of water and soil increased and the algae invaded the major part of the cultivated field. During harvesting period in late September, the growth of the algae started declining.

Of the 9 species of blue-green algae, 4 species were capable of metabolizing elementary nitrogen. Those were ***Anabaena ambigua***, ***Anabaena fertilissima***, ***Cylindrospermum musicola*** and ***Nostoc commune***. The former three did not appear until flowering and graining stage of rice plants were reached in early August. It was interesting to note that they appeared when the pH of water and soil exceeded 6.5. The temperature of water also went down to 28°C or less. ***Nostoc commune*** to the contrary showed reasonable growth since the transplantation of rice seedlings up to the flowering and graining stage of the crop in early August. The growth of this alga was considerable specially when the pH of water and that of soil were low and between 6.2 to 6.5 and 6.0 to 6.2 respectively. When the other $N_2$ fixing algae exhibited their highest growth i.e., from late August to late September, ***Nostoc commune*** disappeared from the field.

## In rice field at Simurali of Nadia district (Table 4.2)

The rice field at Simurali of district Nadia was prepared for cultivation of 'Jaya' variety of rice during the middle of July and was irrigated with water from shallow tube-well. From this field altogether 12 blue-green algae were obtained during the cultivation period of rice of which 6 were capable of fixing atmospheric nitrogen. The species obtained from the field were ***Anabaena fertilissima***, A***nabaena iyengarii,Aphanothece stagnina ,Aulosira fertilissima,Oscillatoria obscura,Oscillatoria princeps, Phormidium anomala,Phormidium molle,Phormidium tenue, Schizothrix arenaria.Spirulina meneghiniana*** and ***Scytonema mirabile***.

At the late stage of field preparation a meagre growth of 2 species of ***Phormidium*** was observed specially near the ridges of earth around the pieces of land. In early August, when the seedlings were transplanted gelatinous free floating thalli of ***Aphanothece stagnina*** were noticed in some places. ***Oscillatoria princeps*** was very transitory in appearance and could be noted only in the middle of August. Other 7 species appeared at the same period when rice seedlings established in the field and continued to grow there.

The broadest spectrum of algal species (11 species) was visible in the field in early September and 10 species were present in late August as well as in late September. Variations in the number of species was only due to addition and elimination of two different species of ***Phormidium*** viz**.,** ***P. anomala*** and ***P. tenue.*** The initial invasion of blue-green algae during transplantation and establishment of rice seedlings (early August to middle of August) is followed by their accelerated growth rate during seedling growth and pre-flowering stage of crop plants (late August to late September) and finally the growth rate declined till harvesting season of rice plants (late November) ensued.

It is interesting to note that during the period of luxurient growth of the algae, the pH of water was 6.0 to 6.2 and that of soil was 6.2. From early October, the pH values of both water and soil increased. When grain setting of rice plants started in early October, three species viz., ***Anabaena iyengarii,Aulosira fertilissima*** and ***Schizothrix arenaria*** disappeared from the field. The other species also suffered a gradual decline and during the harvesting season in late November only ***Anabaena fertilissima***, ***Spirulina meneghiniana*** and ***Scytonema mirabile*** were found growing on the field.

Of the 6 $N_2$ fixing blue-green algae ***Anabaena fertilissima,Anabaena iyengarii*** and ***Spirulina meneghiniana*** came into existence during the stage of establishment of the seedlings in the field (middle of August**).** ***Aulosira fertilissima***, ***Schizothrix arenaria*** and ***Scytonema mirabile*** appeared later in late August, during the growing phase of the seedlings. ***Anabaena fertilissima,Spirulina meneghiniana*** and ***Scytonema mirabile*** were present in the field till the crops were harvested. On the other hand, the rest of the $N_2$ fixers disappeared in late September which was the flowering season for the rice plants.

## In rice field at Guptipara of Hooghly district (Table 4.3)

The succession of different blue-green algae was studied in a field of Guptipara in Hooghly district where 'Ratna' variety of rice was cultivated.

The 'Ratna' seedlings were transplanted in late February and harvested in late June. The flowering stage appeared in late April i.e., 2 months after transplantation of the seedlings and ripening of the grains occurred in the middle of June.

The first appearance of blue-green algae was observed during preparation of field in early February. The broadest spectrum (8 species) of associated algae was observed over a fairly long period during the pre-flowering, flowering and upto grain maturing stage of the crop plants from early April to late May. The most abundant species was ***Oscillatoria formosa.*** The other blue-green algal species noticed were: ***Lyngbya aerugino-coerulea, Nostoc commune, Schizothrix lacustris, Oscillatoria chalybea, Anabaena oscillaroidies, Oscillatoria pseudogeminata*** and ***Oscillatoria princeps*** according to decreased abundance. During grain maturing stage ***Oscillatoria chalybea*** and ***Oscillatoria princeps*** were eliminated, but ***Aulosira fertilissima*** started growing. During grain ripening stage (middle of the June) ***Lyngbya aerugino-coerulea*** and ***Oscillatoria pseudogeminata*** were absent in late June, when the crop was harvested, ***Anabaena oscillarioides*** and ***Schizothrix lacustris*** were not detected.

Of the 10 species growing in the field, 6 were capable of fixing atmospheric nitrogen. Those were ***Anabaena oscillarioides,Anabaena fertilissima,Lyngbya aerugineo-coerulea,Nostoc commune, Scizothrix lacustris*** and ***Scytonema simplex.*** Of these Nitrogen fixing algae ***Nostoc commune*** was present throughout the season of rice cultivation, though it became abundant in season of growth of rice seedlings and remain to till the grains ripened. ***Anabaena oscillarioides*** existed since the season of establishment of rice seedlings. During the growth of the seedlings (late March), ***Lyngbya aerugineo-coerulea*** and ***Schizothrix lacustris*** appeared and they remained in the field till the stages of grain maturation and grain ripening respectively. ***Scytonema simplex*** came into existence slightly later in early May when grain initiation in the rice plant was taking place. ***Aulosira fertilissima*** was last to invade the field in late May i.e., the grain maturing stage in the crop but existed till the crop was harvested.

Among the different abiotic factors prevailing in the rice fields at the three places of the three districts, the pH of soil as well as that of water were considered to find out the correlation with the succession and growth of the algae, if any. The growth of the blue-green algae growing on wet soil is largely influenced by soil pH and those growing in stagnant water on the field is subjected to the pH of water. The pH of water was mostly found a little higher than that of the soil. In all the three fields in the three districts, the soil was slightly acidic and between 6.0-6.8. The pH of water occasionally reached neutrality but never became alkaline except in the field at Hooghly district where it became alkaline (pH 8) at the latter stages of cultivation. The range of soil pH in the fields of Burdwan and Hooghly districts was between 6.0-6.8 and that in Nadia district was between 6.2-6.5. But the changes in soil pH with the progress of growth phases of the cultivated crop plants were extremely different in the three fields of the three districts.

The range of environmental temperature though has a role on the growth and development of rice plants, the water temperature has a significant role for the algal growth. The variations in water temperature in the rice fields were different in the different phases of growth of the rice plants in the 3 districts (Tables 4.1, 4.2 and 4.3).

The data on the relative magnitude of growth of different blue-green algae in the 3 districts show that there was a close relation between the algal growth and the different phases of growth of rice plants in the fields, but it could not be correlated with the water temperature or soil pH of the fields. The magnitude of algal growth gradually increased between the phase of acclimatization followed by establishment of seedlings and flowering stage of the crop plants. At the later stage when the grains mature the magnitude of algal growth gradually decreased. The growth of blue-green algae, thus can only be correlated with the condition of water supply in the field as was necessary for cultivation. Any other abiotic factors prevailing in the fields possibly have no influence on the magnitude of algal growth.

**Fig. A**

**Fig. B**

◈ Fig. A Initiation of colonization of blue-green algae at an early stage of paddy cultivation.

◈ Fig. B A paddy field showing colonies of blue-green algae.

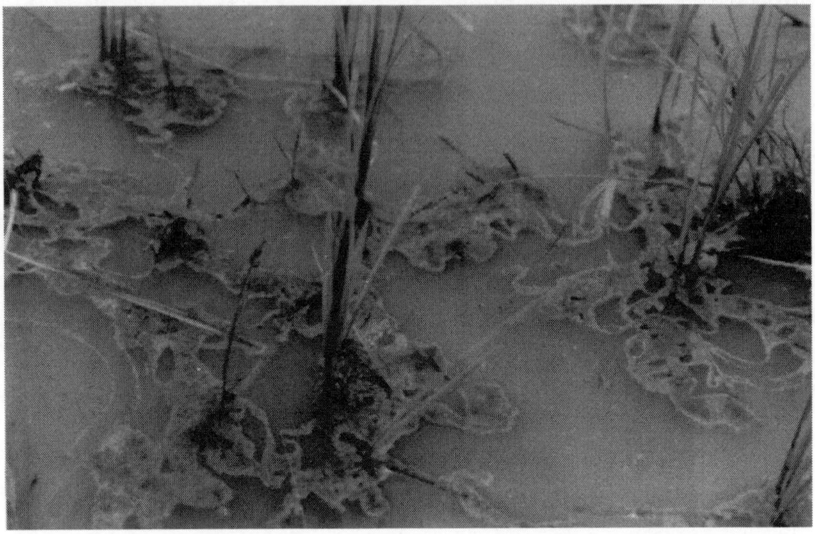

**Fig. C**

**Fig. D**

◈ Fig C. Blue-green algal colonies in a paddy field showing association between rice plant and blue-green algae.

◈ Fig D. Colonies of Gloeotrichia sp. attached to a paddy field weed.

## SUMMARY


- Study of the formation, association and succession of the blue-green algal flora of paddy field with respect to the environmental factors have been undertaken for understanding of the rice field ecosystem.
- Algal flora were collected in different seasons from the rice fields as well as isolated from culture. The abiotic factors e.g., pH, temperature, relative humidity and light/dark cycle have also been considered.
- The different experimental procedures have been given.
- The blue-green algae and other algae of the paddy fields have been enumerated.
- The study of distributional pattern of blue-green algae in the rice field soil at 3 districts has been made. From Nadia district 47 species, from Burdwan district 46 species and from Hooghly district 41 species have been recorded. The effects of environmental conditions for growth of blue-green algae in the three districts under study are almost similar though some genera are not represented in any district. Their effects on the growth, development and distribution of N2 fixing and non N2 fixing are not very different.
- Association, succession and relative frequency of occurrence of blue-green algae in the rice fields have been selected in respect to the environmental factors. Algal growth is significantly influenced by pH and temperature of water in the field as well as pH of the soil. No other abiotic factor could show their importance. The relation between algal growth and different phases of growth of rice plants was exhibited, but such correlation may be due to condition of water supply in the rice field in different phases of growth of rice plants.


# B. EFFECTS OF SOME FUNGICIDES AND PESTICIDES ON BLUE- GREEN ALGAL FLORA OF RICE FIELDS.

## INTRODUCTION

The role of blue-green algae in the nitrogen economy of Indian agriculture has gain attention for the last three decades and has been well documented by several authors (Singh, 1961; Watanabe, 1962; Venkataraman, 1966, 1972, 1978; Watanabe and Yamamoto, 1970). It has been recorded that in case of high yielding dwarf variety of rice, algal inoculation can replace the use of fertilizer (Venkataraman and Goyal, 1969a,b).

The use of fungicides and pesticides in crop field is a common practice and so their effect on the biofertilizer blue-green algae is a matter of interest. Work on the toxic effect of these chemical compounds has been done by different authors (Pillay and Tchan, 1970; Inger, 1970; Venkataraman and Rajya-lakshmi, 1971; Batterton *et al.*, 1971; Ahmed and Venkataraman, 1973; Singh, 1973; Torres *et al.*, 1976; Tarar and Salpekar, 1978, 1979; Anand and Veerappan, 1980; Tarar and Shewale, 1984).

Inger (1970) recorded the effect of herbicides on nitrogen fixation by blue-green algae. The effects of pesticides on blue- green algae have been studied by Venkataraman and Rajyalakshmi (l971a,b), Singh (1973), Torres *et al.* (1976).

Ahmed and Venkataraman (1973) made similar study on ***Aulosira fertilissima.*** The effects of aldrin, dialdrin and endrine on the growth response by these algae have been studied by Batterton and his associates (1971). Tarar and Salpekar (1978, 1979) recorded the effects of insecticides and rodenticides on soil algae. Anand and Veerappan (1980) recorded differential sensitivity of three blue-green algae to the pesticides Dimecron 100 and Ekalux and fungicides Blitox and Dithane. Tarar and Shewale (1984) applied four selected fungicides viz., Vitavax, Benlate, Blue copper 50 and Bavistin on 23 cyanophycean algae, some green algae and diatoms to note the algicidal potential of the chemicals. Various workers (Kolte and Goyal, 1992; Mansour et al., 1994; Rath and adhikary, 1996; Das and Debnath, 2006; Debnath et al., 2012; Dash et al., 2017 a ; Singh et al., 2020) carried out work on the responses of pesticides and herbicides to cyanobacteria.

The present work has been undertaken to note the effect of a few fungicides and pesticides on the survival of the blue-green algae of the paddy field soil of West Bengal. For this purpose, two fungicides were selected, of which one is a systemic fungicide, Bavistin [2-(Methoxycarbamyl)-benzimidazole] and the other Dithane-Z (Zinc ethylene bis di thiocarbamate). One insecticide viz., Dimecron [2-chloro-3(diethylamino)-1-methyl-3-oxo-1-propenyl dimethyl phosphate] and a herbicide 2,4-D (2,4-dichlorophenoxy acetic acid)were also used. The algae selected for the study were both heterocystous nitrogen fixing and non-heterocystous members of cyanophyceae.

## MATERIALS AND METHODS

Samples of surface soil from the paddy fields of Debagram area of Nadia district were collected at random and mixed together. The algicidal action of the fungicides and pesticides was studied on the algal members present in the soil sample. The fungicides (Bavistin and Dithane Z), insecticide (Dimecron) and herbicide (2,4-D) were tested for algicidal action in different concentrations in Chu-10 (modified) medium. The composition of the medium:

- $Ca(NO_3)_2$ 40 mg/l,
- $K_2HPO_4$ 10 mg/l
- $MgSO_4,7H_2O$ 25 mg/l
- $Na_2CO_3$ 20 mg/l
- $Na_2SiO_3$ 25 mg/l
- Fe-citrate 3 mg/l
- Citric acid 3 mg/l

A5 micronutrient solution 1 ml/1 (Boric acid 2.86 g/l, $MnCl_2$ 1.81 g/l, $ZnSO_4$ 0.222 g/l, Sodium molybdate 0.0177 g/l, $CuSO_4$0.079 g/l); pH of the medium was 7.0.

Pesticides were used as freshly prepared solutions. Eight different concentrations viz., 250, 500, 1000, 2000, 3000, 4000, 6000 and 8000 ppm were prepared in sterile distilled water as stock solution. In 90 ml of culture solution,

10 ml from stock solution of pesticidal compound and 10 g of soil sample were added. Thus, the concentrations of pesticides in culture media was 25,50,100,200, 300,400, 600 and 800 ppm.The culture flasks were agitated and kept in diffused light from the north. Each experiment had three replicates and a set of three were maintained as control. The algae were identified between 15 days and 60 days periodically to study their survival.

## RESULTS AND DISCUSSION

In the experiments on the effects of some fungicides and pesticides, altogether 12 species of blue-green algae appeared in culture. Of them 5, viz. ***Chroococcus pallidus, Oscillatoria chalybea.O. obscura, O. amoena*** and ***Phormidium molle*** were non-heterocystous and 7, viz. ***Cylindrospermum musicola,Nostoc commune,N. piscinale. Anabaena variabilis.A. fertilissima,Auloslra fertilissima*** and ***Calothrix brevissima*** were heterocystous members.The days required for the appearance of algal growth in culture and in the media with the fungicidal and pesticidal compounds in varying concentrations are given in Tables 6.

Table 6 indicates that the minimum days required for algal growth is 18 days. In general, all the 4 compounds tested have a inhibitory effect on different algal members at higher concentration. But effects are different in non-heterocystous and heterocystous members. In case of Bavistin inhibitory effects are more in heterocystous forms but in case of other three compounds effects are comparatively more on non-heterocystous members.

Whatever may be the effect on growth of different algae, the effect is similar upto a concentration of 50 ppm except that in Dithane Z where the effect in 25 ppm and 50 ppm are different. It is also noteworthy that in Dithane Z at 25 ppm concentration, the algal growth started earlier than in control experiment. In experiment with Bavistin, Dimecron and 2,4-D the first appearance of algae required longer days than control, but upto 100 ppm concentration, the number of days required were almost same. A generalised conclusion can be made that a total inhibition in algal growth are obtained at the concentration of 600 ppm and above except in Bavistin, where algal growth are obtained even at 800 ppm, though the number of species is very few.

### Effect of Bavistin

In general, the non-heterocystous, non-nitrogen fixing forms are more resistant to Bavistin (Figure 1). ***Chroococcus pallidus*** can tolerate very high concentration (800 ppm). The tolerance of the 3 species of ***Oscillatoria*** tested are different. ***Oscillatoria chalybea*** is very susceptible and can tolerate only up to a concentration of 50 ppm. On the other hand, ***Oscillatoria obscura*** and ***Oscillatoria amoena*** tolerate upto 600 and 400 ppm respectively. It indicates that the species of the same genus have differential sensitivity to this compound. ***Phormidium molle*** is susceptible beyond 400 ppm concentration.

Of the 7 heterocystous algae, ***Cylindrospermum musicola*** and ***Calothrix brevissima*** only can tolerate Bavistin upto 50 ppm concentration. All other species of ***Nostoc,Anabaena*** and ***Aulosira*** are highly susceptible and cannot grow even in 25 ppm of concentration. The effect of this compound is remarkably dissimilar on non-heterocystous and heterocystous form. No heterocystous form can grow beyond 50 ppm concentration.

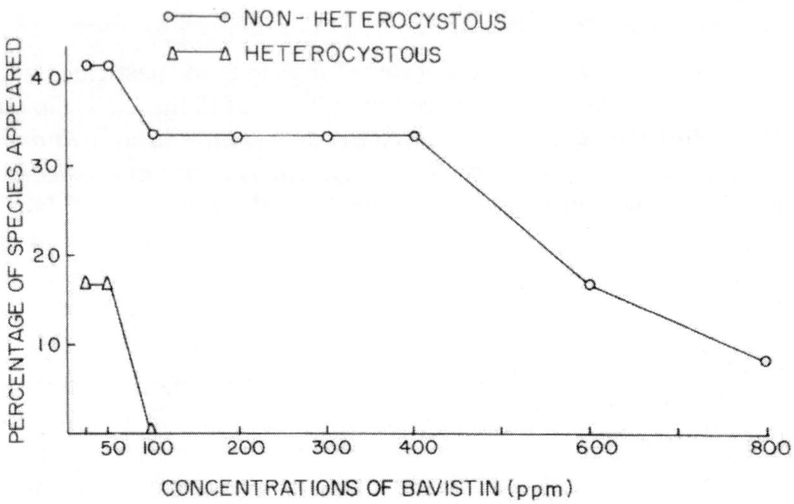


**Fig. 1.** Showing Effect of Bavistin on Blue-Green Algae of Rice Field

## Effect of Dithane Z

Experimentation with Dithane Z at different concentrations shows that this compound has a great effect on the survival of both non-heterocystous and heterocystous forms (Figure 2). No species could tolerate a concentration beyond 100 ppm.

Of the non-heterocystous forms ***Oscillatoria chalybea*** is most susceptible and did not appear even in 25 ppm concentration. ***Chroococcus pallidus*** could tolerate upto 50 ppm concentration. The other members viz. ***Oscillatoriaobscura, Oscillatoria amoena*** and ***Phormodium molle*** did not grow beyond 25 ppm concentration. Among the heterocystous member ***Cylindrospermum musicola*** was more tolerant than others and could survive in a higher concentration of 100 ppm. ***Anabaena fertilissima*** and ***Calothrix brevissima*** could tolerate upto 50 ppm concentration. Other forms viz., ***Nostoc commune, Nostoc piscinale, Anabaena variabilis*** and ***Aulosira fertilissima*** showed high sensitivity and only grew in 25 ppm concentration of the compound.

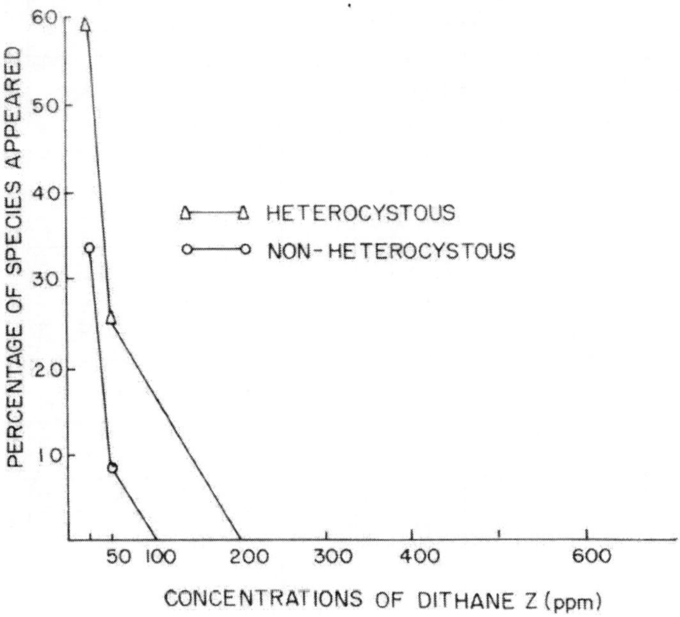


**Fig. 2.** Showing Effect of Dithane Z on Blue-Green Algae of Rice Field

In general, Dithane Z has no differential effect on non-heterocystous and heterocystous blue-green algae. The algal forms tested were highly susceptible to this fungicide for their survival and non-heterocystous form could survive beyond 100 ppm.

## Effect of Dimecron

The effect of Dimecron on the survival of the blue-green algae tested is not remarkably dissimilar in non-heterocystous and heterocystous forms. ***Chroococcus pallidus*** is most resistant and grew at a higher concentration (300 ppm). Of the 3 species of ***Oscillatori****a*, ***Oscillatoria chalybea*** is susceptible and can tolerate a concentration of 50 ppm of Dimecron. Other 2 species viz. ***Oscillatoria obscura*** and ***Oscillatoria amoena*** show growth in 200 ppm.***Phormidlum molle*** can not grow beyond 100 ppm concentration.

Among the heterocystous algae, ***Anabaena variabilis*** is most resistant and can thrive upto 400 ppm. In respect of sensitivity for survival, ***Nostoc piscinale*** and ***Calothrix brevissima*** are similar in response, the maximum limit of concentration in which they can grow being 300 ppm. **Nostoc commune** and ***Aulosira fertilissima***can not grow beyond 200 ppm. ***Cylindrospermum musicola*** and ***Anabaena fertilissima*** can tolerate a concentration of 200 ppm.

Beyond 400 ppm concentration of this compound, no blue-green algae under study can grow. The tolerance varies in different species but no relation could be observed between the occurrence of heterocyst and the survival (Figure 3).

## Table 6: Effect of different fungicides and pesticides on the survival of blue-green algae

| Sl. No. | Algae present in soil culture (control) | Bavistin (ppm) | | | | | | | | Dithane Z (ppm) | | | | | | | |
|---|---|---|---|---|---|---|---|---|---|---|---|---|---|---|---|---|---|
| | | 25 | 50 | 100 | 200 | 300 | 400 | 600 | 800 | 25 | 50 | 100 | 200 | 300 | 400 | 600 | 800 |
| | Non-heterocystous | | | | | | | | | | | | | | | | |
| 1. | Chroococcus pallidus | + | + | + | + | + | + | + | + | + | + | | | | | | |
| 2. | Oscillatoria chalybea | + | + | | | | | | | | | | | | | | |
| 3. | Oscillatoria obscura | + | + | + | + | + | + | + | | + | | | | | | | |
| 4. | Oscillatoria amoena | + | + | + | + | + | + | | | + | | | | | | | |
| 5. | Phormidim molle | + | + | + | + | + | + | | | + | | | | | | | |
| | Number of species appeared | 5 | 5 | 4 | 4 | 4 | 4 | 2 | 1 | 4 | 1 | 0 | 0 | 0 | 0 | 0 | 0 |
| | % of species appeared | 41.6 | 41.6 | 33.3 | 33.3 | 33.3 | 33.3 | 16.6 | 8.3 | 33.3 | 8.3 | 0 | 0 | 0 | 0 | 0 | 0 |
| | Heterocystous | | | | | | | | | | | | | | | | |
| 6. | Cylindrospermum musicola | + | | | | | | | | + | + | + | | | | | |
| 7. | Nostoc commune | | | | | | | | | + | | | | | | | |
| 8. | Nostoc piscinale | | | | | | | | | + | | | | | | | |
| 9. | Anabaena variabilis | | | | | | | | | + | | | | | | | |
| 10. | Anabaena fertillisima | | | | | | | | | + | + | | | | | | |
| 11. | Aulosira fertillisima | | | | | | | | | + | | | | | | | |
| 12. | Calothrix brevissima | + | + | | | | | | | + | + | | | | | | |
| | Number of species appeared | 2 | 2 | 0 | 0 | 0 | 0 | 0 | 0 | 7 | 3 | 1 | 0 | 0 | 0 | 0 | 0 |
| | % of species appeared | 16.6 | 16.6 | 0 | 0 | 0 | 0 | 0 | 0 | 58.3 | 25.0 | 8.3 | 0 | 0 | 0 | 0 | 0 |
| | Total number of algal species present | 7 | 7 | 4 | 4 | 4 | 4 | 2 | 1 | 11 | 4 | 1 | 0 | 0 | 0 | 0 | 0 |
| | Number of days required for the appearance of algae in culture | 26 | 27 | 26 | 27 | 27 | 29 | 31 | 33 | 18 | 25 | 57 | 0 | 0 | 0 | 0 | 0 |

| Sl. No. | Algae present in soil culture (control) | Dimecron (ppm) | | | | | | | | 2, 4-D (ppm) | | | | | | | |
|---|---|---|---|---|---|---|---|---|---|---|---|---|---|---|---|---|---|
| | | 25 | 50 | 100 | 200 | 300 | 400 | 600 | 800 | 25 | 50 | 100 | 200 | 300 | 400 | 600 | 800 |
| | Non-heterocystous | | | | | | | | | | | | | | | | |
| 1. | Chroococcus pallidus | + | + | + | + | + | | | | + | + | + | | | | | |
| 2. | Oscillatoria chalybea | + | + | | | | | | | + | + | | | | | | |
| 3. | Oscillatoria obscura | + | + | + | + | | | | | + | + | + | | | | | |
| 4. | Oscillatoria amoena | + | + | + | + | | | | | + | + | + | + | | | | |
| 5. | Phormidim molle | + | + | + | | | | | | + | + | + | | | | | |
| | Number of species appeared | 5 | 5 | 4 | 3 | 1 | 0 | 0 | 0 | 5 | 5 | 4 | 1 | 0 | 0 | 0 | 0 |
| | % of species appeared | 41.6 | 41.6 | 33.3 | 25.0 | 8.3 | 0 | 0 | 0 | 41.6 | 41.6 | 33.3 | 8.3 | 0 | 0 | 0 | 0 |
| | Heterocystous | | | | | | | | | | | | | | | | |
| 6. | Cylindrospermum musicola | + | + | + | | | | | | + | + | + | | | | | |
| 7. | Nostoc commune | + | + | + | + | | | | | + | + | + | | | | | |
| 8. | Nostoc piscinale | + | + | + | + | + | | | | + | + | + | | | | | |
| 9. | Anabaena variabilis | + | + | + | + | + | + | | | | | | | | | | |
| 10. | Anabaena fertillisima | + | + | + | | | | | | + | + | + | + | | | | |
| 11. | Aulosira fertillisima | + | + | + | + | | | | | + | + | + | | | | | |
| 12. | Calothrix brevissima | + | + | + | + | + | | | | + | + | + | + | + | + | | |
| | Number of species appeared | 7 | 7 | 7 | 5 | 3 | 1 | 0 | 0 | 6 | 6 | 6 | 2 | 1 | 1 | 0 | 0 |
| | % of species appeared | 58.3 | 58.3 | 58.3 | 41.6 | 25.0 | 8.3 | 0 | 0 | 50.0 | 50.0 | 50.0 | 16.6 | 8.3 | 8.3 | 0 | 0 |
| | Total number of algal species present | 12 | 12 | 11 | 8 | 4 | 1 | 0 | 0 | 11 | 11 | 10 | 3 | 1 | 1 | 0 | 0 |
| | Number of days required for the appearance of algae in culture | 26 | 27 | 26 | 25 | 30 | 82 | 0 | 0 | 24 | 24 | 26 | 57 | 76 | 74 | 0 | 0 |

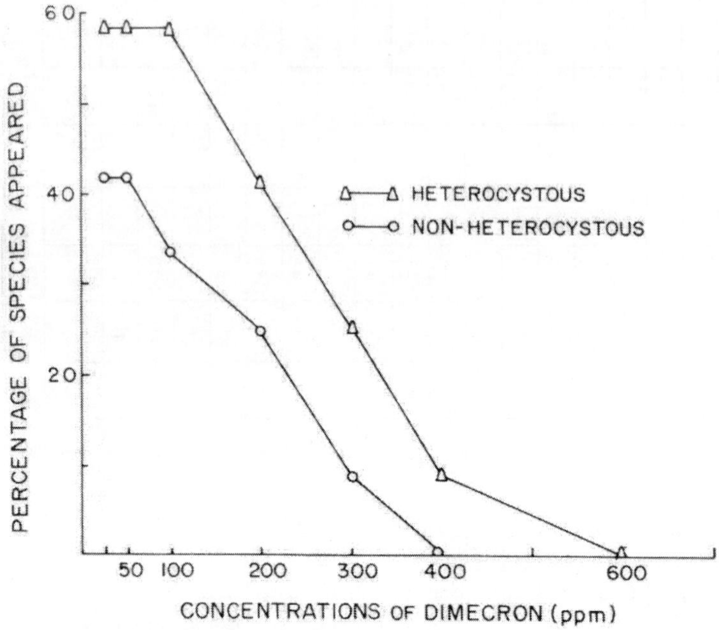


**Fig. 3.** Showing Effect of Dimecron on Blue-Green Algae of Rice Field

## Effect of 2,4-D

Both non-heterocystous and heterocystous members show almost similar response to this herbicide and none could thrive beyond 200 ppm concentration (Figure 4) except ***Calothrix bravissima*** which can tolerate a concentration of 400 ppm and ***Anabaena variabilis*** which is very susceptible and shows no growth even in 25 ppm concentration. Among non-heterocystous algae ***Oscillatoria amoena*** can grow upto 200 ppm concentration and ***Chroococcus pallidus,Oscillatoria obscura*** and ***Phormidium molle*** show growth at a concentration of 100 ppm. But the maximum limit of concentration for survival in ***Oscillatoria chalybea*** is only 50 ppm. Among the heterocystous members, ***Cylindrospermum musicola, Nostoc commune, Nostoc piscinale,*** and ***Aulosira fertilissim*a** show similar tolerance and show growth upto 100 ppm. ***Anabaena fertilissima*** is more resistant and can grow in 200 ppm concentration of 2,4-D.

In general, no distinction can be made between the non-heterocystous and heterocystous forms regarding the effect of this compound.It is highly lethal to ***Anabaena varibilis***, but ***Calothrix brevissima*** can tolerate a higher concentration (400 ppm), both being heterocystous form.

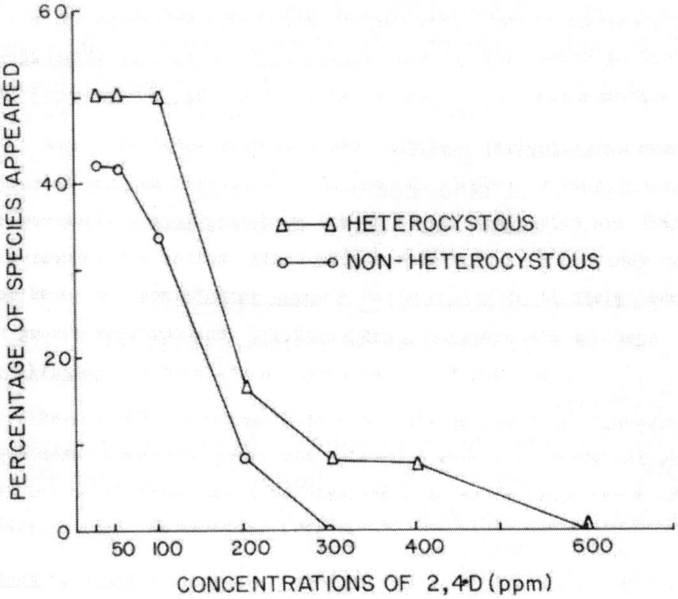


**Fig. 4.** Showing Effect of 2,4-D on Blue-Green Algae of Rice Field

## Table 6

## SUMMARY


- Two fungicides viz. Bavistin (systemic) and Dithane Z, and two pesticides viz.,Dimecron (insecticide) and 2,4-D (herbicides) were tested for their effects on the survival of blue- green algae of the rice fields. The effects on heterocystous and non-heterocystous blue-green algae were noted separately.
- The experimental procedures have been given in Materials and Methods chapter.
- Observations have been made on 12 species of blue-green algae of which 7 were heterocystous.
- A higher concentration (600 ppm) of the 4 compounds generally have an inhibitory effect on the growth of different blue-green algae, and more so in case of the non-heterocystous members except Bavistin. Bavistin inhibits the growth of heterocystous forms more, but growth of a non-heterocystous alga is obtained even in 800 ppm.
- Non-heterocystous forms are more resistant to Bavistin and some can tolerate a very high concentration (400-800 ppm). Species of the some genus have different sensitivity. But heterocystous forms can not grow beyond 50 ppm concentration.

- No species can tolerate Dithane Z beyond 100 ppm concentration and the effects are not differential for non-heterocystous and heterocystous forms. Non-heterocystous members can not grow beyond 50 ppm concentration.
- The effect of Dimecron is not remarkably dissimilar for non-heterocystous and heterocystous forms. No blue-green algae under study could grow beyond 400 ppm concentration.
- Except 2 species, all other non-heterocystous and heterocystous members show similar response to 2,4-D and mostly could not thrive beyond 200 ppm concentration.

## C. EFFECTS OF INORGANIC FERTILIZERS AND ORGANIC MANURE ON CYANOBACTERIA OF PADDY FIELDS

## INTRODUCTION

The modern day intensive crop cultivation requires the use of chemical fertilizers, but fertilizers are not only in short supply but they are expensive in developing countries like India. Therefore, the current trend is to explore the possibility of supplementing chemical fertilizers with organic one, more particularly biofertilizers of microbial origin. It has been well established that cyanobacterial inoculation can replace the use of inorganic fertilizers in case of different high yielding varieties of rice. Besides the nitrogen fixing ones, other cyanobacteria in the rice fields are also important because of their effects through the addition of different organic compounds to the soil by their death and decay, by extracellular secretions, by adding certain substances which are stimulatory or inhibitory to the existing algae, fungi and other micro-organisms. The regular use of inorganic fertilizers in agriculture posed the problem of their effect on microorganisms. The consumption of fertilizers has increased with the development of intensive agriculture. It was considered important to evaluate their effect in field conditions rather than evaluation in the laboratory in pure cultures.

During recent years, emphasis is given on the use of cyanobacteria to enrich paddy fields, which fixes nitrogen. Paddy fields of Purba Bardhaman district in West Bengal are known to provide favourable environment for the luxuriant growth of cyanobacteria. The use of nitrogen fertilizers in paddy fields is a common practice and so their effect on cyanobacteria is a matter of great interest. The current trend is to explore the possibility of supplementing chemical fertilizers with organic one, more particularly biofertilizers of microbial origin.

A number of workers from time to time have reported saving or less application of nitrogenous fertilizers with the application of blue-green algae in rice (Pereira *et al.*, 2009; Begum *et al.*, 2008, 2011; Misra and Pabbi 2004; Pabbi 2015). The Govt. of India made concerted efforts to promote the biofertilizers by initiating and supporting National and State Projects on Development and use of Biofertilizers (NPDB). Biofertilizers have, therefore emerged as a potential alternative to chemical fertilizers due to their eco-friendly, easy to apply, non-toxic and cost-effective nature.

Also, they provide nutrients to the plants and act as supplements to agrochemicals (Majid and Khan, 2014).

In the present study an attempt has been made to know the effects of inorganic fertilizers like ammonium sulphate, urea, potash and superphosphate and organic manure specially Farmyard manure on cyanobacteria of paddy fields. In some selected paddy fields of Purba Bardhaman district the appearance and decline of cyanobacteria during the different period of growth of the rice plants have been observed. Tolerance of different heterocystous and non-heterocystous species was studied in presence of inorganic fertilizers, organic manure, 1:2 ratio of inorganic fertilizers and organic manure. Altogether 29 species of cyanobacteria were recorded from the field collections as well as in soil culture. A correlation could be established between such observation, which indicates cyanobacteria as a biofertilizer actually come into benefit of the rice plants in different growth periods and its role suffers from the application of inorganic fertilizers but not so from organic manure. In the present observation a healthy growth of cyanobacteria was observed in the control field and in the field where 1:2 ratio of inorganic fertilizers and organic manure was given.

An effort has been made to understand the succession of cyanobacteria in rice fields in relation to the growth stages of rice plants. Ecological parameters like water pH and water temperature were also considered to understand the survival and appearance of cyanobacteria in paddy fields. Observations have been made on the fields near the river Ganga at the eastern part of the Purba Bardhaman district.

## MATERIAL AND METHODS

Cyanobacteria were collected between June to October, for consecutive three years 2011 to 2014 from the surface growth as well as water on the paddy field soil where Kharif crop (summer rice) Autumn or Aus varieties were cultivated. Collections were made from fields in different places of eastern Bardhaman district, four rice fields were selected to study the interaction of cyanobacteria with inorganic fertilizers, organic manure, 1:2 ratio of inorganic fertilizers and organic manure and one control field where no fertilizers or manure were given.

**Application of inorganic fertilizers on experimental field:** In one field Phosphorus and potash were applied as 40 kg/ha each as single superphosphate and mureate of potash after puddling as basal dressing. Nitrogen as ammonium sulphate and urea super granule were applied at the rate of 50 kg/ha in three splits (50% + 25% + 25%). First doses were applied as basal while the second and third doses were applied at maximum tillering and panicle initiation stages of rice crop.

**Application of organic manure on experimental field:** In one experimental field Farm Yard Manure (FYM) was applied at the rate of 90 kg/ha basally.

**Application of inorganic fertilizers and organic manure (1:2 ratio) on experimental field:** In one experimental field inorganic fertilizers like phosphorus and potash were applied 20kg/ha each as single super phosphate and mureate of potash after puddlig as basal dressing. Ammonium sulphate and urea super granule were applied at the rate of 25 kg/ha in three splits (50% + 25% + 25%). Farm Yard Manure (FYM) was applied at the rate of 90 kg/ha during preparation of the field.

Collected cyanobacteria were fixed in FAA solution (Formalin- 5 ml., Ethyl alcohol-70 ml., Glacial acetic acid- 2ml., Distilled water-20 ml.) and brought to the laboratory for identification. Soil samples were also collected in plastic bags for raising cultures in Chu-10 (modified) agar medium (with or without nitrogen). Monograph on Cyanophyta by Desikachary (1959) was consulted principally for the purpose of identification.

## RESULTS

### Rice field with inorganic fertilizers:

The field was prepared in the latter half of June and transplantation done in early July. Fertilizers were given accordingly. No growth of cyanobacteria was found before transplantation. In the middle of July, when the first crown root started emerging, detectable growth of cyanobacteria was observed for the first time. The cyanobacterial growth was not increased by the middle of August, the number of species became scanty in the field during flowering and grain setting stage. No species were found in grain maturing and grain ripening stages. (**Table 7**). Altogether 12 species appeared in the field.

### Rice field with FYM

Organic manure FYM was applied during preparation of the field. Altogether 20 species appeared in the field, among which most of them were capable of fixing atmospheric nitrogen. Maximum number of species appeared in the field during pre-flowering stage and shown their presence up to grain maturing stage (**Table 8**)

### Rice field with 1:2 ratio of inorganic fertilizers and organic manure:

Recommended dozes of fertilizers and organic manure were given as mentioned earlier. Maximum number of species appeared during tillering and pre-flowering stage. Total number of species appeared in this field was 16 only. Excellent growth of cyanobacteria was observed during Flowering stage. (**Table 9**)

### Rice field with no fertilizers and organic manure (control field):

Luxuriant growth of cyanobacteria was observed in this field. Altogether 25 species appeared in the field among which 13 species were nitrogen fixers. At field preparation stage, sparse growth of species was observed specially along bunds. It was interesting to note that during the period of luxuriant growth of cyanobacteria, the pH of water in the field was 6.2 to 6.5 (**Table 10**)

Following members of cyanobacteria were collected from paddy fields and appeared in the culture medium:

- ***Chroococcus minimus*** (Keissler) Lemm.
- ***Gloeocapsa rupestris*** Kutz.
- ***Aphanocapsa biforsmis***A. Br.
- ***Aphanothece pallida*** (Kutz.) Rabenh.
- ***Oscillatoria formosa*** Bory ex Gomont
- ***O. agardhii*** Gomont
- ***O. animalis*** Ag.ex Gomont
- ***O. chalybea*** (Mertens) Gomont
- ***O. obscura*** Bruhl et Biswas
- ***O. sancta*** (Kutz.) Gomont
- ***O. splendida*** Grev. ex Gomont
- ***Phormidium ambiguum*** Gomont
- ***P. tenue*** (Menegh.) Gomont
- ***Lyngbya putealis*** Mont.ex Gomont
- ***L. major*** Menegh. ex Gomont
- ***Cylindrospermum musicola*** Kutzing ex Born. et Flah.
- ***Nostoc palludosum*** Kutzing ex Born. Et Flah.
- ***N. piscinale*** Kutzing ex Born. et Flah.
- ***N. commune*** Voucher ex Born. et Flah.
- ***N. punctiforme*** Ag. ex Born. et Flah.
- ***Anabaena fertillisima*** Rao, C.B.
- ***A. oryzae*** Fritsch
- ***A. sphaerica*** Bornet et Flahault
- ***A. naviculoides*** Fritsch
- ***Aulosira fertilissima*** Ghose
- ***Scytonema simplex*** Bharadwaja
- ***Tolypothrix campylonemoides*** Ghose
- ***Calothrix braunii*** (A. Br.) Bornet et Flahault
- ***Gloeotrichia ghosei*** Singh, R.N.

## DISCUSSION

Majority of the cyanobacterial species were collected during rainy season and in autumn, few forms were recorded during summer months which indicate heat tolerance. Among all forms ***Oscillatoria*** was the most dominant genus followed by ***Aabaena*** and ***Nostoc.*** The paddy fields were composed of alluvium and pH ranges from 6.2 to 8.0, water temperature of the fields varies from 28°C to 35°C.

**Table 7. Growth of cyanobacteria at different growth stages of rice crop in presence of inorganic fertilizers**

| | Field preparation (Late June) | Transplantation (Early July) | Crown root (Mid July) | Tillering (Early August) | Pre-flowering (Mid August) | Flowering and grain setting (Early September) | Grain maturing (Late September) | Grain ripening (Early October) | (Late October) |
|---|---|---|---|---|---|---|---|---|---|
| Water pH | 6.5 | 6.5 | 6.1 | 6.2 | 6.2 | 6.5 | 6.6 | 7.0 | 6.8 |
| Water temperature (◦C) | 33 | 34 | 34 | 33 | 31 | 30 | 29 | 28 | 28 |
| Chroococcus minimus | - | - | - | - | - | + | - | - | - |
| Gloeocapsa rupestris | - | - | + | + | + | + | - | - | - |
| Aphanothece pallida | + | + | ++ | - | - | - | - | - | - |
| Oscillatoria formosa | - | + | + | - | - | - | - | - | - |
| O. animalis | - | - | + | + | - | - | - | - | - |
| O. chalybea | - | ++ | ++ | + | - | - | - | - | - |
| O.obscura | - | + | + | - | - | - | - | - | - |
| O. splendida | - | - | - | - | + | + | - | - | - |
| Lyngbya major | - | - | - | + | + | - | - | - | - |
| Cylindrospermum musicola | - | + | - | - | - | - | - | - | - |
| Nostoc palludosum | - | - | - | + | + | - | - | - | - |
| N. commune | - | - | + | + | - | - | - | - | - |

-Absence of the species, + moderate growth, ++ significant growth

**Table 8. Growth of cyanobacteria at different growth stages of rice crop in presence of FYM (Organic manure)**

| | Field preparation (Late June) | Transplantation (Early July) | Crown root (Mid July) | Tillering (Early August) | Pre-flowering (Mid August) | Flowering and grain setting (Early September) | Grain maturing (Late September) | Grain ripening (Early October) | (Late October) |
|---|---|---|---|---|---|---|---|---|---|
| Water pH | 6.5 | 6.5 | 6.4 | 6.4 | 6.5 | 6.7 | 6.6 | 6.8 | 7.0 |
| Water temperature (◦C) | 33 | 34 | 35 | 35 | 34 | 33 | 32 | 33 | 30 |
| Chroococcus minimus | - | - | + | + | ++ | +++ | +++ | + | - |
| Gloeocapsa rupestris | - | - | - | + | + | ++ | ++ | + | - |
| Aphanothece pallida | - | - | - | + | + | ++ | ++ | ++ | + |
| Oscillatoria formosa | - | - | - | - | + | + | ++ | + | - |
| O. agardhii | - | + | ++ | ++ | ++ | +++ | ++ | + | - |
| O. animalis | - | + | + | ++ | ++ | +++ | +++ | ++ | + |
| O. obscura | - | - | - | + | + | ++ | ++ | ++ | - |
| O. sancta | - | - | - | + | ++ | ++ | +++ | ++ | + |
| O. splendida | - | - | - | - | + | + | ++ | + | - |
| Phormidium ambigum | - | - | + | + | ++ | +++ | ++ | + | - |
| P. tenue | - | + | + | ++ | ++ | +++ | ++ | + | - |
| Lyngbya putealis | - | - | + | + | + | ++ | ++ | + | - |
| Cylindrospermum musicola | - | - | - | - | + | + | ++ | + | - |
| Nostoc palludosum | - | - | - | + | ++ | ++ | +++ | + | + |
| N. punctiforme | - | + | + | + | ++ | ++ | +++ | + | - |
| Anabaena fertilissima | + | + | + | + | ++ | +++ | ++ | + | - |
| A. oryzae | - | - | - | + | + | ++ | ++ | + | + |
| A. sphaerica | + | + | ++ | ++ | ++ | +++ | ++ | + | + |
| Tolypothrix campylonemoides | _ | - | - | + | + | ++ | +++ | + | + |
| Glocotrichia ghosei | - | - | - | + | ++ | +++ | ++ | + | - |

-Absence of the species, + moderate growth, ++ significant growth, +++ excellent growth

**Table 9. Growth of cyanobacteria at different growth stages of rice crop in presence of 1:2 ratio of inorganic fertilizers and organic manure (FYM)**

| | Field preparation (Late June) | Transplantation (Early July) | Crown root (Mid July) | Tillering (Early August) | Pre-flowering (Mid August) | Flowering and grain setting (Early September) | Grain maturing (Late September) | Grain ripening (Early October) | (Late October) |
|---|---|---|---|---|---|---|---|---|---|
| Water pH | 6.5 | 6.5 | 6.2 | 6.2 | 6.2 | 6.5 | 6.8 | 6.8 | 6.9 |
| Water temperature (◦C) | 33 | 33 | 34 | 34 | 34 | 35 | 33 | 33 | 29 |
| Aphanocapsa biformis | - | - | + | + | ++ | ++ | + | + | - |
| Oscillatoria agardhii | + | + | + | ++ | +++ | +++ | ++ | + | + |
| O. animalis | - | - | - | + | ++ | ++ | ++ | + | - |
| O. obscura | - | - | + | ++ | +++ | +++ | ++ | + | + |
| Phormidium tenue | - | - | + | + | ++ | ++ | + | + | - |
| Lyngbya putealis | + | + | + | ++ | ++ | ++ | + | + | - |
| L. major | + | - | - | + | ++ | ++ | + | + | - |
| Nostoc palludosum | - | - | + | + | + | ++ | + | + | - |
| N. piscinale | - | + | + | ++ | ++ | +++ | ++ | + | - |
| N. punctiforme | - | - | + | + | ++ | + | + | + | - |
| Anabaena oryzae | - | - | + | ++ | +++ | +++ | ++ | + | + |
| A. naviculoides | - | - | + | + | ++ | + | + | - | - |
| Scytonema simplex | - | - | - | - | + | + | + | - | - |
| Tolypothrix campyloemoides | - | - | + | + | ++ | + | + | - | - |
| Calothrix braunii | - | + | + | ++ | ++ | ++ | + | + | + |
| Gloeotrichia ghosei | + | + | + | ++ | +++ | ++ | + | + | - |

-Absence of the species, + moderate growth, ++ significant growth, +++ excellent growth

**Table 10. Growth of cyanobacteria at different growth stages of rice in absence of inorganic fertilizers and organic manure (control field)**

| | Field preparation (Late June) | Transplantation (Early July) | Crown root (Mid July) | Tillering (Early August) | Pre-flowering (Mid August) | Flowering and grain setting (Early September) | Grain maturing (Late September) | Grain ripening (Early October) | (Late October) |
|---|---|---|---|---|---|---|---|---|---|
| Water pH | 6.5 | 6.5 | 6.4 | 6.5 | 6.5 | 6.8 | 6.9 | 7.0 | 8 |
| Water temperature (◦C) | 32 | 32 | 33 | 34 | 34 | 35 | 31 | 30 | 29 |
| Chroococcus minimus | - | + | + | + | ++ | +++ | ++ | + | - |
| Gloeocapsa rupestris | + | + | + | ++ | ++ | +++ | ++ | + | + |
| Aphanocapsa biformis | - | - | - | - | + | ++ | ++ | + | - |
| Aphanothece pallida | - | - | + | - | ++ | +++ | ++ | + | - |
| Oscillatoria formosa | + | + | ++ | ++ | +++ | ++ | + | + | + |
| O. agardhii | - | - | - | + | ++ | ++ | ++ | + | + |
| O. animalis | - | - | + | + | ++ | +++ | ++ | + | + |
| O. chalybea | - | - | + | + | ++ | ++ | ++ | + | + |
| O. sancta | + | + | ++ | +++ | +++ | +++ | ++ | + | + |
| O. splendida | - | - | - | + | ++ | ++ | + | + | - |
| Phormidium tenue | - | - | - | + | + | ++ | ++ | + | - |
| Lyngbya major | - | - | + | + | ++ | ++ | ++ | + | - |
| Cylindrospermum musicola | - | + | + | +- | +++ | ++ | ++ | + | - |
| Nostoc piscinale | + | + | ++ | +++ | +++ | ++ | ++ | + | - |
| Anabaena fertilissima | - | - | - | + | ++ | ++ | + | + | - |
| A. oryzae | - | - | + | + | +++ | +++ | ++ | - | - |
| A. naviculoides | - | - | - | + | ++ | ++ | + | - | - |
| A. sphaerica | - | + | ++ | +++ | +++ | ++ | + | + | - |
| Scytonema simplex | - | - | ++ | ++ | ++ | + | + | - | - |
| Tolypothrix campylonemoides | - | - | - | + | ++ | ++ | + | + | - |
| Calothrix braunii | - | - | + | ++ | +++ | ++ | ++ | + | + |
| Glocotrichia ghosei | - | - | - | + | ++ | ++ | + | - | - |
| Nostoe commune | - | + | + | ++ | +++ | ++ | ++ | + | - |
| O. obscura | - | - | + | ++ | ++ | ++ | ++ | + | + |
| L. putealis | - | - | + | ++ | +++ | ++ | + | + | - |

-Absence of the species, + moderate growth, ++ significant growth, +++ excellent growth

A consistent decrease in cyanobacterial growth with inorganic fertilizers was marked during rice cultivation cycle in an experimental field which is attributed to the inhibitory effect of ammonium sulphate and urea super granules (Table 7). The experimental field with FYM showed excellent growth of cyanobacteria because of the following reasons:

- They increase the nutrient holding power and cation exchange capacity of the soil and provide conditions for multiplication of soil microorganisms;
- Improve soil aeration and percolation and downward movement of water and thus reduces the risk of surface run off and soil erosion;
- Organic manures are carriers of nutrients including number of micronutrients;
- They help in binding soil fine particles together into soil aggregates and help to maintain the soil in a loose and open granular structure and thus improving physical condition;
- They produce organic acids which help to dissolve soil nutrients and also act as chelating agents.

The experimental field with 1:2 ratio of inorganic fertilizers and organic manure (FYM) showed luxuriant growth of cyanobacteria from crown root stage to grain maturing stage. Although organic manures contain more or less the essential elements for crop growth which are released slowly throughout the growth period but the amounts of nutrients are not so high to keep pace with crop requirement, particularly for high yielding varieties. Therefore a judicious application of inorganic fertilisers and organic manures together in right proportion according to soil and climatic condition and type of crop is considered as a right approach for increasing soil fertility and crop productivity.

In control field maximum no. of species (25) appeared during different stages of rice cultivation. It was possibly due to absence of inorganic fertilizers (specially nitrogen fertilizers) biofertilizers cyanobacteria had shown hyper activity in the soil and perhaps different ecological factors like field water temperature, alkaline pH of the water, humidity etc. favours cyanobacterial growth in this field.

In situations where inorganic nitrogen fertilizers are not used or used as low doses due to economical or ecological reasons, biological nitrogen fixation by free-living or symbiotic cyanobacteria can contribute a considerable amount of nitrogen to soil. The decrease in biomass and $N_2$-fixation could perhaps be due to reduction in nitrogenase activity which was inhibited by nitrogenous compounds (Stewart 1973). Effect of inorganic fertilizers and organic manures on paddy field soil of eastern part of Bardhaman district in West Bengal was worked out by Sen (1987). It was noted that $N_2$ fixing alga ***Anabaena*** was more abundant in unfertilized plots than in fertilized plots (Okuda and Yamaguchi 1952). Rice fields are known to provide congenial environment for the luxuriant growth of BGA in tropical countries which play an important role in its nitrogen economy (Singh 1961, Venkataraman 1981, Singh 1981a, b, 1982, 1985).

These results indicated that the rice field soils under study had favourable environment for the vigorous growth of cyanobacteria. It was observed that biofertilizer cyanobacteria prefers organic manure than that of inorganic fertilizers for their appearance and growth in the rice fields. This study also suggests that reckless applications of inorganic fertilizers should be avoided and a judicious application of inorganic fertilizers and organic manures together with biofertilizer cyanobacteria is considered as a right approach for increasing crop productivity and soil fertility.

## D. CYANOBACTERIA AS BIOFERTILIZERS

During the last three decades agricultural production has increased both in developed and certain developing countries due to use of high yielding varieties and enhanced consumption of chemical fertilizers and water. Nitrogenous fertilizers produced in industry by Haber-Bosh process consume high energy (about 13,500 k cal/kg N fixed). In such industries, fossil fuel is the source of energy. In recent years the cost of crude oil increased about seven fold. Therefore, fossil fuel (oil and coal) based method of farming has become more expensive accordingly. To combat with this situation, however, it is necessary to develop an alternative method of supplying nutrients to plants.

The current world population of 7.6 billion is expected to reach 8.6 billion in 2030, 9.8 billion in 2050 and 11.2 billion in 2100, according to a United Nations report being launched on 21 June 2017. With roughly 83 million people being added to the world's population every year, the upward trend in population size is expected to continue. Since the global population is expected to touch more than 8 billion by 2030, by which time humanity would require large amount of cereals as their main food. To meet these targets, the fertilizer consumption has to be increased as early as possible. Some of the developing countries are historically and culturally the oldest one in the world and hence their soils have been so intensely cultivated for generations that their nutrient content has been depleted. These countries are also handicapped by their low purchasing power, hence they can not use their own soft currencies to import fertilizer or food. It is therefore, essential for us to evolve and adopt a strategy of integrated nutrient supply by using a judicious combination of chemical fertilizer, organic manure and biofertilizers.

In recent years, use of microbial inoculants as a source of biofertilizers has become a hope for most of countries, as far as economical and environmental view points are concerned. Biological systems able to fix atmospheric nitrogen are called as diazotrophs. There are three different forms of $N_2$-fixation i) Physical, ii) Industrial, and iii) Biological and the amount fixed by these processes are 0.5, 49.0, and 50.5% of total fixation (about $139 x 10^9$ kg). Biologically fixed nitrogen is such a source which can supply an adequate amount of nitrogen to plants and other nutrients to some extent. It is a non-hazardous way of fertilization of field. Moreover biologically fixed nitrogen consumes about 25% to 30% less energy than normally done by chemical process. Therefore, in developing countries like India, it can solve the problem of high cost of fertilizers and help in saving the economy of the country.

Cyanobacteria are popular biofertilizers because of their innumerable advantages to the agro-ecosystem. Chemical fertilizers can damage the environment but the cyanobacterial biofertilizers are eco-friendly. They are cheap alternatives as they require low cost inputs. Many cyanobacterial species can fix atmospheric nitrogen, enriches soils with organic matter and reduces the C:N ratio of the soil. They are responsible for the enhanced solubilization of immobile phosphates and producing growth promoting substances in the paddy fields. They improve the physical chemical and biological nature of the paddy field soil and are responsible for long term soil fertility. Cyanobacteria can reduce the pH of soil and helps the soil to retain exchangeable calcium. Prof. G.S. Venkataraman introduced the term "algalization" for application of blue-green algae as a fertilizer. BGA can contribute about 20-30 kig N $ha^{-1}$ along with organic matter to the soil, which is quite significant for the economically weak farmers unable to invest costly inorganic nitrogen fertilizers. Many Asian countries like India, China, Vietnam, Pakistan, Indonesia etc., have been using BGA in paddy cultivation as an alternative to chemical fertilizers. Another reason for cyanobacteria becoming popular as biofertilizer is the presence of environmental factors favourable for the growth of cyanobactrial forms. Cyanobacteria occurred at densities from: $1.0x10^{-2}$ to $8.0x10^{-6}$ CFU/$cm^{-2}$and their abundance was correlated to pH and available 'P' content of soil.

## Governments Initiative for BGA Mass Production

The awareness of utilizing algal biofertilizer to supplement the nitrogen requirement of rice crop was introduced as early as 1939. In the later half of the $20^{th}$ century considerable attention has been focussed on nitrogen fixation by cyanobaeteria as well as transferring the acquired knowledge effectively for application. Cyanobacteria are now components of biofertilizer programmes supporting the chemical fertilizer. A new technology was developed with the financial assistance provided by Department of Biotechnology, in which cyanobacteria were grown in a formulated medium in glass/polyethylene house and soil was replaced with lighter carrier (straw). In 1972 at IARI, New Delhi the first ever production of Cyanobacterial Biofertilizer in India was reported. Since then, significant progress has been made, mainly in the technology of production, identifying efficient strains, newer techniques of inoculation for better establishment of inoculated organism in various soils and in demonstrating their beneficial effect on the crop yield in field trials on different crops. Most commonly used species in production of biofertilizers are ***Anabaena variabile*, Nostoc *muscorum*, *Aulosira fertilissima*** and ***Tolypothrix tenuis*.**

For acceleration and extension of works on algal biofertilizers, department of Biotechnology, Govt. of India, established four major centres in different paddy growing areas of the country. The programme was launched in Lucknow (UP), Madurai (Tamil Nadu) West Bengal (Kolkata) and New Delhi. As a result, cyanobacterial biofertilizer was found very useful, especially for small and marginal farmers of our country.

## Production Methods for BGA

5 g of top soil is takeN from a selected paddy field and added in to a 100 ml of Fogg's medium in a flask. The flask is shaken well and incubated at room temperature. Illumination of 1500 lux is provided to the culture to increase the cyanobacterial growth. Now the culture is transferred to 10ml of water in a tube with the help of a loop. Tube has to be shaken well so that algal filaments can be separated. The content is serially diluted. Every dilution drop is inoculated into Fogg's medium to establish algal growth. This process is executed in a pteridish.

**(Fogg's Medium:** $KH_2PO_2$-0.2g; $MgSO_4.7H_2O$-0.2g; $CaCl_2$-0.1g; $Na_2MoO_4$-0.1mg; $MgCl_2$-0.1mg; $H_3BO_3$-0.1mg; $CuSO_4$-0.1mg; $ZnSO_4$-0.1mg; Fe-EDTA-1.0 ml; Distilled water – 100 ml; pH-7.)

Each culture is examined properly with microscope. If it is supposed to be a pure culture for a single species then only, it is used further. In case of more than one species, the sample is diluted till the isolation of a pure culture is attained. The pure cultures of BGA are transferred to culture flasks having Fogg's medium for growth. Adequate light is provided for the growth. Now the algal cultures are used as starter culture to initiate the mass culture of BGA.

In general, there are four methods of algal production that have been reported viz,: i) Trough or tank method ii) Pit Method iii) Field Method and iv) Nursery method. The former two methods are essentially for individual farmers and latter two are for bulk production on a commercial basis.

## Trough or tank method:

- Preparation of shallow trays (2m x1m x 23cm) of galvanised iron sheet or permanent tank. The size of the tank can be increased if more material is to be produced.
- Spreading of 4 to 5 kg of river soil and mixing well with 100g of superphosphate and 2g Sodium molybdate.
- 5 to 15 cm of water poured in the trays. This will depend upon local conditions (rate of evaporation). Ingredients were mixed properly.
- 10 to 15g Furadan granules or Malathion was added o protect from insects.
- Soil and water mixture was allowed to settle for 10-11 hours. During this time mother culture (200-250g) of BGA was added to the surface water without disturbing the water.
- The reaction of the soil should be neutral. CaCO was added in order to bring the pH of the soil to neutral, if the soil is acidic.
- If sunlight and temperature are normal then within 10-15 days the growth of the blue-green algae will appear and hard flakes are formed on the surface of the water/soil.

- Water in the tray or pit is allowed to evaporate and the growth of the cyanobacterial flakes is allowed to dry.
- Dried algal flakes are separated from the soil and collected in plastic bags. From one sq.m tray or pit about half tonnes kg blue-green algal flakes are obtained.
- Again water was added to trays and stared the soil well and allowed the algae to grow. Now it is not necessary to add mother culture or superphosphate. By this process one can harvest and grow algae 2-3 times.

**Pit method:**This method is easy and less expensive to operate by small farmers. By this method BGA production is very much similar to the trough or tank method. Here, pits are dug in the ground and layered with thick polythene sheet to hold the water. The soil is filled in pit for 20 cm and watered for 10cm height. pH is maintained and carbofuran is added to the pit. Then saw dust is spread over the soil and is sprinkled with the starter culture. The pits are watered for BGA growth. Thereafter, the soil is allowed to dry.

**Field Method:** It is an ideal method to produce BGA biofertilizer on a commercial scale.

- The area in the field for production of BGA was demarcated, the suggested area is 40sq.m. After crop harvest, the stubble is removed and algal production may be started.
- The area is covered with water to a depth of 2.5 cm. Flooding is needed repeatedly to keep the water standing.
- Superphosphate 12 kg/40 sq.m was applied.
- To control insects-pests attack, carbofuran (3% granules) is applied.
- No fresh algal application is required for at least two consecutive cropping seasons if the field has received algal application previously. Otherwise the composite algal culture of 5 kg/40m is applied.
- In clear sunny weather good growth of BGA takes place within weeks in clayey soils, while in loamy soils it takes three to four weeks.
- When BGA have grown properly and formed floating mats, they are allowed to dry in the sun and dried algal flakes are collected in sunny bags for further use.
- By reflooding the same area one can continually harvest algal mat by applying superphoshate and pesticides. Addition of algal inoculums is not necessary.
- During summer (April-June) the average yield of cyanobacteria per harvest ranges from 18-30 kg/40m.

**Nursery Method:** Farmers can produce algae in their nurseries along with rice seedlings. If 320m$^2$ of land is allotted to form a nursery and an additional 40m$^2$ of plot can be prepared alongside to produce algae as described above. By the time rice seedlings are ready for transplantation about 15-20 kg of algae material will be available.

This much quantity of algal mass will be sufficient to inoculate one and half hectares of area.

If every farmer produces the algal material required to inoculate his own land then he will reduce the cost of algal inoculum required to be purchased. So, also one can cut the cost of chemical fertilizers to be applied as recommended.

## Method of Application of BGA Biofertilizer

One packet (500 g) of ready to use multani mitti based BGA biofertilizer is recommended for one acre of rice field. The packet is opened and mixed with 4 kg dried and sieved farm soil. The mixture is broadcast on standing water 4-6 days after transplantation. Use of excess BGA material is not harmful; instead it accelerates the multiplication process and early establishment of algae in the field.

The field should be kept water-logged for about 10-14 days after inoculation to allow good growth of BGA.

When nitrogenous chemical fertilizers are used, reduce the dose by one third and supplement with BGA.

To get cumulative effect, apply BGA for at least four consecutive seasons.

One may not need to apply BGA further as these will establish in the field and reappear as and when the condition becomes favourable.

A gap of 3-4 days should be maintained in between application of fertilizers and pesticides and the BGA application in the field.To accelerate BGA multiplication a small dose of phosphate fertilizer may be given after BGA inoculation.

## Benefits of using biofertilizers

Some of the advantages associated with biofertilizers include:

- They are comparatively low on cost inputs and are light on the pockets of the farmers.
- Environment friendly, unlike chemical fertilizers,that cause damage to the ricefield as well as environment.
- Application of biofertilizers leads to soil enrichment and the quality of the soil improves with time.
- Do not show immediate effect/results, but over time are extremely beneficial to the soil and ecosystem.
- Complex organic materials can be converted to simple compounds by inoculants, as a result plant can easily absorb the nutrients through water.
- They increase the phosphorous content of the soil by solubilising and releasing unavailable phosphorous.
- Biofertilizers can release growth promoting chemicals which causes root proliferation.
- Responsible for increasing the crop yield by 12-30%.

## Disadvantages of using chemical fertilizer

- It causes a disturbance to the plant-microbe associations.

- Softening of plant tissues ensuing in an augmented vulnerability to disease and pests.
- Loss of nutrients from soils through leaching or greenhouse gas emission.
- It enhances the decomposition and depletion of soil, leading to the acidification or alkalization of soil, which reduces soil fertility.
- Excess use of chemical fertilizers causes weed problems in the fields.
- Chemical fertilizers contain nitrates and phosphates when they are leached or removed from the soil enter into ground or surface water causing water contamination, leading to eutrophication of surface water resulting growth of aquatic plants and algae.
- Over use of inorganic fertilizers more than the recommended amounts for continuous monoculture cropping caused formation, accumulation of mineral soil of fertilizers that lead to compaction layer, compaction and soil degradation in long-term. Soil degradation affects significantly the soil structure and nutrient uptake.

## Ability of Cyanobacteria as Biofertilizer for Rice

Biofertilizers are living microorganisms which can add nutrients to the soil, improves soil properties, solubilizes insoluble phosphatic compounds and produces plant growth promoters in the soil and thus plays a vital role in the overall growth of the plants. Cyanobacteria can be used as biofertilizer in paddy fields owing to its nitrogen fixing potential and at the same time it releases different bioactive compounds into the field which can stimulate the growth of rice plants. In wetland paddy fields cyanobacteria usually appears as floating mass on the surface of water. Cyanobacteria produces a wide range of bioactive compounds such as amino acids, carbohydrates, proteins, vitamins and growth hormones like auxins, gibberellins and cytokinin which are necessary for the growth of paddy plants Kaushik, 2014; Sood *et al.*, 2011). Decomposition of cyanobacteria gives organic matter rich in nitrogen and phosphate to the soil. They can increase soil fertility by fixing atmospheric nitrogen and it attaches with paddy field soil particles, helps to retain moisture of the soil, and reduces erosion of the soil.

Many ecological as well as environmental factors influence the cyanobacterial colonization in the rhizosphere of paddy plants. The roots of the paddy plants release sugars, amino acids and a variety of other substances which can attract or repel microbial communities in the field. pH of the soil (acidic or alkaline), temperature, light, humidity and crop canopy and nutritional status of the soil may have important impact on colonization. The pH of the soil is most critical parameters for cyanobacterial abundance and growth. They require a neutral to slightly alkaline pH for optimal growth. Most paddy soils have a natural population of cyanobacteria which provides a potential source of nitrogen fixation at no cost. Ammonia can be taken up by cyanobacteria through passive diffusion or as ammonium ($NH_4^+$) by a specific uptake system. Nitrate and Nitrite are important sources, which later reduce into ammonia. Large number of cyanobacteria are able of using dinitrogen ($N_2$) of the air, as the source of nitrogen, and this is termed as nitrogen fixation. A high molecular weight, oxygen sensitive metalloprotein enzyme called as nitrogenase is responsible for nitrogen fixation. In presence of hydrogen, nitrogenase reduces molecular nitrogen to ammonia. The dried cyanobacteria are given to the fields as inoculum, for better yield of rice. 'Algalization' is now applied to the use of a defined mixture of cyanobacterial species to inoculate soil. Cyanobacteria introduced as a result of algalization can establish themselves permanently in the fields if inoculation is done consecutively for 3-4 cropping seasons.

The basic advantage of mass production technology is that farmers after getting the soil based starter culture can produce the biofertilizer on their own with minimum additional inputs. For one hectare of paddy field 10-12 g of inoculum is considered enough. The open-air algal biofertilizer production has few limitations, which are as follows:

- Mass cultivation in open-air is possible during summer months only, but in rainy season and in winter production has to be stopped.
- High chance of contamination due to open-air type of production.
- Production rate is very slow.
- Require heavy inoculum per hectare as population density is very low.

Now a days in many places instead of open-air system, indore production of algal biomass under controlled conditions is followed. In this process, the growth of algae takes place in a unit that may be a polyhouse or glass house. The individual unit in the polyhouses can be of either RCC, brick and mortar, or even polythene lined pits in the ground. The algae are grown by inoculating separate tanks with pure culture grown in the laboratory. After fully grown, the culture is harvested, mixed with the carrier material, pre-soaked over-night in water and multani mitti (1:1 ratio) and then sun dried. The dried material is packed in small polythene bags for future use. The final product contains 10,000 to 1,00,000 units or propagules per g of carrier material and, therefore, 500g material is sufficient to inoculate one acre of rice growing area (Venkataraman, 1972). The best results are obtained when mixed inocula are produced from local stocks, and the biofertilizers are used in combination with a low level of nitrogen fertilizer. Cyanobacteria form a major component of the flora as long as nitrogen content of the paddy field is not very high. If high doze of nitrogen fertilizer is applied in the field, green algae tend to dominate the soil flora.

During $N_2$ fixation, $N_2$ from the atmosphere is fixed by cyanobacteria with the help of enzyme nitrogenase into ammonium using ATP as the energy source. The ammonia is added to 2-oxoglutarate by an enzyme glutamate dehydrogenose to form glutamate. By the addition of a second ammonium group to glutamate, glutamine is produced. Glutamine molecule is transferred from one cyanobacterial cell to another. Nitrogenase is sensitive to inactivation by oxygen. To exclude oxygen from the area of cells containing nitrogenase different mechanisms are observed in cyanobacteria.

### Heterocystous Cyanobacteria

***Nostoc, Anabaena, Aulosira, Calothryx, Tolypothrix, Scytonema, Stigonema, Rivularia* etc.** are heterocystous genera capable to fix atmospheric nitrogen. Heterocyst is the site of fixing Nitrogen. A glycolipid layer encloses heterocyst and this layer is impermeable to $O_2$. It is photosynthetically inactive, i.e. neither it fixes $CO_2$ nor it produces $O_2$ but exhibits a high rate of respiratory $O_2$ consumption.

### Non-heterocystous Filamentous Cyanobacteria

Non -heteterocystous filamentous cyanobacteria like ***Oscillatoria, Spirulina, Phormidium, Lyngbya, Trichodesmium, Pleurocapsa, Hydrocoleus*** and ***Katagnymene*** etc. are reported to fix nitrogen from the atmosphere.

Two marine genera ***Trichodesmium* and *Katagnymene*** are devoid of heterocysts, but able to fix nitrogen under aerobic conditions (Bergman *et al*., 1997). They have few specialized cells (10-15%) in the filament which are capable to fix nitrogen, these $N_2$ fixing specialized cells are called as diazocytes (Lundgren *et al*., 2001). Diazocytes ae locted adjacent to one another.

### Non-heterocystous, Non-filamentous Cyanobacteria

Some genera like ***Microcystis, Gloeothece, Synechococcus, Synechocystis, Prochlorococcus*** are reported to fix atmospheric nitrogen and they are non-heterocystous and non-filamentous. These genera are able to fix nitrogen in the dark when nitrogenase inhibiting oxygen is produced in the cell during photosynthesis.

### Use of BGA as biofertilizers in paddy fields

- Algal culture is applied as dried flakes at 10 kg/ha over the standing water in field rice.
- This is done two days after transplanting in loamy soils and six days after planting in clayey soils.
- The field is kept water logged for few days immediately after algal application.
- The biofertilizer is to be applied for 3-4 consecutive seasons in the same field.

## Precautions to take while using BGA biofertilizers:

- Packets of biofertilizers need to be stored in cool and dry place away from direct sunlight and heat.
- Right combinations of biofertilizers have to be used.
- Other chemicals should not be mixed with the biofertilizers.
- While purchasing one should ensure that each packet is provided with following information:
- Name of the product, name of the crop for which intended, name and address of the manufacturer, date of manufacture, date of expiry, batch number and instructions for use.
- Biofertilizers are live products and require proper care during storage.

- Both nitrogenous and phosphatic biofertilizers are to be used to get the best results.
- It is important to use biofertilizers long with chemical fertilizers and organic manures. Biofertilizers are not replacement of fertilizers but can supplement plant nutrient requirements.

# PART – IV
# ANALYSIS & ASSESSMENT

## A. A REVIEW ON MODERN CYANOBACTERIAL TAXONOMY

Cyanobacteria occupy a broad range of habitats throughout the world being widespread in lakes, ponds, springs, wetlands, streams, rivers, paddy fields, berk of the trees etc. and they play a major role in the nitrogen, carbon and oxygen dynamics in different types of water bodies. They are the autotrophic prokaryotic organisms appeared in the plant earth before long days, the exact timing of appearance of this prokaryotic microbes on earth is still not clear because of controversy over the elucidation of Precambrian fossils. They play a major role in the accumulation of oxygen in the earth's early atmosphere. According to Fogg *et al.*, (1973) cyanobacteria are thought to have originated during Precambrian Eon and can tolerate wide range of environmental stresses like heat and cold shock, anaerobiosis and oxygeny, photo oxidation, nitrogen starvation, salinity and osmoticum. They are such organisms which can utilize the elemental nitrogen of the atmosphere by the process of nitrogen fixation. It is a wonderful prokaryotic group that are able to conduct oxygenic photosynthesis process and can fix nitrogen by the process of nitrogen fixation which is a process sensitive to oxygen: These two processes are accomplished by the same cyanobacterial cell and they have the capacity to protect nitrogenase (the key enzyme) activity from oxygen. Due to this inherent modalities, this particular group (cyanobacteria) is the main attraction of present day research in biology.

BGA have been classified in the Kingdom Plantae and in the Kingdom Monera (Procaryotae) due to the resembleness of this algal group with bacteria. In Kingdom Plantae BGA are included in division Cyanophyta or class Cyanophyceae or Myxophyceae while in Kingdom Monera, they are included in division Cyanobacteria (Murray, 1968) or order Cyanobacteriales (Gibbons and Murray, 1978). So, the classification of BGA into taxonomic categories at the levels of Kingdom, division, class or order has not agreed upon. That is why, the problems encountered in identification and classification of BGA require new approaches which are being explored to overcome taxonomic studies.

Traditional classifications are based on some morphological features (Geitler, 1932; Fritsch, 1945; Drouet, 1951; Desikachay, 1959; Bourrelly, 1970a) because no sexual reproduction has been reported among BGA. The main morphological criteria used in taxonomy of BGA are 1) Shape and compactness of the colonies 2) Nature of the thallus: unicellular, colonial or filamentous 3) Shape of the filaments 4) Presence or absence of the sheath, if present how is the shape 5) Size and shape of vegetative cells, heterocysts and akinetes 6) Presence or Absence of akinetes 7) Shape of the terminal cell of the trichome or filament 8) Branching of the filament: presence or absence, false or true etc. 9) Nature of branching: Uniseriate or multiseriate. Combinations of above mentioned characters are considered during classification of BGA.

Several classifications were proposed emphasizing different morphological characters. Elenkin (1936) considered 12 orders and 47 families, whereas Drouet (1951) recognized no distinct orders but only 8 families. The classification proposed by Geitler (1942). Fritsch (1945) and Desikachary (1959, 1973) fall in between these two types. Geitler recognized four orders of the class Cyanophyceae: which are as follows: (a) Chroococcales, (b) Dermocarpales, (c) Pleurocapsales, and (d) Hormogoniales. Fritsch divided Myxophyceae on the basis of hormogones into two categories, five orders and 17 families.

Category 1 (without hormogones): Comprises of three orders – Chroococcales, Chamaesiphonales and Pleurocapsales.

Category 2 (with hormogones) comprises of two orders – Nostocales and Stigonematales.

According to Desikachary the phylum Cyanophyta contains single class Cyanophyceae which has five orders – a) Chroococcales, b) Chamaesiphonales, c) Pleurocapsales, d) Nostocales and e) Stigonematales.

After the introduction of electron microscopy and applications of molecular and genetic methods of characterization of cyanobacterial taxa, the taxonomic system of cyanobacteria was radically changed. The taxonomy of cyanobacteria was continually being changed / revised since the work of Francis Drouet; different proposals being made over the last fifty years by microbiologists and phycologists. A group of workers, typified by Drouet, preferred to simplify the classification by substantial reducing the number of taxa (Drouet & Daily 1956, Drouet 1968, 1973, 1978, 1981; Bourrelly 1970; Otsuka *et al.*, 2001). A second group of researchers proposed splitting of species and genera and all higher level taxa to achieve monophyly in all taxonomic groups (Anaganstidis & Komarek, 1985; Casamatta *et al.*, 2005; Johansen & Casamatta, 2005; Rehakova *et al.*, 2007; Siegesmund *et al.*, 2008; Perkerson *et al.*, 2011). Another group (3rd group) recommended caution and advice a prohibition on taxonomic revision until considerable molecular evidence (Hoffmann 2005). This approach effectively discarded the definition of orders, families, genera and species and restore them with subsections, "families" and "form genera" though it do not reflect evolutionary history but lay out a temporary, artificial and nomenclatural invalid system which is convenient and stable method of referring to cyanobacterial strains (Castenholz, 2001). So, controversies in traditional and modern approaches to cyanobacterial taxonomy still persist. This is a time when we are going through fairly radical and rapid systematic and taxonomic development. The 20th IAC symposium in 2016 at University of Innsbruck, special emphasis was given on functional molecular biology and phylogeny, biogeography, systematics and taxonomy with functional ecology and biodiversity of cyanobacteria (Rott, 2016). The symposium provided forum for discussion on merits and value of molecular vs. classical taxonomy of cyanobacteria. Up to late 20th century to describe a particular cyanobacterial species we had to depend on morphological characters. It has been observed that when cyanobacteria brought into culture often changed their morphological characters and this was largely ignored by taxonomists. To overcome this problem, it has been proposed to base the identification and taxonomy of cyanobacteria on standardized and reputed stable genetic markers (Edelman *et al.*, 1967).

## Disadvantage of classifying BGA on the basis of morphological characters

Satisfactory classification of unicellular BGA belonging to the order Chroococcoles is not possible in terms of traditional taxonomic features, which has been recognized. With the help of microscope very few useful characters are determined for identification due to structural simplicity of unicellular forms (Zehnder, 1973). Some forms of BGA exhibit morphological variations (polymorphism) in response to environmental changes (Nikitina *etal.*, 1979; Stam and Holleman, 1979; Rosenchuk ad Mikhailo, 1980; Seki *et al.*, 1981). It has been observed that the ranges of variations of morphological characters of natural populations of the same genus (e.g. *Aphanizomenon*) were reported to overlap, forming no clear differences among the members (Baker, 1981). Following problems encountered in utilizing morphological characters to describe BGA.

### Shape of the Colony

Stainer (1971) stated that *Aphanocapsa*, in culture, forms colonies like that of *Merismopedia*. According to them *Merismopedia* may be growth form of *Aphanocapsa*. Forest (1968) compared *Nostoc* and *Anabaena* and concluded that *Nostoc* can be differentiated by a firm sheath and a macroscopic colony form was unjustifiable and that the usual practice of calling all free filaments as *Anabaena* was not viable taxonomically. So, to distinguish these two genera developmental morphology must be followed.

### False branching

Fritsch (1945) and Desikachary (1973) included the genus *Plectonema* under Scytonemataceae on the basis of false branching, where as closely, related genera *Lyngbya–Phormidium – Plectonema* were placed in Oscillatoriaceae as they do not form false branching.This character of false branching was overe-stimated by them. Stainer and Cohen-Bazire(1977) recognized *Lyngbya-Phormidium-Plectonema as LPP group.*

### Sheath

Presence or absence of sheath, its morphological characters etc. were used as taxonomic characters. But Schwabe (1962) pointed out that presence or absence of sheath is influenced by environmental factors. Bourrelly (1970b) proposed unification of two genera *Lyngbya* and *Phormidium* in one genus and differentiate them from *Oscillatoria* by the presence of sheath. Stam (1980) suggested that *Lyngbya*, *Phormidium* and *Plectonema* should be merged into one genus on the basis of characters like false branching and sheath morphology.

### Non-filamentous colonies

Non-filamentous colonies resulted from a persistence or a confluence of the gelatinous envelopes surrounding the individual cells. Genera with evident sheaths about the individual cells show a strong tendency toward colonial dissociation and have smaller colonies than do genera in which the cell sheaths are fused to form a homogeneous matrix. The shape of non-filamentous colonies dependent upon the planes in which the cells divide. When divisions are restricted to two planes, resulted the layer one cell in thickness and either a flat plate or a hollow sphere may be formed. When divisions are in three planes, a cubical colony is resulted due to regular sequence of divisions.

### Filamentous colonies

Repeated divisions in a single plane produce a filamentous colony. Cells of filamentous colonies may be held together solely by walls common to two adjacent cells; but usually there is also a cylindrical sheath of gelatinous material enveloping the file of cells. A filamentous colony with single row of cells is called a trichome, and a trichome with enclosed sheath is called a filament. Depending on genera, a filament may contain a single or several trichomes. In most genera trichomes are unbranched, but in some genera they may be branched.

### Apical cell

Apical cell of the trichome has been used as a taxonomic character to identify different species of *Oscillatoria*. However form and size of the apical cells of *O. redekei* was reported to change depending on age, light intensity and growth stage of the strains (Meffert *et al.*, 1981).

### Size of the cells

In the taxonomy of unicellular BGA, cell size plays an important role to identify a species, but it has been observed that cell size of a particular species is influenced by environmental conditions.Variation in size of the cells of filamentous BGA in response to variation in environmental conditions have also been reported.

### Mutation

Heterocyst is consider as one of most significant characters in taxonomy of filamentous BGA. Position of akinetes whether adjacent to or away from heterocysts, was considered as one of the most important characters in identifying or determining many species belongs to the Nostocaceae. However, a number of mutants of BGA unable to form akinetes have been reported which occurred spontaneously (Singh & Sinha, 1965). Rippka *etal.*, 1979 reported several spontaneous heterocyst minus (Het-) mutants when the cultures of heterocystous BGA were maintained for a long time in the presence of a source of combined nitrogen. Such spontaneous Het-mutants may create confusion in identification of some heterocystous BGA, because the mutant sometimes outgrows the wild type after several transfers.

The nomenclature of heterotrophs and Archaea follows the International Code of Nomenclature of Bacteria ("bacteriological code", ICNB) from 1980 onwards, cyanobacterial nomenclature is treated both by the botanical code and the bacteriological code. Several solutions have been proposed to solve this confusion but none found agreeable or amicable acceptance (Stanier *et al.*, 1978).

In the year 1985, the "Subcommittee on the Taxonomy of Phototrophic Bacteria" proposed to consider species validity published under botanical code as valid species in the sense of the bacteriological code, but this proposal given by the subcommittee was not accepted and the debate is perpetual (Oren & Ventura, 2017). Two opposite proposals were made to regulate cyanobacterial nomenclature, one proposed for deletion of this prokaryotic group from the groups of organisms whose nomenclature is regulated by the International code of Nomenclature of Prokaryotes (ICNP, earlier known as ICNB), the second proposal was in favour of application of the rules of ICNP to all cyanobacteria.

Following the botanical code, the number of species is continuously increasing. However, following the bacterial code, only few number of cyanobacterial species are considered. During classification of cyanobacteria in a taxonomic system, phylogenetic relationship is very important criteria that can reflect the grouping of organisms in hierarchical taxa. Now a days, biochemical and molecular methods have been used increasingly in microbial taxonomy. On the basis of molecular data, a number of cyanophycean taxa have been renamed after revision – and there will be more revisions in future (Table 11).

**Table 11. Cyanobacterial species names that underwent revision in recent years**

| *New name* | *Old name (basionym)* | *Reference* |
|---|---|---|
| Chrysosporum bergii | Anabaena bergii | Zapomelova et al. (2012) |
| Chrysosporum ovalisporum | Aphanizomenon ovalisporum | Zapomelova et al. (2012) |
| Cuspidothryx issatchenkoi | Aphanizomenon issatchenkoi | Rajaniemi et al. (2005b) |
| Raphidiopsis raciborskii | Cylindrospermopsis raciborskii | Aguilera et al. (2018) |
| Dolichospermum flosaquae | Anabaena flos-aquae | Wacklin et al. (2009) |
| Dolichospermum circinale | Anabaena circinalis | Wacklin et al. (2009) |
| Dolichospermum crassum | Anabaena crassa | Wacklin et al. (2009) |
| Dolichospermum lemmermannii | Anabaena lemmermannii | Wacklin et al. (2009) |
| Dolichospermum planctonicum | Anabaena planctonica | Wacklin et al. (2009) |
| Dolichospermum smithii | Anabaena smithii | Wacklin et al. (2009) |
| Dolichospermum solitarium | Anabaena solitaria | Wacklin et al. (2009) |
| Dolichospermum spiroides | Anabaena spiroides | Wacklin et al. (2009) |
| Dolichospermum viguieri | Anabaena viguieri | Wacklin et al. (2009) |
| Kamptonema formosum | Oscillatoria formosum | Strunecky et al. (2014) |
| Moorea producens | Lyngbya majuscula | Engene et al. (2012) |

| New name | Old name (basionym) | Reference |
|---|---|---|
| Microcoleus autumnalis | Phormidium autumnale | Strunecky et al. (2013) |
| Microseira wollei | Lyngbya wollei | Mcgregor & Sendall (2015) |
| Planktothricoides raciborskii | Planktothrix raciborskii | Suda et al. (2002) |
| Sphaerospermopsis aphanizomenoides | Aphanizomenon aphanizomenoides | Zapomelova et al. (2011) |
| Sphaerospermopsis reniformis | Anabaena reniformis | Zapomelova et al. (2011) |
| Wilmottia murrayi | Phormidim murrayi | Strunecky et al. (2011) |

The list is not complete but comprises primarily those species that are potentially toxigenic or are closely related to toxigenic species. Note that a new name not necessarily comprises all forms previously published under the old name, and therefore, a retrospective renaming may be critical. (Source: Vidal *et al.* 2021).

Now, most of the cyanobacterial species have been described according to the botanical code of nomenclature based on some morphological criteria which are mainly quantitative. Descriptions of the species are based on drawings and microphotographs only because the botanical code does not require the deposition of a type strain. Some workers like Woese, 1987 and Pace, 1997 were of opinion that the description of cyanobacterial taxa according to the bacteriological code would be biologically more appropriate as there is no doubt that cyanobacteria are a monophyletic branch in the global bacterial phylogenetic tree. Molecular approaches to cyanobacterial taxonomy are very significant to understand the phylogenetic relationships. Molecular approaches involve sequencing of marker genes (16SrDNA, phycocyanin operon), DNA-DNA hybridisation, genome sequencing and biochemical characteristics like fatty acid profiles or immunological procedure (Wilmotte, 1994; Whitton & Potts, 2000). Table 12 summarises taxonomic schemes for cyanobacteria proposed by Geitler (1932) to Komárek *etal.*(2014). Many system avoid the use of nomenclatural categories and instead use groups like "sections" (Rippka *etal.*, 1979) or "subsections" (Castenholz *etal.*, 2001) instead of orders visible by the suffix –"ales" in order to high light that at least some of these groups do not characterized by monophyletic origin but are defined based on shared morphological features (Ishida *et al.*, 2001; Gugger & Hoffmann, 2004).

**Table 12. Overview on taxonomic classification systems of cyanobacteria, following either the International Code of Nomenclature for algae, fungi and plants (ICN) or the International Code of Nomenclature of Bacteria (ICNB)**

| Reference | Criteria | Classification and morphology | Observations |
|---|---|---|---|
| Komarek *et al.* (2014) (ICN) | Polyphasic, including ultrastructure, molecular, genomic | Order Gloeobacterales (U)<br>Order Synechococcales (U/Fi,C)<br>Order Spirulinales (coiled Fi)<br>Order Chroococcales (U/pFi,C)<br>Order Pleurocapsales (U/pFi,Bc)<br>Order Oscillatoriales (U/Fi,fB)<br>Order Chroococcidiopsidales (U/Bc)<br>Order Nostocales (Fi,fB/tB,Ho,Hc,Ak) | New orders: Spirulinales, Chroococcidiopsidales |
| Hoffmann *et al.* (2005) (ICN) | Polyphasic, including morphology, ultrastructure, molecular, | Order Gloeobacterales (U)<br>Order Synechococcales (U/Fi,C)<br>Order Pseudoanabaenales (Fi)<br>Order Chroococcales (U,C)<br>Order Oscillatoriales (Fi,fB,Ho)<br>Order Nostocales (Fi,fB/tB,Ho,Hc,Ak) | New orders: Pseudoanabaenales, Synechococcales;<br>Some orders with both filamentous and colonial forms;<br>No partitioning of true and false branching in forms with heterocytes;<br>Order Stigonematales is dispersed and former members moved to other orders |
| Cavalier-Smith (2002) (ICNB, not valid) | Morphology | Order Gloeobacterales (U)<br>Order Chroococcales (U,C)<br>Order Pleurocapsales (U,C,Bc)<br>Order Oscillatoriales (Fi,fB,Ho)<br>Order Nostocales (Fi,fB,Ho,Hc,Ak)<br>Order Stigonematales (Fi,tB,Ho,Hc,Ak) | Classification of cyanobacteria in global bacterial taxonomy;<br>Gloeobacterales (with genus Gloeobacter) as sister group of all other cyanobacteria;<br>No details on classification of genera |

| Reference | Criteria | Classification and morphology | Observations |
|---|---|---|---|
| Castenholz*et al.* (2001) (ICNB, not valid) | Morphology | Subsection 1 (U,C)<br>Subsection 2 (U,C,Bc)<br>Subsection 3 (Fi,fB,Ho)<br>Subsection 4 (Fi,fB,Ho,Hc,Ak)<br>Subsection 5 (Fi,tB,Ho,Hc,Ak) | Largely based on morphological characteristics; Genera known at the time as "form genus"; Largely in agreement with Rippka*et al.* (1979) |
| Anagnostidis&Komarek (1985) (ICN) | Morphology | Order Chroococcales (U,C,Bc)<br>Order Oscillatoriales (Fi,fB,Ho)<br>Order Nostocales (Fi,fB,Ho,Hc,Ak)<br>Order Stigonematales (Fi,tB,Ho,Hc,Ak) | All nonfilamentous taxa are unified in Chroococcales |
| Rippka*et al.* (1979) (ICN) | Morphology | Section I (U,C)<br>Section II (U,C,Bc)<br>Section III (Fi,fB,Ho)<br>Subsection IV (Fi,fB,Ho,Hc,Ak)<br>Subsection V (Fi,tB,Ho,Hc,Ak) | Based on morphology, cell organisation and types of cell division |
| Geitler (1932) (ICN) | Morphology | Order Chroococcales (U,C)<br>Order Chamaesiphonales (U,C,Bc)<br>Order Hormogonales (Fi,fB/fB,Ho,Hc,Ak) | |

For each system, the main groups are given and a brief description of applied criteria and a summary of observations.

Morphological characteristics: U: unicellular; Fi: filaments; C: colonies; pFi: pseudofilaments; Bc: baeocytes; fB: false branching; tB: true branching; Ho: hormogonia; Hc: heterocytes; Ak: akinetes; "Not valid" refers to the fact that despite a description following the rules of ICNB a formal recognition was not granted by taxonomic committees. (Adapted from Vidal *et al.*, 2021)

**Table 13. Major groups of cyanobacteria in the taxonomic schemes prposed by Castenholz et al. (2001) and Cavalier-Smith (2002).**

| *Group* | *Morphological characteristics* | *Genera (selection)* |
|---|---|---|
| Subsection 1 "Chroococcales" | • Unicellular<br>• Colonies with regular or irregular cell arrangement<br>• Embedded in extracellular mucilage | Aphanocapsa, Gomphospheria, Merismopedia, Microcystis, Synechococcus, Synechocystis, Woronichinia |
| Subsection 2 "Pleurocapsales" | • Colonial or filamentous<br>• Reproduction through baeocytes | Pleurocapsa, Chroococcidiopsis, Cyanocytis |
| Subsection 3 "Oscillatoriales" | • Multiplication by hormogonia<br>• Unbranched, linear filaments<br>• No heterocytes or akinetes<br>• Cells typically shorter than broad | Leptolyngbya, Lyngbya, Microcoleus, Oscillatoria, Phormidium, Planktothrix, Pseudoanabaena, Tychonema |
| Subsection 4 "Nostocales" | • Multiplication by hormogonia<br>• Nonbranching or flase branching<br>• Heterocytes (can be absent in individual filaments<br>• Akinetes | Anabaena, Aphanizomenon, Raphidiopsis (Cylindrospermopsis), Cuspidothrix, Chrysosporum, Dolichospermum, Nostoc, Sphaerospermopsis |
| Subsection 5 "Stigonematales" | • Multiplication by hormogonia<br>• True branching<br>• Heterocytes (can be absent in individual filaments)<br>• Akinetes | Chlorogloeopsis, Fischerella, Stigonema |

The morphological characteristics are based on microscopic observation. Exemplary genera are given for subsections. (Source:Vidal *et al.*, 2021)

Komárek *et al.* (2014) emphasized on whole genome sequences and on ultrastructural features (specially distribution of thylakoids in the cells) in their classification. This classification system is not helpful for the routine analysis of the field samples because these features are not observed under light microscope in the laboratory. Not only that, characteristics like presence of sheath and formation of filaments in the thallus etc. are not so important. It has been observed that a filamentous genus *Pseudoanabaena* is grouped together with the unicellular form *Synechococcus* in a new order Synechoccales. For the analysis and examination of large number of field samples, traditional taxonomic systems are considered as more suitable.

The taxonomic scheme as proposed by Castenholz *et al.* (2001) may be considered for systematics and identification because it is primarily based on morphological features observed by light microscopy and largely have a close similarity to the scheme given by Rippka *et al.* (1979). This scheme is considered as invalid because here nomenclature definition of categories like orders and families were not used and replaced by "subsections", "families" and "form genera" which do not speculate monophyletic origin. So, the scheme proposed by Castenholz *et al.* (2001) does not reflect phylogenetic relationships, however, it is convenient practical and stable method for the identification of different strains and samples of cyanobacteria with the help of microscope. Traditionally cyanobacteria were classified according to the morphological characteristics. Later with the onset of techniques in molecular biology and bioinformatics many phycologists tend to use molecular data to classify cyanobacteria. Due to this reason, some taxa which were previously assembled on the basis of some morphological characteristics appeared phylogenetically not related (Gugger & Hoffmann, 2004; Schirrmeister *et al.*, 2011). These taxa formed polyphyletic species, genera and higher taxomic categories (Komárek *et al.* (2014). Polyphyletic forms are resulted due to taxonomic mislabelling of many taxa and it requires revision of early taxonomic grouping. So, it is very important to classify cyanobacteria through a polyphasic approach which could involve molecular data and morphological features.

The use of rRNA gene sequences in prokaryotic classification is very significant. Among the rRNAs, 16S rRNA sequence identity is widely accepted because it is present in all prokaryotic genomes and has highly conserved 5' and 3' end regions. This 16S rRNA sequence identity has also been used in classifying cyanobacteria. Lane *et al.* (1985) stated that due to low sequence homology in 16S rRNA between bacteria and cyanobacteria, applicability of universal primers designed for bacterial 16S rRNA does not always give positive results for cyanobacteria. That is why, cyanobacteria – specific primers have been specially designed for the sequencing of 16S rRNA (Nubel *et al.*, 1997). According to Shirani and Gamage (2020) the 16S rRNA gene sequence is the basis of modern classification of cyanobacteria. It is considered as one of the major achievement in taxonomy of cyanobacteria and significantly replaced traditional approach of cyanobacterial taxonomy, which is morphology based and time consuming. However morphological variation and diversification is the key feature in taxonomy. Available Molecular data can only give information on genetic variability and also phylogenetic affiliations but could not recognize ecological importance of different phenotypes, on-going diversification of strains etc. They also pointed out that, in addition to the traditional morphological data and molecular data, ultrastructural morphological features, ecophysiological and biochemical properties that are combined in a polyphasic approach would support more accurate taxonomic classification of cyanobacteria .

Cyanobacteria do not have the capacity to perform sexual reproduction and the evolution of this prokaryotic group is a constant process which is based on the horizontal transfer of nucleic acids (DNA) between cyanobacterial strains and within populations (Rudi *et al.*, 1998, 2000; Barker *et al.* 1999; Hayes *et al.*, 2006). In addition to horizontal transfer of DNA rapid acclimation and adaptation are also significant in the continual process of evolution (Erdmann & Hagemann, 2001; Hagemann, 2002; Komárek and Kaśtovsky', 2003). Continual rapid development of morphotypes and ecotypes may be resulted due to flexibility of cyanobacterial genomes to adopt to a wide span of environmental conditions, breakdown of countless cellular functions due to stress and induction of new proteins to cope with such conditions. So, it may be concluded that, the situation of cyanobacteria and their phylogenetic relationships are regulated by distinct and determined metabolic and genetic processes. Early taxonomic classifications of blue-green algae were totally based on few particular/distinct morphological features. In the last decades of 20$^{th}$ century modern methods had been introduced and considerably changed our understanding of these prokaryotes. Modern cyanobacterial taxonomy is influenced by the electron microscope, modern ecological investigations, establishment of large numbers of cyanobacterial types/strains in cultures, molecular methods, ideas regarding toxicity etc. stimulate the growth and development of cyanobacterial research and dominate their taxonomy also. Designation of cyanobacterial taxa is a problem, because the random and arbitrary use of botanical or bacteriological nomenclature rules produces many uncertainty and misinterpretations. Microbiologists and biotechnologists do not use formal recommendations, and designate experimental strains with taxonomic binomials based on incorrect identifications. Strain designation is often based on arbitrary names selected from old, unrevised literature. Modern criteria must be used in cyanobacterial taxonomy and the classification must be revised and based on studies that include adequate molecular data.

Komárek (2006) reported that nucleic acid sequencing of strains and populations is without doubt the basis for modern classification of cyanobacteria, but the criteria form the classical botanical system (traditional taxonomy) can not be omitted. Without the complementary morphological approaches, molecular data alone have limited capacity to recognize the ecological importance of different genotypes, the morphological variability in situ, the ongoing adaptional processes and the continual origin of new cynobacterial eco – and morphotypes, Komárek stated that combination of both molecular & morphological approaches for modern cyanobacterial taxonomy is therefore, essential. He suggested following approaches should be used for evaluation of cyanobacterial diversity:

- Molecular analyses, mainly those related to phylogenetic relationships, diversity of genotypes, diversification processes and speciation.
- Morphological diversity including variation in nature as well as in culture;
- Ecological, ecophysiological and biogeographical limits;
- Ultrastructural studies;
- Biochemical features and information about special metabolic processes like production of secondary metabolites, adaptation processes, etc.

- Accurate approved designation of taxa that regards Botanical and Bacteriological rules of nomenclature.

It may be concluded that the taxonomy and systematics of cyanobacteria is a big challenge. It is very clear that, molecular techniques have facilitated in describing and elucidating biodiversity of cyanobacteria. Though, morphology sometimes conflict with molecular markers and it is observed that, cryptic species and polyphyletic genera represent problematic phenomena resulting uncertainty of proper morphological identification. So, more attention should be given to the use of molecular markers in taxonomy of cyanobacteria. It is to be mentioned that ecological and morphological characters are very important, which should be taken into consideration. Gen Bank and other databases should be completed with adequate data and they should be accessible very easily for the future workers. All these data may be a source for a barcoding database which would resolve cryptic taxa and polyphyletic genera problems by providing correct and quick identification technique (system). We are now in a period of very significant and necessary changes in cyanobacterial taxonomy and we hope our knowledge on cyanobacterial systematics would be increased and amplified many times within few years.

## B. BIOACTIVE METABOLITES OF CYANOBACTERIA WITH PHARMACEUTICAL APPLICATIONS

The term "bioactive" is an alternative term for "biologically active". A bioactive compound is simply a substance that has a biological activity. A compound or a substance having biological activity, if it has a direct effect on a living organism. These effects may be positive or negative depending on the substance, the dose or the bioavailability. Bioactive compounds are of high aesthetic value in pharmaceutical industries. These compounds are derived from plants, fungi, bacteria and cyanobacteria also. This review focuses on the significance of bioactive metabolites produce by cyanobacteria like alkaloids, terpenoids, non-ribosomal peptides, polyketides, ribosomal peptides, mycosporine-like aminoacids, vitamins, phenolics etc.

Cyanobacteria are the most abundant oxygenic photosynthetic organisms that possess various cellular strategies and physiological capacities to facilitate their adaptations for colonization in various ecosystems on earth. They can thrive in extreme environments like deserts, polar zones, geothermal springs, saline lakes, and in soils with high metal concentrations. According to their habitats cyanobacteria can be classified as alkaliphilies (prefer high pH), oligotrophics (grow under low nutrients), endolithics (beneath the rock), halophiles (prefer to grow in high salinity), thermophilies (in high temperature), psychrophilies (low temperature and radiophilies (which grow under high radiation).

Cyanobacteria have the capacity to produce biologically active compounds that are capable of meeting the global demand as an alternative source of drugs. Different environmental factors influence the metabolic activities of cyanobacteria, which results in their adaptation ability to survive in diverse ecosystems. According to WHO report 80% of the world population gamble on plants for traditional medicines and 40% of the industries depend on plants to produce drugs (Jain, 2017). Cyanobacteria and algae are the good sources of many bioactive metabolites like as alkaloids, carbohydrates, flavonoids, pigments, phenols, saponins, steroids, tannins, terpenes, and vitamins which can be utilized in biotechnology and industrial fields (Guiheneuf *et al.*, 2016). Microorganisms forming microbial mats in extreme environment have been identified as a good source of bioactive compounds for different biotechnological application (Bhatnagar and Kim, 2010). Cyanobacteria have been identified as a new and rich source of bioactive compounds (Abed *etal*, 2009; Shalaby, 2010;Nowruzi *etal.*, 2017). Few compounds like hassallic hapalindole and Y-lactone isolated from *Nostoc* sp. of CENA 219KP701037 and *Nostoc calicola* 6 sf Cale KP701034, *Fischerella* sp. and *Scytonema* sp. (Shishido *et al.*, 2015; Singh *et al.*, 2016) have attained importance for their antimicrobial activity in the field of biotechnology and Microbiology. Cyanobacteria can produce many pharmacologically active compounds that have antibacterial, anticancerous, antifungal, antiplasmodial, antiviral and immuno suppressive activities (Malathi *et al.*, 2014; Mukud and Sivasubramanian, 2014; Semary and Fouda, 2015; Rath and Priyadarshani, 2013; Papendorf *et al.*; Abdo *et al.*, 1998, 2012; Vijayakumar and Menakha, 2015).

**Table 14. Different classes of cyanobacterial species based on their physiological characteristics. (After Nandagopal*et al.*, 2021).**

| Class | Habitats | Cyanobacteria | References |
|---|---|---|---|
| Alkaliphiles | Hypersaline swamps, alkaline-saline lake or ponds, hot spring, alkaline hot spring, alkaline-saline volcanic lake, soda deserts | *Microcoleus* sp., *Pleurocapsa* sp., *Synechocccus* sp., *Cyanobacterium* sp., *Spirulina subsalsa, Spirulina platensis, Spirulina maxima,* and *Arthrospira* sp. | Cho *et al.*, 2015; Herbert & Codd, 1986; Shih *et al.*, 2013; Bhaya *et al.*, [1-10] 2007; Cheevadhanarak, *et al.*, 2012; Fujisawa *et al.*, 2010; Lefort *et al.*, 2014 Carrieriet al., 2011; Janssen *et al.*, 2010; Dog *et al.*, 2015 |
| Acidophiles | Sulfuric pools and acid mine drainage | Cyaobcteria cannot survive under this condition | Hirooka *etal.*, 2017; Gross 2000; Ferrisetal., 2005. |
| Endolithic | Rocks, granites and quartzites in desert, freshwater | *Chroococcidiopsis*-like Cyanobacterium | Khomutovska *et al.*, 2021. |
| Halophilic | Hypersaline lakes, coastal hypersaline lagoons, saline, springs, salt flats and ponds | *Synechocccus* sp., *Leptolyngbya* sp., *Nodosilinea* sp., *and Geitlerinema* sp. | Ramos *et al.*, 2017 |
| Oligotrophics | Coastal regions of marine and freshwater | Dolichospermum lemmermanii | Sterner *et al.*, 2020; Reinl *et al.*, 2020 |
| Psychrophilic | Alpines and polar regions | *Nostoc* sp., *Leptolyngba* sp., *Oscillatoria* sp., *and Phormidium* sp. | Nadeau & Castenhol, 2000; Singh *et al.*, 2007; Thangaraj *et al.*, 2017 |
| Thermophilic | Thermal springs and soil crusts f deserted area | *Synechococcus* sp., *Thermosynechococcus vulcanus*, *Leptolyngbya* sp., *Thermosynechococcus elongates*, and *Phormidium* sp. | Carrieri*et al.*, 2011; Janssen *et al.*, 2010; Dong *et al.*, 2015 |
| Radiophiles | Marine, freshwater and desert | *Synechocystis* sp., *Chroococcus minutus*, *Leptolyngbya* sp., *Trichodesmium* and *Crocosphaera* | |

**Table 15. List of cyanobacteria producing diverse class of indole alkaloids (After Nandagopal *et al.*, 2021)**

| Cyanobacteria species | Habitat | Compounds | Bioactivities |
|---|---|---|---|
| *Hapalosiphon* sp. CBT1235 | Terrestrial | Hapalindoles | Inhibit T Cell Proliferation |
| Hapalosiphon fontinalis | Soil | Hapalindoles | Antibacterial ad atimycotic |
| Hapalosiphon fontinalis | Soil | Hapalindoles | Antialgal |
| *Westiellopsis* sp. (SAB 20.93) and *Fischerella muscicola* (UTEX LB1829) | Freshwater and terrestrial | Hapalindoles | Antibacterial |
| *Fischerella ambigua* UTEX1903 | Terrestrial | Ambiguine | Unknown |
| *Hapalosiphon welwitschii* UTEX B1830 | Freshwater | Welwitindolinone | Unknown |
| *Westiella intricate* UH strain HT-29-1 | Freshwater | Welwitindolinone | Unknown |
| *Fischerellaambigual* (UTEX 1903), *Westiellopsisprolifica* and *Hapalosiphon hibernicus* BZ-3-1 | Terrestrial | AmbiguineIsonitriles | Fungicidal |
| Fischerella muscicola | Terrestrial | Fischerindole | Antifungal |
| *Fischerella ambigua* (UTEX 1903) | Terrestrial | Fischambiguines and ambiguines | Antibacterial |
| *Fischerella* sp. | Terrestrial | Welwitindolinones | Multi-drug resistance reversing activity |
| *Hapalosiphon welwitschia* and *Westiella intricata* | Soil | Welwitindolinones | Multi-drug resistance re versing activity and insecticidal activity |

Thus far, 80 variants of indole alkaloids have been identified exclusively produced by the genera *Wesiella*, *Westiellopsis*, *Fischerella* and *Hapalosiphon* (belonging to subsection V formerly order Stigonematales) The variants belong to nine different groups.

# BIOACTIVE METABOLITES FOUND IN CYANOBAETERIA

## Indole Alkaloids

Indole alkaloids have a bicyclic structure, consisting of a six-membered benzene ring fused to a five-membered nitrogen-containing pyrrole ring. The pyrrole ring with nitrogen atom gives rise to the basic properties of indole alkaloids that make them particularly pharmacologically active (El-Sayed and Verpoorte, 2007). In plants, alkaloids are produced as secondary metabolites in response to biotic or abiotic stresses (Taha *et al.*, 2009). Different types of indole alkaloids synthesized by cyanobacteria have been reported as the bioactive secondary metabolites. These bioactive substances have unique biological and pharmacological properties (**Table 15**) reported by Nandogopal *et al.,* 2021. So far, 80 variants of indole alkaloids have been reported (Richter *et al.*, 2008; Demay *et al.*, 2019; Walton & Berry, 2016) by the genera *Westiella, Westiellopsis, Fischerella* and *Hapalosiphon*. Walton & Berry, 2016 stated that the variants belong to nine different groups based on their carbon skeletons. The groups are Hepalindoles (it is the largest group produced by cyanobacteria). Group I tetracyclic hapalindoles, Group 2 tricyclic hapalindoles, Group 3 hapalinololes, Group 4 tetracyclic Ambiguines, Group 5 pentacyclic Ambiguines, Group 6 Fischambiguines, Group 7 Fischerindoles and Group 8 and Group 9 are Welwitindolinones (Welwitinolinones).

Endole alkaloids are identified mainly from cyanobacterial genera *Westiella, Westiellopsis, Fischerella* and *Hapalosiphon*. Scytonemin is a indole alkaloid UV-filtering pigment produce by cyanobacteria (Castenholz and Garcia-Pichel, 2012; Nageli and Gattungen, 1849). Scytonemin found in the *Lyngbya* sp. CU 2555, showed high resistance towards UV-B and heat and can protect the cells against adverse conditions. It has been established that this oxidising agent can act as photoprotective substance against UV radiation and also protect living cells against harmful short-wave-length radiation (Dillion *et al.*, 2002). Another indole alkaloid compound β-carotene has been detected in cyanobacteria. Norparmane (9H-pyrido (3, 4-b)indole) excreted by a cyanobacterium *Nodularia harveyana* is responsible for causing high algicidal activity (Volk, 2005). Becher et al., 2009 reported that nostocarboline from *Nostoc* 78-12A could be used as an acetylcholinesterase inhibitor to treat neuronal diseases.

**Terpenoids:** Terpenoids are isoprene derivatives and are, therefore, sometimes referred to as isoprenoids. The terpenes are hydrocarbons derived from combination of several isoprene units. The derivatives of these terpenes where the methyl group has been removed or oxygen atoms have been added to produce different functional groups, are called terpenoids. These are the group of organic compounds widely distributed in cyanobacteria and algae (Keeling and Bohlmann, 2012). Terpenoids can be classified as hemiterpenoid, monoterpenoids, sesquiterpenoids, diterpenoids, sesterpenoids, triterpenoids (steroids) and tetraterpenoids (carotenoids) based on the number of isoprene units (**Fig. A**) [Kandi *et al.*, 2015].

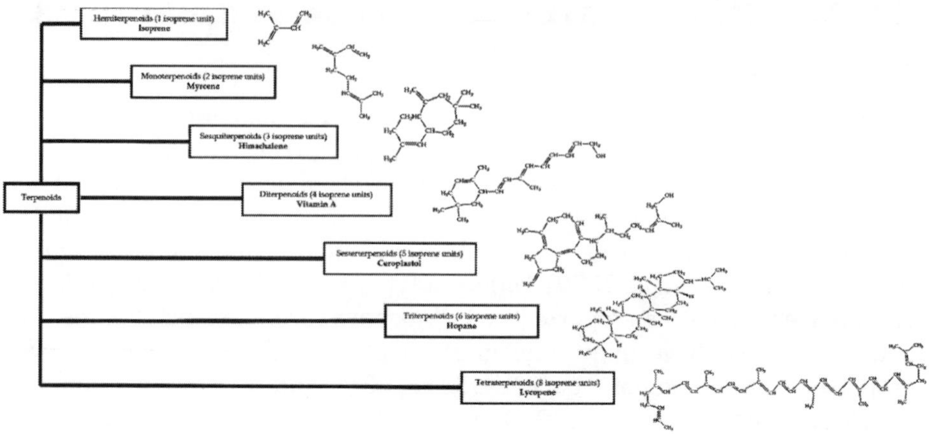


**Fig. A: Classes of terpenoids based on their isoprene units (Abdallah & Quax, 2017) [Adopted fron Nandagopal** *et al.*, **2021]**

It has been reported that terpenes can be supplemented in products for their fragrances and flavours (Kirby and Keasling, 2009. Pattanaik and Lindberg, 2015) Terpenoids present in *Cylindrospermum musicola* HPUSD12 and *Phormidium* sp. HPUSD13 play roles in providing protection against free radical oxidative damage to the cells that might be due to high – salinity condition of the Drang salt mine in India (Devi and Rani, 2020). It has been reported that 2-methylhopanoids (2-Me BH Ps), were present in both laboratory cyanobacterial cultures and in natural cyanobacteria dominated microbial mats (Jahnke *et al.*, 2004; Summons *et al.*, 1999). This pentacyclic triterpenoids (2-Me BHPs) play the role of bio- mark for cyanobacteria in some environmental settings (Summons *et al.*, 1999). Terpenoids, also known as carotenoids, present in most photosynthetic organisms are also found in cyanobacteria. Common carotenoids produced by cyanobacteria are β-carotene, zeaxanthin and echinenone. Carotenoids are necessary to facilitate cyanobacteria against, direct UV light exposure and photooxidative damage while conducting photosynthesis. Two common carotenoids echinenone and zeaxanthin protect PSII against singlet oxygen (Kusama *et al.*, 2015). Carotenoids also play significant roles in the survival of endolithic cyanobacteria, *Chroococcidiopsis* sp. at the Atacama Desert (Vitek *etal.*, 2017). Lopes *et al.*, 2020 reported that many strains of Cyanobacteria like fresh water *Alkalinema* aff.*pantanalense* LEGE15481, *Cyanobium gracile* LEGE12431 *Cuspidothrix issatschenkoi* LEGEO3282, the terrestrial *Nodosilinea* (*Leptolyngbya*) *antartica* LEGE13457 and the marine *Leptolyngbya*-like sp. LEGE13412, have been characterized for their high content of carotenoids. Scytoscalarol present in *Scytonema* sp. (UTEX1163) culture showed growth inhibition against *Bacillus anthracis*, *Staphylococcus aureus*, *Escherichia coli*, *Candida albicans* and *Mycobacterium tuberculosis* (Mo *et al.*, 2009). The function of pure monoterpenes has been suggested to be aniparasitic (Goulart *et al.*, 2004; Bedoux *et al.*, 2014).

**Non-Ribosomal Peptides and Polyketides:** Polyketides are compounds containing alternating carbonyl and methylene groups (β-poly ketonese). Polyketides are synthesized by sequential reactions which one catalysed by a set of enzymes known as polyketide synthases (PKS). Three types of PKS have been recognized – Type I, The most common example is polyketide synthase (deoxyerythronolide β-synthase) which is later modified as erythromycin. Type II, the important example of this type is Doxorubicin polyketide synthase, Type III-common example is the chalcone synthase. In bacteria, PKS Type I are widely found to be responsible for polyketide chain clongation, processing and termination (Dutta *et al.*, 2014).

Non-ribosomal peptides (NRPs) are a kind of secondary metabolites which are synthesized by multidomain mega-enzymes called nonribosomal peptide synthetases (NRPs), without the need for the cell ribosomal machinery and messenger RNAs (Evans *et al.*, 2011).

The gene cluster of NRPS and PKS were more frequently found in bacteria, including cyanobacteria, than archaea and eukarya (Wang *et al.*, 2014). Gene clusters have been observed in the genera like *Lyngbya, Microcystis, Planktothrix, Nodlularia, Nostoc, Pleurocapsa* and *Aabaena* (Wang *et al.*, 2014; Welker and Döhren, 2006). Among other cyanobacteria *Nostoc* and *Pleurocapsa* can produce NRP and PK (Wang *et al.*, 2014). According to Welker & Döhren, 2006; NRPS comprises modules, each of which integrates proteinogenic amino acids with non-proteinogenic amino acids, fatty acids, carbohydrates and other building blocks into peptide chains. It has been observed that in bacteria, PKS Type I are widely found which is responsible for elongation, processing and termination of polyketide chains. (Dutta *et al.*, 2014).

**Ribosomal Peptides:** Peptides and small proteins that are genetically encoded and synthesized by Ribosomes act as important regulators in many fundamental cellular processes, including gene expression, signalling, development as well as metabolic processes. These small proteins can play a very significant role in prokaryotic and eukaryotic defence mechanism against many microorganisms. Protenogenic amino-acids form peptide chains on ribosomes are called as Ribosomal peptides (RPs). Only 20 proteinogenic amino acids are used as the building blocks during the biosynthesis of RPs (Kehr *et al.*, 2011. Cyanobactin, microviridins and lantipeptide are the three major RP families. Cyanobactin is multifariously present in symbiotic and planktonic cyanobacteria (Sivonen *et al.*, 2010). Largest RPs are the Microviridins consist of 12-20 amino acids, which are classified into four classes (GroupI-IV) and these are found in freshwater cyanobacteria (do Amaral *et al.*, 2021). Four different kinds of lantipeptides are observed which are produced by four different classes of lantipeptidases. Non ribosomal peptides, polyketides and ribosomal peptides are post-translational modified compounds which are well known for their antibacterial properties, which could be important when in competition with other microbes in the ecosystem (Li *et al.*, 2010).

**Table 16. Bioactivities of non-ribosomal peptides (NRPS) and polyketides (PKS) produced by cyanobacteria. (Adapted from Nandagopal *et al.*, 2021)**

| Cyanobacteria species | Habitat | Compounds | Bioactivities | References |
|---|---|---|---|---|
| Microcystis aeruginosa | Freshwater | Microcystins | Inhibit eukaryotic types 1 and 2A phosphates, cytoskeletal collapse, massive hepatic bleeding, potential tumour promoters and carcinogens | Nishizawa *et al.*, 1999, 2000; Tillet *et al.*, 2000) |
| *Planktothrix agardhii* NIVA-CYA 126 | Freshwater | Aeruginosin | Inhibit serine proteases | Ishida et al., 2007 |
| *Cylindrospermopsis raciborskii*, *Aphanizomenon ovalisporum* and *Aphanizomenon flos-aquae* | Freshwater | Cylindrospermopsin | Cytotoxic,neurotoxic effects and carcinogen | Mihali *et al.*, 2008 |
| *Anabaena* sp. 90 | Freshwater | Anabaenopeptin | Inhibit proteases | Rouhiainen *et al.*, 2010 |
| Lyngbya bouillonii | Marine | Apratoxin | Reversible inhibition of several cancer-associated receptors | Tidgewell *et al.*, 2010; Grindberg, 2011 |
| Lyngbya majuscula | Marine | Lyngbyatoxin | Potent skin irritant | Edwards and Gerwick, 2004 |
| *Lyngbya majuscule* JHB | Marine | Hectochlorin | Antifungal and anticancer activity | Ramaswamy *et al.*, 2007 |
| *Lyngbya majuscule* 19L | Marine | Barbamide | Anti-molluscidal | Chang *et al.*, 2002 |
| *Lyngbya majuscule* 19L | Marine | Curacin A | Antiproliferative and cytotoxic activities | Chang *et al.*, 2004 |
| *Lyngbya majuscule* JHB | Marine | Jamaicamide | Block sodium-channel | Edwards *et al.*, 2004 |
| *Nostoc* sp. GSV 224 | Terrestrial | Nostopeptolide | No cytotoxic, antifungal and inhibit protease activities | Hoffmann *et al.*, 2003 |
| *Nostoc* sp. ATCC 53789 | Terrestrial | Nostocyclopeptide | Antitoxin activity | Becker *et al.*, 2004 |
| *Nostoc* sp. ATCC 53789 | Terrestrial | Cryptophycins | Tubulin-destabilizing compound | Magarvey*et al.*, 2006 |
| *Cylindrospermum alatosporum* CCALA 988 | Terrestrial | Puwainaphycins | Cytotoxic | Maris *et al.*, 2014 |
| Nostoc calcicola | Wasterater | Nostophycin | Antibacterial and antifungal | Gupta &Vyas, 2021 |
| *Nadularia spumigena* NSOR 10 | Freshwater | Nodularin | Inhibits phosphatase type 1 and 2A, cytoskeletal collapse, massive hepatic bleeding, potential tumour promoters and carcinogens | Moffit&Neilan, 2004 |

Benthic or sessile cyanobacteria may produce NRPs, PK, and RPs for cell signalling in order to form epilithic/endolithic biofilms in water bodies, soil coating of desert, saline soil surface and polar regions. Many organisms like sponges, eukaryotic algae, plants etc. can use different types of peptides produced by cyanobacteria. Actually these peptides are mainly oligopeptides and used as precursors for the metabolic activities of many living organisms of the ecosystem.

**Table 17. Bioactivities of ribosomal peptides (RPs) produced by cyanobacteria. (Adapted from Nandogopal *et al.*, 2021)**

| Cyanobacteria | Habitat | Compounds | Bioactivities |
|---|---|---|---|
| Mirocystisaeruginosa | Freshwater | Aerucyclamide A,B,C and D | Cytotoxic and antimalarial |
| Stigonemadendroideum | Terrestrial | Dendroamide A | Multidrug-resistance reversing activity |
| Trichodesmiumerythraeum | Marine | Trichamide | No biological effects found |
| *Prochlorondidemnid* (symbioant) | Marine | Patellamide A and C | Cytotoxic |
| *Anabaena* sp. 90 | Freshwater | Anacyclamide | Cytotoxic |
| *Microcystisaeruginosa* PCC 7806 | Freshwater | Microcyclamide | No biological effects found |
| | | Microviridin | |
| *Mirocystisaeruginosa* NIES-298 | Freshwater | Miroviridios B and C | Inhibits elastase |
| | | Lantipeptides | |
| *Prochlorococcus* MIT9313 | Marine | Prochlorosins | Bacteriocidal and act as signaling molecule |

**Mycosporine** – Like Amino acids (MAAs): Mycosporine like amino acids (MAAs) are a family of intracellular compounds engaged in the protection of aquatic organisms against solar radiation. UV radiation can induce the production of MAAs and osmotic stresses can also induce its production, though the mechanism is not clearly understood (Pathak et al., 2019). Antioxidant activity is a very significant property of MAAs. Many members of Synechoccales, Chroococcales, Osillaoriales and Nostocales are able to produce MAAs for adaptations (Geraldes et al., 2020; Jain et al., 2017). MAAs also provides protection against the damaging effects of solar UVR to the cyanobacterial mats in the Arctic (Muller et al., 2005). An usual mycosporine-glycine-alanine (MGA, a derivative of MAA) was found in *Sphaerospermopsis torques-regine* ITEP–024 (Geraldes et al., 2020). Studies have shown that MAAs originate from Shikimate pathway, but the exact route of their biosynthesis is unknown. Favre-Bonvin et al., (1987) were of opinion that an intermediate of the Shikimate pathway (3-dehydroquinate) can protect human skin and it can be commercially explored for the protection of skin as suncare products.

Carreto *et al.* (1990) reported that there was a close relationship between the photosynthetic process and synthesis of MAAs, they have shown that after application of 3-(3,4-dichlorophenyl)-1, 1-dimethylurea (DCMU), the biosynthesis of MAAs was inhibited in an alga *Alexandrium excavatum*. MAAs are very much attractive to cosmetic industries as an active ingredients for sunscreens and antiaging products due to their characteristics (Chrapusta *et al.*, 2017; Kageyama & Waditee-Sirisatha, 2018). Ishihara *et al.*, (2017) reported that a novel MAA

(13-o-β-galactosyl-porphyra- 334) from *Nostoc sphaericum* can protect human keratinocytes against UV radiation.

**Phenolics:** Polyphenols are the group of secondary metabolites like phenolic acids, flavonoids (flavonols), tannins, and lignins (Thomas and Kim, 2011). Structurally, phenolic compounds have at least one phenolic ring and show strong biological activities when halogenated (Cabrita *et al.*, 2010).

**Flavonoids:** are polyphenolic compounds that are widely distributed in plants (Singh, *et al.*, 2017). The flavonoids form a large group of secondary metabolites which have two six carbon, aromatic rings linked by three carbon chain. Both the aromatic rings contain a variable number of hydroxyl groups. Flavonoids are also antioxidants that are important for the survival of cyanobcteria. Singh *et al.* (2014) reported that two flavonoid molecules (quercertin and lutin), might facilitate that salt acclimation mechanisms in *Plectonema boryanum, Haplosiphon intricatus, Anabaena doliolum* and *Oscillatoria acuta*. By charomtographic analysis it has been identified that *Leptolyngbya* sp. (thermophilic) produces a high quantity of luteolin-7-glucoside and naringenin, these two compounds could protect the cells from oxidative damage due to high temperatures. Nephroprotective, neuroprotective, anticancer and antiatherosclerotic properties of cyanobacterial flavonoids have been reported by many workers (Hernández-Aquino & Muriel, 2018; Sugumar *et al.*, 2019; Choi *et al.*, 2020; Mulvihill *et al.*, 2010; Assini *et al.*, 2013).

**Phenolic acids:** Phenolic acids consist of one carboxyl group and one or more hydroxyl groups joined to the aromatic ring. When a high concentration of NaCl was supplied into the cultures of *Plectonema boryanum, Haplosiphon intricatus, Anabaena doliolum* and *Oscillatoria acuta* the accumulation of gallic acid, caffeic acid, chlorogenic, ferulic acid and vanillic acid was detected (Singh *et al.*, 2014) and it was suggested that the above mentioned phenolic acids play vital roles in the scavenging of free radicals under salt stress conditions. Ferulic acids can produce skin whitening and anti-wrinkle effect (Park *et al.*, 2018). *Monteiro Silva et al., (2017) reported antiaging property of gallic acid.* Cyanobacterial phenolic acids might have these therapeutic effects due to their abilities to detoxify Reactive oxygen species and scavenge free radicals (Singh *et al.*, 2014; Patipong *et al.*, 2019). Cyanobacterial phenolic compounds are able to prevent vascular damage as well as cardiovascular diseases progression.

**Vitamins:** Plants, bacteria and cyanobacteria are generally synthesize different types of vitamins. These are diverse group of organic molecules which are required in small amounts in the diet of higher animals. They are classified by their biological functions. Based on their solubility they may be referred to as fat soluble and water soluble vitamins.Water soluble vitamins are Vitamin B complex, and C. Vitamin B complex includes B1(thiamin), B2 (riboflavin), B3 (niacin), B5 (pantothenic acid), B6 (pyridoxine), B7 (biotin), B9 (folic acid), B12 (cobalamin), B10 (para-amino benzoic acid), B11 (pteryl-heptaglutamic acid), B13 (orotic acid), B14(pterin phosphate), B15 (pangamic acid), B16 (dimethyl glycine or lipoic acid), B17(amygdalin) etc. Fat soluble vitamins are – A (retinol),

D (calciferol), E (tocopherol) and K (phyton adione). To mitigate environmental stresses plants produce vitamin A,B,C,E and K in most organ parts (Asensi-Fabado & Munné-Bosch, 2010). Microalgae and cyanobacteria can produce vitamin D, K and B12, which are absent in higher plants (Del Mondo et al., 2020). Cyanobacteria like *Arthrospira maxima, Anabaena cylindrica* and *Synechococcus* produce much pro-vitamin A (*β-carotene*) than some fruits like carrots and oranges, B-Vitamins might be necessary in the metabolic pathways of cyanobacteria (Del Mondo et al., 2020). From Fresh water *Anabaena cylindrica* high content of B2 , B5 and B6 was detected, but the marine forms of the same alga was found to produce a high amount of B12 (Aaronson et al., 1977). It was stated by Aaronson et al., that Anabaena cylindrica is an excellent source of vitamin C. Marine form of *Anabaena cylindrica* can produce vitamin K1 higher than spinach and parsley (Tarento et al., 2018). A very low amount of vitamin D was detected in *Arthrospira* sp. (Ljubic et al., 2020). This lipid-soluble vitamin might be important to mitigate the damage or degradation of cyanobacterial cell membranes caused by UV radiation. Goiris et al. (2015) were of opinion that Arthrospira sp. and Oscillatoria sp. synthesize high amounts of vitamin E in response to nitrogen deficiency during logarithmic growth phase. It has been also reported by them that microalgae also produce vitamin E in response to nutrient limitation. Many cyanobacteria like *Anabaena variabilis, Mastigocladus laminosus, Nostoc muscorum, Prochlorococcus* sp., Anacysti nidulans and Synechocystis sp. PCC6803 can synthesize vitamin K1 (Phylloquinone). *Spirulina* sp. is used as a supplementary diet in many countries. Different studies on *Spirulina* have showed that this alga is a rich source of Vitamin B12, β-carotene and Vitamin E. Cyanobacteria have the ability to overcome extreme environmental conditions and possess the capacity to produce biologically active molecules which have multifunctional applications in nutraceuticals as well as in pharmaceuticals. Cyanobacterial and algal secondary metabolites have several pharmaceutical applications like anticancer, antiviral and antimicrobial activities. We have already entered in the flourishing and blooming period of cyanobacteria and it is the time to expose the mystery of cyanobacterial bioactive metabolites. Actually these bioactive compounds are produced by cyanobacteria as a response to environmental stresses. Bioactive metabolites produced by higher plants are very low in amounts. To obtain the industrial yield large amount of secondary metabolites are required. It is expected that cyanobacteria and algal metabolites can fulfil the demand of industry for the production of various pharmaceutical and nutraceutical products.

## C. NITROGEN FIXATION BY CYANOBCTERIA

Nitrogen is an important element for plants and it comes only next to carbon, hydrogen and oxygen. It is a major component of chlorophyll (the pigment needed for photosynthesis) as well as amino acids, the main building units of proteins. It figures in the composition of important biomolecules, such as nucleic acids, ATP, hormones, vitamins etc. The atmosphere contains vast quantities (about 77% by volume) of molecular nitrogen ($N_2$). Plants can not absorb this element directly from the atmosphere and they have to depend for its to different prokaryotic micro-organisms mainly bacteria and cyanobacteria. Acquirement of nitrogen form the atmosphere requires the breaking of an exceptionally stable triple covalent bond between two nitrogen atoms (N≡N) to produce ammonia ($NH_3$) or nitrate ($NO^-_3$). So, nitrogen fixation can be defined as the phenomenon of conversion of free di-nitrogen of the atmosphere into nitrogenous salts to make it available for absorption by plants. Plants can only utilize reduced forms of this element by – 1) the addition of ammonia and/or nitrate fertilizers into the soil, 2) During decomposition of organic matter these compounds are released and mixed into the soil, 3) the conversion of atmospheric nitrogen into the compounds by natural processes, such as lighting and photochemical reactions and 4) Biological nitrogen fixation.

**Lightning:** Lighting is responsible for about 8% of the nitrogen fixed. Nitrogen can be fixed through the electrical discharge that occur during lightning first to oxide (NO) and then to peroxide ($NO_2$) forms by reacting with atmospheric oxygen.

Lightning

$N_2 + O_2$--------------------------> 2NO (Nitric oxide)

2NO + $O_2$-------------------------------> $2NO_2$ (Nitrogen peroxide)

The nitrogen peroxide combines with water during rain to form nitrous and nitric acid which falls on the ground with rain water. The acids then react with Ca or K salts present in the soil and forms corresponding nitrates which are directly absorbed by the plants.

$2NO_2 + H_2O$——-—> $HNO_2 + HNO_3$

$HNO_3$ + Ca(or K) salts——> Ca (or K) nitrates

**Photochemical reactions:** Approximately 2% of the nitrogen fixed derives from photochemical reactions between gaseous nitric oxide (NO) and ozone ($O_3$) that produce nitric acid ($HNO_3$).

**Biological nitrogen fixation:** The remaining 90% results from biological nitrogen fixation, in which bacteria and cyanobacteria (blue-green algae) fix $N_2$ into ammonia ($NH_3$). This ammonia dissolves in water to form $NH_4^+$ (ammonium):

$$NH_3 + H_2O \rightarrow NH_4^+ + OH^-$$

This biological nitrogen fixation may be categorised into following two types:

a) Symbiotic nitrogen fixation and b) Asymbiotic nitrogen fixation.

Biological nitrogen fixation is carried out by specialized group of prokaryotes. These prokaryotes include different types of bacteria (free living soil bacteria and symbiotic) & cyanobacteria, presented in the Fig. B

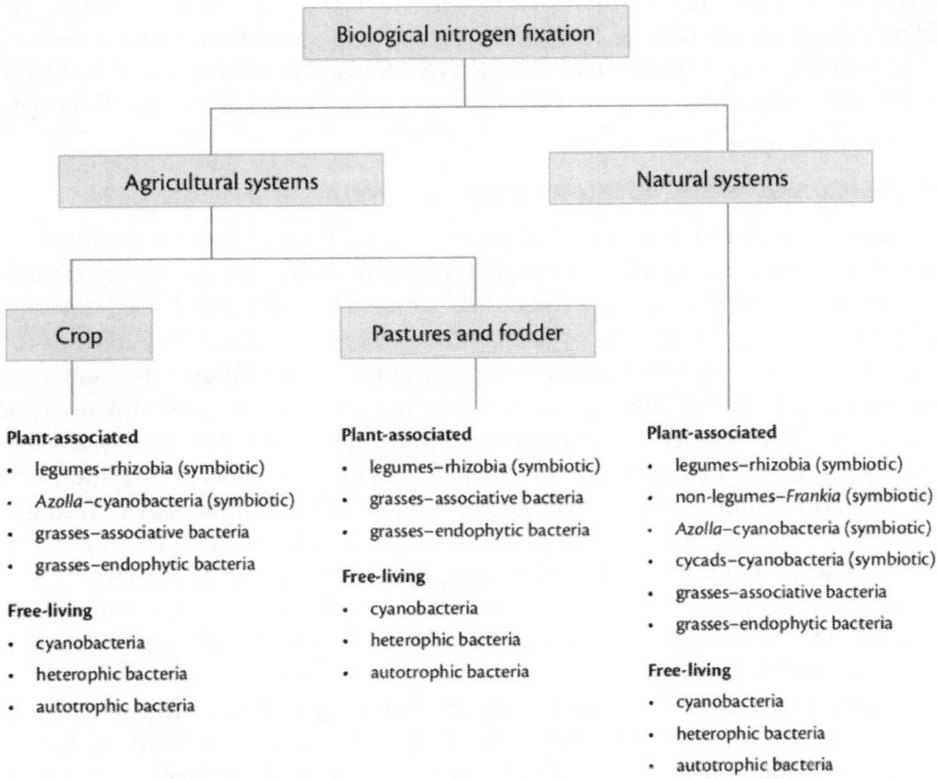


**Fig. B: Nitrogen-fixing organisms found in agricultural and natural systems**

Biological nitrogen fixation is the conversion of inert $N_2$ of the atmosphere into a combined form by living organisms through a biochemical process in which nitrogenase is the key enzyme. $N_2$ fixing organisms could be autotrophic or heterotrophic, aerobic, anaerobic or micro-aerobic, and among them the cyanobacteria are the only prokaryotes that carry out oxygenic photosynthesis and oxygen-sensitive nitrogenase mediated $N_2$-fixation. The cyanobacteria which do not possess a compartmentalised cellular organisation with and cell organelles carry out these two incompatible processes of photosynthesis and $N_2$ fixation, has drawn a variety of explanation by scientists but this enigma is not yet completely resolved. To avoid the destructive effects of oxygen on nitrogenase enzyme, heterocystous cyanobacteria have spatial separation of nitrogenase in heterocysts and photosynthesis in vegetative cells only. But in non heterocystous filamentous forms and in unicellular forms, the two processes are separated temporally, here both photosynthesis process and nitrogenase activity are taking place in the same cell but function alternately. The enzyme nitrogenase responsible for fixation of nitrogen is oxygen labile but the cyanobacteria have their innate mode to avoid and combat the situation.Nitrogen fixing cyanobacteria can be divided into many categories depending on conditions under which they fix nitrogen. These are as follows: 1. Aerobic $N_2$–fixers, 2. Anaerobic and micro-aerophilic nitrogen fixers, 3. $N_2$ –fixers with spatial separation of nitrogen fixation and photosynthesis, 4. Nitrogen fixers with temporal separation of the two processes and 5. Those exhibiting dual modulation of nitrogen fixation.(Anand, 1998).

## $N_2$-fixation by Heterocystous cyanobacteria

The interrelation between $N_2$ fixation and $O_2$ have fascinated scientist ever since it was understood that the enzyme nitrogenase was inactivated by exposure to oxygen. The mechanism to protect nitrogenase from oxygen is well developed in cyanobacteria. A majority of filamentous cyanobacteria that exhibit aerobic nitrogen fixation possess heterocysts. Compared to vegetative cells heterocysts are generally larger in size, lighter in colour, have thicker cell wall and have polar nodules at the point of attachment to adjacent cells (Fogg *et al.*, 1973). A mature heterocyst has a thick envelope that consists of an outer polysaccharide fibrous layer, a middle homogeneous layer and inner glycolipid laminated layer (Lang and Fay, 1971). Heterocysts protect nitrogenase from inactivation by $O_2$ by several mechanisms, including a high rate of respiration and decreased permeability to $O_2$. Heterocysts are dependent to adjacent vegetative cells for carbon source as they do not evolve $O_2$ and can not fix $CO_2$. The cell envelope of heterocysts limits the ingress of $O_2$ into the heterocysts (Murry *et al.*, 1984; Murry & Wolk, 1989). Kulassoriya *et al.* (1972) reported on the heterocysts differentiation and also stated how it coincided with the restoration of nitrogenase activity in *Anbaena cylindrica*. Studies on heterocyst development are being greatly facilitated by recent advances in the genetics of *Anabaena*. Genes involved in heterocyst development can now be tagged by transposon mutagenesis. Isolated heterocysts of *A. cylindrica* showed higher rates of respiration than vegetative cells and contained very little chlorophyll a and were devoid of phycocyanin and phycoerythrin pigments, which are associated with the oxygenic photosynthesis (Thomas, 1970). Heterocysts and vegetative cells are dependent on each other. Photosystem-II and carbon fixation are not observed in heterocysts. That is why they are dependent on vegetative cells for supply of carbon, which was probably supplied as sucrose to some extent (Cumino *et al.*, 2007; Marcozzi *et al.*, 2009). On the other hand vegetative cells procure fixed nitrogen in the form of amino acids from the heterocysts (Meeks & Elhai, 2002). In some species, vegetative cells supply glutamate to heterocysts, where glutamate is converted to glutamine and other amino acids (Martin-Figueroa *et al.*, 2000). The exchange of metabolites and intercellular signals that control regulated spacing of the heterocysts require movement of molecules between cells along a filament, possibly through a continuous periplasm (Flores *et al.*, 2006). The green fluorescent protein (GFP) expressed from the pat S promoter, which is expressed at low basal level in vegetative cells but strongly over-expressed in heterocysts and targeted on the periplasm by a cleavable twin – arginine signal peptide, could diffuse through the heterocyst periplasm to nearby vegetative cells. GFP attached to the cytoplasmic membrane was visible only in heterocysts but not in adjacent vegetative cells. Zhang *et al.* (2008) reported that, GFP targeted to the periplasm of vegetative cells or heterocysts showed no intercellular diffusion from one cell to the next and the exported GFP diffused around individual cells but not beyond the cell borders. Difference of opinions may be because of the different signal sequences they used (Kumar *et al.*, 2009).It may be concluded that although heterocysts in these cyanobacteria are the sites of nitrogen fixation but their continued functioning depends upon intimate connection and interaction with the vegetative cells.The genetic regulation of nitrogen fixation in cyanobacteria differs from that in other $N_2$- fixers. In presence of combined nitrogen like nitrate or ammonium in the environment generally prevents cyanobacteria from fixing $N_2$. This is due to reason that assimilation of reduced forms of nitrogen is less expensive. Nitrogenase is not synthesized when cyanobacteria are grown in presence of nitrate or ammonium, that is why, the genetic regulation of nitrogen metabolism in cyanobacteria is different.

Nitrogenase consists of two proteins: a) MoFe-protein and (b)Fe-protein. In some other nitrogen fixing organisms, the Fe-protein undergoes reversible covalent modification (attachment of ADP-ribose) rendering the Fe-protein inactive. This inactivation of nitrogenase is stimulated by ammonium or sources, by transfer of cultures to the dark or exposure to oxygen. In a number of cyanobacteria the Fe-protein can also be resolved into two components, but there is no evidence as yet that it can be covalently modified to inactivate. Some heterocystous cyanobacteria contains two molybdenum-based nitrogenase systems, nif1 and nif2, with different expression patterns. The vegetative cell nitrogenase (nif2) has its counterpart in non-heterocystous cyanobacteria that fix $N_2$ only anaerobically. Not only that, some cyanobacteria may produce Mo-independent 'alternative' (V or Fe) nitrogenases, encoded by the Vnf and anf genes.

## $N_2$ fixation by Non-hetrocystous cyanobacteria

Unicellular and non-heterocystous filamentous cyanobacteria are considered unique since two processes (Photosynthesis and $N_2$ fixation) are accomplished in the same cell. The first report on filamentous non-heterocystous forms fixing nitrogen was that of *Trichodesmium* sp. In open sea natural population (Dugdale *et al.*, 1961). nitrogenase activity of *Oscillatoria*, *Lynbgya* and *Plectonema* was demonstrated by Kenyan *et al.*, (1972). Nitrogenase activity of unicellular cyaobacteriaum *Gloeocapsa* was reported by Wyatt and Sylvey (1969). Mistsui *et al.*, (1986) stated activity of this enzyme on *Synechococcus* and *Aphanothece*. Rippka *et al.*, (1971) surveyed nitrogenase activity in 46 strains and they suggested the inclusion of all nitrogen fixing unicellular species into a single genus *Cyanothece*. Huang and Chow (1991) suggested that $N_2$ fixation in *Cyanothece* takes place during dark and in preceding light period ATP and carbon substrates are produced. In filamentous, non-heterocystous cyanobacterium *Trichodesmium*, $N_2$ fixation is observed under aerobic conditions in presence of sun light. *Trichodesmium* forms colonies in Tropical seas and are estimated for fixing one quarter of the total nitrogen in the World's ocean (Bergman and Carpenter, 1991). Two types of cells are observed in this alga – some cells are specialized to fix nitrogen, while others do not fix nitrogen. Nitrogen fixing cells are found adjacent to one another and these $N_2$-fixing cells have a dense thylakoid network with fewer gas vacuoles and cyanophycin granules (Fredrikson and Bergman, 1997). Issa *et al.* (2014) explained how the two incompatible process of oxygenic photosynthesis and nitrogen fixation exist in unicellular and filamentous non-heterocystous cyanobacteria. They observed $N_2$–fixation was confined to work phases of growth, followed by a cyclic pattern of continuous light period, higher rate of respiration causing micro-aerobic conditions in certain cells and $O_2$ evolution regulated by phycobiliproteins, to a possible switching off mechanism of photosynthesis in *Oscillatoria*. They were of opinion that *Trichodesmium* should be treated separately as this cyanobacterium regulate and conduct $N_2$–fixation under aerobic conditions in presence of sunlight. They even question regarding incorporation of this genus as cyanobacteria. Nitrogen fixation by *Lyngbya aestuarii* appears unique among non-heterocystous cyanobacteria. Pearl *et al.*(1991) reported that active nirogenase present at the terminal region of the trichomes where photosynthesis takes place at very low rate or it is totally lacking. This cyanobacterium fixes nitrogen in the dark, possible this organism is able to separate photosynthesis and nitrogen fixation both spatially and temporally. Among non-heterocystous filamentous cyanobacteria different strategies are observed to protect the nitrogenase enzyme from oxygen damage. *Pseudoanabaena*, *Phormidium* and *Plectonema* prefer sulphide rich anaerobic environments. Padan and Cohen (1982) reported that *Oscillatoria limentica* conduct anoxygenic, type of photosynthesis and avoids evolution of oxygen to maintain anaerobic atmosphere within cell to perform nitrogen fixation. It has been observed in *Trichodesmium* that to prevent entry of sun light for effecting photosynthesis trichomes are clumped together (Bryceson and Fay, 1981). Many unicellular and non-heterocystous filamentous cyanobacteria (**Table 18**) have been reported those are capable to fix atmospheric nitrogen.

**Table 18. Nitrogen fixing non-heterocystous cyanobacteria (Adapted from Anand, 1998)**

| *Sl. No.* | *Organism* | *Reference* |
|---|---|---|
| 1. | Aphanothece sp. | Singh, 1973; Roger, 1985; Chan et al., 1988. |
| 2. | Gloeocapsa sp. | Wyatt and Sylvey, 1969; Wang and Wang, 1988. |
| 3. | Gloeothece sp. | Rippka et al., 1979; Reynaud and Fracnche. 1986. Khamees et al., 1987; Haung ad Chow. 1988; Maryan et al. 1986; Kallas et al., 1985; Gallon et al., 1973; Chaplin et al., 1984; Peschek et al. 1991. |
| 4. | Lyngbya sp. | Paerl et al., 1991; Anand and Murugesan, 1995. |
| 5. | Microcoleus chthonoplastes | Pearson et al., 1981. |
| 6. | Oscillatoria sp. | Stal and Krumbein, 1981. 1985; Kumazawa and Mitsui, 1985; Khamees et al. 1987; Parameswaran, 1993. |

| Sl. No. | Organism | Reference |
|---|---|---|
| 7. | Phormidium sp. | Weisshar and Boger, 1983; Anand and Murugesan, 1995. |
| 8. | Plectoema sp. | Stewart and Lex, 1970; Smoker and Barnum, 1986; Giani and Krumbein. 1986; Apte and Thomas, 1987. |
| 9. | Pseudoanabaena sp. | Singh et al., 1987. |
| 10. | Synechococcus sp. | Mitsui et al., 1986; Grobbelaar et al., 1986; Spiller and Shanmugam. 1987; Huang and Chow, 1988. |
| 11. | Trichodesmium sp. | Dugdale et al., 1961; Carpenter and Price, 1976; Saito and Hattori, 1982; Carpenter et al. 1990; Capone et al, 1990; Bergman and Carpenter, 1991. |

## REFERENCES

Aaronson, S., Dhawale, S.W., Patni, N.J., DeAngelis, B., Frank, O. and Baker, H. 1977. The cell content and secretion of water-soluble vitamins by several freshwater algae. *Arch. Microbiol.* 112: 57–59.

Abdallah, I.I. and Quax, W.J. 2017. A Glimpse into the Biosynthesis of Terpenoids. In Proceedings of the International Conference on Natural Resources and Life Sciences (2016), Surabaya, Indonesia, 20–21 October 2016, KnE Life Sciences: Dubai, United Arab Emirates, pp. 81–98.

Abdos, S.M. Hetta, M.H., Samhan, F.A., EI Din, R.A.S. and Ali, G.H. 2013 phytochemical and antibacterial study of five freshwater algal species. Asian J. Plant Sci. 11 (3):109-116

Abed, R.M.M., Dobretsov, S., Sudesh, K. 2009. Applications of cyanobacteria in biotechnology. J. App-Id. Microbiology.106 (1):1-12.

Aguilera, A. Berrendero G.E., Kastovsky, J. Echenique, R.O., and Salerno, G.L. 2018. The polyphasic analysis of two native Raphidiopsis isolates supports the unification of the genera Raphidiopsis and Cylindrospermopsis (Nostocales, Cyanobacteria). Phycologia.57(2):130-146.

Ahmed, M.H. and Venkataraman, G.S. 1973. Tolerance of ***Aulosira fertilissima*** to pesticides. ***Curr. Sci.*** 42: 108.

Ahmed, O.M., Mahmoud, A.M., Abdel-Moneim, A. and Ashour, M.B. 2012. Antidiabetic effects of hesperidin and naringin in type 2 diabetic rats. *Diabetol. Croat.* 41: 53–67.

Aitsui, A., Kumazawa, S., Takahashi, A., Ikemoto, H., Cao, S. and Arai, T. 1986. Strategy by which nitrogen fixing unicellular cyanobacteria grow photoautotrophically. *Nature* (London). 323: 720722.

Aiyer, R.S. 1965. Comparative Algological studies in rice fields in Kerala State. *Agric. Res. J. Kerala* 3: 100–104.

Aiyer, R.S. 1965. Comparative algological studies in rice-fields of Kerala State. ***Agri. Res. J. Kerala*** 3: 100–104.

Allen, M.B., Arnon, D.I. and Peters, R.F. 1957. Progr. Rep. Interagency. Agreem. No. 1236.

Allen, M.M. 1984. Cyanobacterial cell inclusions. *Annu. Rev. Microbiol.* 38: 1–25.

Anagonostidis, K. and Komárek K. 1985. Modern approach to the classification system of the Cyanophytes 1: Introduction. *Algol. Stud.* 38/39: 291–302.

Anand, N. 1998. Nitrogen Fixation by Non-heterocystous Cyanobacteria. Advances in Phycology. APC Publications Pvt. Ltd., New Delhi. Pp. 209–219.

Anand, N. and Revathi, G. 1987. Blue-green algae from rice-fields of Tamil Nadu. *Phykos* 26: 17–21.

Anand, N., Murugesan, V.S.R. 1995. Nitrogen fixation by bon-heterocystous Cyanobacteria. Abst. National Seminar on Biology and Biochemistry and Biotechnology of Cyanobacteria. Madurai Kamraj University, Madurai.

Anand, V.K. 1979. Blue-green algae of Gadigarh (Miran Sahib, Jammu) - I. ***Phykos*** 18(1 & 2): 21–24.

A Nonymous. 1983. The Agrochemicals hand book (The Royal Soc. of Chem. The University of Nottingham, England). P. 326.

Apte, S.K. and Thomas, J. 1987. Nitrogen fixation genes (nif K,D,H) in the filamentous non-heterocystous cyanobacterium *Plectonema boryanum* do not rearrange. *J. Genet.* 66: 101–110.

Asensi-Fabado, M.A. and Munné-Bosch, S. 2010. Vitamins in plants: Occurrence, biosynthesis and antioxidant function. *Trends Plant Sci.* 15: 582–592.

Ashtekar, P.V. and Kamat, N.D. 1980. Nostocales of Marathwada, Maharashtra. *Phykos* 19(1): 89–93.

Assini, J.M., Mulvihill, E.E. and Huff, M.W. 2013. Citrus flavonoids and lipid metabolism. *Curr. Opin. Lipidol.* 24: 34–40.

Attridge, E.M. and Rowell, P. 1997. Growth, heterocyst differentiation and nitrogenase activity in the cyanobacteria *Anabaena variabilis* and *Anabaena cylindrica* in response to molybdenum and vanadium. *New Phytol.* 135: 517–526.

Bakers, K.K. 1981. Ecology and taxonomy of 5 natural populations of the genus Aphanizomenon (Cyanophyceae). Arch. Hydrobial.92(2):222-251.

Baker, A.F. and Bold, H.C. 1970. Phycological studies. X. Taxonomic studies in the Oscillatoriaceae. ***Univ. Texas Publ.*** 7004: 105.

Barker, G.L.A., Hayes, P.K., Ómahony, S.L., Vacharapiyasophon, P. and Walsby, A.E. 1999. A molecular and phenotypic analysis of *Nodularia* (cyanobacteria) from the Baltic Sea. *J. Phycol.* 35: 931–937.

Bandaranayake, W.M. 1998. Mycosporines: are they nature's sunscreens? *Nat. Prod. Rep*. 15: 159–172. doi: 10.1039/a815159y

Banerjee, S., Singh, A. and Pal, R. 2021. A taxonomic investigation on heterocystous cyanobacteria of West Bengal, India. Phytomorphology 70(3 & 4): 53–70.

Banerji, J.C. 1936. Studies on the Myxophyceae of lower Bengal-I. ***J. Indian Bot. Soc*.** 285–302.

Banerji, J.C. 1938. Studies on the Myxophyceae of lower Bengal-II. ***J. Dept. Sci. Kalcutta Univ*.** 1: 95–109.

Barhate, V.P. & Tarar, J.L. 1983. Algae of Maharashtra: I. Additions to Cyanophyceae of Khandesh. ***Phykos*** 22: 67–72.

Barman, N., Satpati, G.G and Pal, R. 2015. A Morphotaxonomic account of cyanobacterial diversity of Indian Sundarbans, *J. Algal Biomass Utln*. 6(3): 39–46.

Basavaraja, B. and Naik, T.P. 2018. Diversity of Blue-green algae of Paddy field, Hassan, Karnataka. *Plant Archives* 18(1): 721–722.

Batterton, J.C., Boush, G.M. & Matsumura, F. 1971. Growth response of blue-green algae to aldrin, dialdrin, endrine and their metabolites. *Bull. Environ. Contam. and Toxicol*. 6: 589–594.

Baumgartel, O. 1920. Das problem der Cyanophyceenzelle. *Arch. Protistenk*. 41: 50–148.

Becher, P.G., Baumann, H.I., Gademann, K. and Jüttner, F. 2009. The cyanobacterial alkaloid nostocarboline: An inhibitor of acetylcholinesterase and trypsin. *J. Appl. Phycol*. 21: 103–110.

Beck, G. De & Zahlbruchner, A. 1897. Shedae ad'Kryptogames exsiccatae editae a Museo Palatine Vindobonensis. Ann. de K.K. Naturhist. Hofmauseum 12: 75-98.

Becker, J.E., Moore, R.E. and Moore, B.S. 2004. Cloning, sequencing, and biochemical characterization of the nostocyclopeptide biosynthetic gene cluster: Molecular basis for imine macrocyclization. *Gene* 325: 35–42.

Bedoux, G., Hardouin, K., Burlot, A.S. and Bourgougnon, N. 2014. Bioactive components from seaweeds cosmetic applications and future development. In: N. Bourgougnon (eds.), *Advances in Botanical Research, Sea Plants Book*, Vol. 71, pp. 346–367.

Begum, Z.N., Mandal, R. and Amin, F.B. 2008. Quantification and nitrogen fixation of cyanobacteria in rice field soils of Bangladesh, *Bangladesh J. Bot*. 37(2): 183–188.

Begum, Z.N., Mandal, R. and Islam, S. 2011. Effect of Cyanobacterial biofertilizer on the growth and yield components of two HYV of rice. *J. Algal Biomass Utln*. 2(1): 1–9.

Bendre, A.M. and Kumar, S. 1975. Cyanophyceae of Meerut. *Phykos* 14(1 & 2): 1–7.

Bergman, B. and Carpenter E.J. 1991. Nitrogenase confined to randomly distributed trichomes in the marine cyanobacterium *Trichodesmium thiebauti. J. Phycol.* 27: 158–165.

Bergman, B. and Carpenter, E.J. 1991. Nitrogenase confined to randomly distributed trichomes in the marine cyanobacterium *Trichodesmium. J. Phycol.* 27: 158–165.

Bergman, B., Gallon, J.R., Rai, A.N. and Stal, L.J. 1997. $N_2$ Fixation by non-heterocystous cyanobacteria. *FEMS Microbiology Reviews.* 19(3): 139–185. https://doi.org/10.111/j.1574-6976.1997. Tboo296.x

Bergman, B., Gallon, J.R., Rai. A.N. and Stal, L.J. 1997. $N_2$ fixation by non-heterocystous cyanobacteria. *FEMS Microbiol. Rev.* 19 (3): 139–185.

Bharadwaja, Y. 1934. The Taxonomy of *Scytonema* and *Tolypothrix* including some new records and new species from India and Ceylon. *Rev. algol., Paris* 7: 149–178.

Bharadwaja, Y. 1935. The Myxophyceae of the United Provinces, India - I, *Proc. Indian Acad. Sci.* B. 2(l): 95–107.

Bhatnagar, I., and Kim, S.K. 2010. Immense essence of excellence: marine microbial bioactive compounds. Mar. Drugs. 8 (10):2673-2701.

Bhaya, D., Grossman, A.R., Steunou, A.-S., Khuri, N., Cohan, F.M., Hamamura, N., Melendrez, M.C., Bateson, M.M., Ward, D.M. and Heidelberg, J.F. 2007. Population level functional diversity in a microbial community revealed by comparative genomic and metagenomic analyses. *ISME J.* 1: 703–713.

Bhoge, O.N. and Ragothaman, G. 1986. Studies on the Cyanophyceae from Jalgaon region, Maharashtra. *Phykos* 25: 129–131.

Biswas, K. 1924. The Subaerial algae of the Berkuda Island, in the Chilka Lake, Ganjam District, Madras Presidency. *J. Proc. Asiatic Soc. Bengal n.s.* 20: 359–365.

Biswas, K. 1925. Road slimes of Kalcutta. *J. Dept. Sci. Kalcutta Univ.* 7: 1–10.

Biswas, K. 1926. Flora of the Salt Lakes, Kalcutta. *J. Dept. Sci. Kalcutta Univ.* 8: 1–47.

Biswas, K. 1927. Aquatic vegetation of Bengal in relation to supply of oxygen to the water. *J. Dept. Sci. Kalcutta Univ.* 8: 49–56.

Biswas, K. 1930. Contributions to the knowledge of the Freshwater Algae of Manipur, Assam. *J. Bombay Nat. Hist. Soc.* 34: 189–192.

Biswas, K. 1932a. Role of aerophilous algae in producing colour effects on the bark of *Oredoxa regia* of the *Oredoxa avenue* of the Royal Botanic Garden, Kalcutta. *Hedwigia* 72: 31–41.

Biswas, K. 1932b . Algal flora of the Chilka Lake. Mem. Asiatic Soc., Bengal 11|(5): 165–198.

Biswas, K. 1934. Observations on the algal collections from the Khasia and Jaintia Hills, Assam, India. *Hedwigia* 74: 1-28.

Biswas, K. 1942. The role of common algal communities of the river Hoogly on the drinking water, Kalcutta. 150th Ann. Vol. Roy, Bot. Gdn.: 189–206.

Boergesen, F. 1935. A list of marine algae from Bombay. *K. Danske Vidensk. Biol. Medd.* 12: 1–64.

Bongale, U.D. and Bharati, S.G. 1980. On the algal flora of cultivated soils of Karnataka State, India. *Phykos* 19(1): 95–109.

Bourelly, P. 1979. Les Cyanophycees, algues ou bactenes? *Rev. Algol.. N.S.* 14: 5–9.

Bourrelly, P. 1970. Les algues d'eau douce III. – N. Boubée & Cié., Paris.

Bruhl, P. and Biswas, K. 1922. Algae of Bengal filter beds. *J. Dept. Sci. Kalcutta Univ.* 4: 1–17.

Bruhl, P. and Biswas, K. 1924. Algae Epiphyticae Epiphloiae Indicae or Indian Bark Algae. *J. Dept. Sci. Kalcutta Univ.* 5: 1–22.

Bruhl, P. and Biswas, K. 1926. Algae of the Loktak Lake. Mem. Asiatic Soc. Bengal 8(5): 257–315.

Bryceson, I. and Fay, P. 1981. Nitrogen-fixation in *Oscillatoria* (*Trichodesmium*) *erythraea* in relation to bundle formation and trichome differentiation. *Mar. Biol.* 61: 159–166.

Cabrita, M.T., vale, C., Rauter, A.P. 2010. Halogenated compounds from marine alage. Mar Drugs. 8(8):2301-2317.

Carpenter, E.J. and Price, C.C. 1976. Marine *Oscillatoria* (*Trichodesmium*)Explanation for nitrogen fixation without heterocysts.*Science*(Washington).191:1278-1280..

Carpenter, E.J. and Price, C.C. 1977. Nitrogen fixation, distribution and production of *Ocillatoria* (*Trichodesmium*) spp. in the Western Sargasso and Caribbean seas. *Limnol. Oceanogr.* 22: 60–72.

Carpenter, E.J., Chang, J., Cottrel, M., Schubuer, J., Pearl, H.W., Bebout, B.M. and Capone, D.G. 1990. Revaluation of nitrogenase oxygen-protective mechanisms in the planktonic marine cyanobacterium *Trichodesmium*. Marine Ecology, Progress series. 65: 151–158.

Capone, D.G., Zehr, J.P., Paerl, H.W., Bergman, B. and Carpenter E.J. 1997. *Trichodesmium*, a globally significant marine cyanobacterium. *Science*, 276: 1221–1229.

Carreto, J.I., carignan, M.O. Daleo, G. and de Marco, S.G. 1990. Occurrence of mycosporine like amino acids in the red tide dinoflagellate Alexandrium exeavatum: UV Photo-protective compounds. JPlankton Res 12:921.

Carrieri, D., Ananyev, G., Lenz, O., Bryant, D.A. and Dismukes, G.C. 2011. Contribution of a sodium ion gradient to energy conservation during fermentation in the cyanobacterium *Arthrospira* (*Spirulina*) maxima CS-328. *Appl. Environ. Microbiol.* 77: 7185–7194.

Casmatta,D.A.,Johansen,J.R.,Vis,M.L and Broadwater,S.T.2005:Molecular and morphological characterization of ten polar and near-polar strains within the Oscillatoriales (Cyanobacteria)-----J.Phycol.41: 421-438.

Castenholz, R.W. 1989c. Order Stigonematales, p. 1794–1799. In: J.T. Staley, M.P. Bryant, N. Pfennig and J.G. Holt (eds.), *Bergey's Manual of Systematic Bacteriology*, Vol. 3, Williams and Wilkins, Baltimore.

Castenholz, R.W. 1989a. Order Oscillatoriales, p. 1771–1780. In: J.T. Staley, M.P. Bryant, N. Pfennig and J.G. Holt (eds.), *Bergey's Manual of Systematic Bacteriology*, Vol. 3. Williams and Wilkins, Baltimore.

Castenholz, R.W. 1989b. Order Nastocales, p. 1780–1798. In: J.T. Staley, M.P. Bryant, N. Pfennig and J.G. Holt (eds.), *Bergey's Manual of Systematic Bacteriology*, Vol. 3, Williams and Wilkins, Baltimore.

Castenholz, R.W.,Wilmotte,A.,Herdman,M.,Rippka.,Waterbury,J.B.,Iteman,et. al 2001. Phylum BX. Cyanobacteria. In: Boone D.R. & Castenholz R.W. (eds.). Bergey's manual of systematic bacteriology, p. 473–599, Springer, New Work.

Castenholz, R.W. and Garcia-Pichel, F. 2012. Cyanobacterial responses to UV radiation. In *Ecology of Cyanobacteria II*, Whitton, B.A., Ed., Springer: Dordrecht, The Netherlands, pp. 481–499.

Castenholz, R.W., Wilmotte, A., Herdman, M., Rippka, R., Waterbury, J.B., Iteman, I. *et al.* 2001. Phylum BX. Cyanobacteria. In: Garrity, G., Boone, D.R., Castenholz, D.W. (eds.). Bergey's Manual of systematic bacteriology. New York: Springer. 473–599.

Chadha, A. and Pandey, D.C. 1982. Distribution of algae on the walls of Buildings. *Phykos* 21: 122–123.

Chakraborty, T., Mukhopodhyay, A. and Pal, R. 2010. Microalgal diversity of Kolkata, West Bengal, India, *Indian Hydrobiology* 12(2): 204–224.

Chang, Z., Flatt, P., Gerwick, W.H., Nguyen, V.-A., Willis, C.L. and Sherman, D.H. 2002. The barbamide biosynthetic gene cluster: A novel marine cyanobacterial system of mixed polyketide synthase (PKS)-non-ribosomal peptide synthetase (NRPS) origin involving an unusual trichloroleucyl starter unit. *Gene* 296: 235–247.

Chang, Z., Sitachitta, N., Rossi, J.V., Roberts, M.A., Flatt, P.M., Jia, J., Sherman, D.H. and Gerwick, W.H. 2004. Biosynthetic Pathway and Gene Cluster Analysis of Curacin A, an Antitubulin Natural Product from the Tropical Marine Cyanobacterium Lyngbya m ajuscula. *J. Nat. Prod.* 67: 1356–1367.

Chaporkar, C.B. & Gangawane, L.V. 1984. Blue green algae of some cultivated soils of Marathwada, Maharashtra-I. *Phykos* 23(1 & 2): 55–58.

Chatterjee, M. & Chatterjee, S.P. 1983. Nitrogen fixing cyanobacteria from the paddy field soils of Burdwan district. *Phykos* 22: 64–66.

Chaturvedi, U.K. & Pandey, U.G. 1976. A list of blue-green and green algae from Rohilkhand division, U.P., India - IV. *Phykos* 15(1 & 2): 127–131.

Cheevadhanarak, S., Paithoonrangsarid, K., Prommeenate, P., Kaewngam, W., Musigkain, A., Tragoonrung, S., Tabata, S., Kaneko, T., Chaijaruwanich, J. and Sangsrakru, D. 2012. Draft genome sequence of *Arthrospira platensis* C1 (PCC 9438). *Stand. Genom. Sci.* 6: 43–53.

Chen, T.H., Huang, T.C. and Chow, T.J. 1988. Calcium requirement in nitrogen fixation in the cyanobacterium *Synechococcus* R.F-1. *Planta* (Berlin). 173: 253–256.

Chilczuk, T., Steinborn, C., Breinlinger, S., Zimmermann-Klemd, A.M., Huber, R., Enke, H., Enke, D., Niedermeyer, T.H.J. and Gründemann, C. 2020. Hapalindoles from the cyanobacterium *Hapalosiphon* sp. inhibit T cell proliferation. *Planta Med.* 86: 96–103.

Cho, S.M., Jeoung, S.C., Song, J.-Y., Kupriyanova, E.V., Pronina, N.A., Lee, B.-W., Jo, S.-W., Park, B.-S., Choi, S.-B. and Song, J.-J. 2015. Genomic survey and biochemical analysis of recombinant candidate cyanobacteriochromes reveals enrichment for near UV/violet sensors in the halotolerant and alkaliphilic cyanobacterium Microcoleus IPPAS B353. *J. Biol. Chem.* 290: 28502–28514.

Choi, J., Lee, D.-H., Jang, H., Park, S.-Y. and Seol, J.-W. 2020. Naringenin exerts anticancer effects by inducing tumor cell death and inhibiting angiogenesis in malignant melanoma. *Int. J. Med. Sci.* 17: 3049.

Chrapusta, E., Kaminski, A., Duchnik, K., Bober, B., Adamski, M. and Bialczyk, J. 2017. Mycosporine-like amino acids: Potential health and beauty ingredients. *Mar. Drugs* 15: 326.

Cosmatta, D.A., Johansen, J.R., Vis, M.L. and Broadwater, S.T. 2005. Molecular and morphological characterization of ten polar and near – polar strains within the Oscillatoriales (Cyanobacteria). *J. Phycol.* 41: 421–438.

Cumino, A.C., Marcozzic., Barreiro, R. and Salerno, G.L. 2007. Carbon cycling in *Anabaena* sp. PCC 7120. Scucrose synthesis in the heterocysts and possible role in nitrogen fixation. *Plant physiology* 143: 1385–1397 DOI: http://dx.doi.org/10.1104/pp.106.091736

Das, A.C. and Debnath, A. 2006. Effect of systemic herbicides on $N_2$-fixing and phosphate solubilizing microorganisms in relation to availability of nitrogen and phosphorus in paddy soils of Wet Bengal. *Chemosphese* 65: 1082–1086.

Dasgupta, M. and Ahmed, S.U. 2013. Some potential rice field BGA isolates from Sonitpur district. Assam, Northeast India. *J. Nat. Prod. Plant Resour.* 3(4): 17–23.

Dash, N.P., Kumar, A., Kaushik, M.S., Abraham, G. and Singh, P.K. 2017a. Nitrogenous agrochemicals inhibiting native diazotrophic cyanobacterial contribution in Wet land rice ecosystem. *J. Appl. Phycol.* 29: 929–939.

Davidson, F.F. 1961. Antibacterial activity of *Oscillatoria formosa* extract. *Amer. J. Bot.* 48: 542.

De, P.K. 1936. The problem of the nitrogen supply of rice. I. Fixation of nitrogen in the rice soils under water logged conditions. *Indian J. Agri. Sci.* 6: 1237–1245.

De, P.K. 1939. The role of Blue-green Algae in Nitrogen fixation in rice fields. *Proc. Roy. Soc.. Lond.* B. 127: 121–139.

Debnath, M., Mandal, N.C. and Ray, S. 2012. Effect of fungicides and insecticides on growth and enzyme activity of four cyanobacteria. *Indian J. Microbiol.* 52: 275–280.

Del Mondo, A., Smerilli, A., Sané, E., Sansone, C. and Brunet, C. 2020. Challenging microalgal vitamins for human health. *Microb. Cell Factories* 19: 201.

Delaney, S.F., Herdman, M. and Carr, N.G. 1976. Genetics of blue-green algae. In: The Genetics of Algae, Lewin, R.A. (ed.), Black well, Oxford, pp. 7–28.

Demay, J., Bernard, C., Reinhardt, A. and Marie, B. 2019. Natural products from cyanobacteria: Focus on beneficial activities. *Mar. Drugs* 17: 320.

Desikachary, T.V. & Ramkrishnan, K.R. 1953. On an interesting aegagrophilous *Schizothrix* from Madras Province forming balls. *Proc. Indian Sci. Congr.*, Part III, 1953.

Desikachary, T.V. 1959. *Cyanophyta*. Indian Council of Agricultural Research, New Delhi.

Desikachary, T.V. 1973. *Status of Classical Taxonomy*. In: N.G. Carr and B.A. Whitton (eds.), The biology of blue-green algae. Pp. 473–481. Blackwell Publications, Oxford.

Devi, S., Rani, N. and Sagar, A. 2020. GC-MS Analysis and antioxidant activity of two species of cyanobacteria isolated from Drang salt mine of district Mandi, Himachal Pradesh, India. *Plant Arch.* 20: 7505–7510.

Dey, H. and Chakraborty, S. 2020. Diversity of non-hetrocystous filamentous cyanobacteria from freshwater bodies of Bidhannagar, North 24 Parganas, West Bengal. Asian J. Biol. Life Sci.. 9,( 3):386-390.

Dey, H.S., Tayung, K. and Basita, A.K. 2010. Occurrence of nitrogen fixing cyanobacteria in local rice fields of Orissa, India. *Ecoprint* 17: 77–85.

Dhanya, V. and Ray, J.G. 2015. Ecology and Diversity of cyanobacteria in Kuttanadu paddy Wetlands, Kerala, India. *Amer. J. Pl. Sci.* 6: 2924–2938.

Dillion, J.G., Tatsumi, C.M., Tandingan, P.G. and Castenholz, R.W. 2002. Effect of environmental factors on the synthesis of scytonemin, a UV-screening pigment, in a cyanobacterium (*Chroococcidiopsis* sp.). *Arch. Microbiol.* 177: 322–331.

Dixit, S.C. 1936. The Myxophyceae of the Bombay Presidency, India, I. *Proc. Indian Acad. Sci.B.3:93-106.*

do Amaral, S.C., Monteiro, P.R., Neto, J.d.S.P., Serra, G.M., Gonçalves, E.C., Xavier, L.P. and Santos, A.V. 2021. Current knowledge on microviridin from cyanobacteria. *Mar. Drugs* 19: 17.

Dong, S., Chen, J., Wang, S., Wu, Y., Hou, H., Li, M. and Yan, C. 2015. Draft genome sequence of cyanobacteria *Arthrospira* sp. TJSD091 isolated from seaside wetland. *Mar. Genom.* 24: 197–198.

Drouet, F. 1968; Revision of the classification of Oscillatoriaceae. *Monogr. Acad. Nat. Sci. Phila.* 15: 370.

Drouet, F. 1969. Revision of the classification of the Oscillatoriaceae: additions and corrections. *Monogr. Nat. Acad. Sci. Phila.* 422: 1–5.

Drouet, F. 1973. *Revision of the Nostocaceae with cylindrical trichomes.* Hafner Press, Macmillan, New York, 292 pp.

Drouet, F. 1978. Revision of the Nostocaceae with constricted trichomes. *Beih. Nova Hedw.* 57: 1–258.

Drouet, F. 1981. Revision of the Stigonemataceae with a summary of the classification of the Blue-green Algae. *Beih. Nova Hedw.* 66: 221.

Drouet, F and Daily, W.A. 1956. Revision of the Coccoid Myxophyceae. *Butler Univ. Bot. Stud.* 12: 1–218.

Dugdale, R.C., Menzel, D.W. and Rythr, J.H. 1961. Nitrogen fixation in the Sargasso Sea. *Deep-sea Research.* 7: 297–300.

Dutta, S., Whicher, J.R., Hansen, D.A., Hale, W.A., Chemler, J.A., Congdon, G.R., Narayan, A.R.H., Håkansson, K., Sherman, D.H. and Smith, J.L. 2014. Structure of a modular polyketide synthase. *Nature* 510: 512–517.

Edelman, M.D., Swinton, J.A., Schiff, H.D., Epstein and Zelden, B. 1967. Decoxyribonucleic acid of the blue-green algae (Cyanophyta). *Bacteriological Review* 31: 315–331.

Edwards, D.J. and Gerwick, W.H. 2004. Lyngbyatoxin biosynthesis: Sequence of biosynthetic gene cluster and identification of a novel aromatic prenyltransferase. *J. Am. Chem. Soc.* 126: 11432–11433.

Edwards, D.J., Marquez, B.L., Nogle, L.M., McPhail, K., Goeger, D.E., Roberts, M.A. and Gerwick, W.H. 2004. Structure and biosynthesis of the jamaicamides, new mixed polyketide-peptide neurotoxins from the marine cyanobacterium *Lyngbya majuscula*. *Chem. Biol.* 11: 817–833.

El-Sayed M. and Verpoorte, R. 2007. Catharan thus terpenoide indole alkaloids: biosynthesis and regulations 6:277-305.

El-Mohsnawy, E. and Abu-Khudir, R. 2020. A highly purified C-phycocyanin from thermophilic cyanobacterium Thermo *Synechococcus elongatus* and its cytotoxic activity assessment using an in vitro cell-based approach. *J. Taibah Univ. Sci.* 14: 1218–1225.

Elenkin, A.A., 1936, 1938, 1949. Monographia algarum cyanophcearum aguidulcium et terrestrium in finibus USSR inventarum. [Sinezelenye vodorosli SSSR]. 1,2(1-2):1-1908, Izd. ANSSSr. MosKva-Leningrod.

Engene, N., Tottaeker, E.C., Kastovsky, J., Byrum, T., Choi, H., Ellisman, M.H., Komarek, J. and Gerwick, W.H. 2012. *Moorea producta* ge. nov., sp. *Moorea bouillonii* comb. nov. tropical marine cyanobacteria rich in bioactive secondary metabolites Int. J. Syst. Evol. Microbial 62:117-1178.

Erdmann, N. and Hagemann, M. 2001. Salt accumulation of algae and cyanobacteria: a comparison, In. Rai, L.C. and Gaur J.P. (eds), Algal Adaptation to Environmental Stresses, Springer, Heidelberg, pp. 323–362.

Ecans, C. Thomson, P.G., Davidson, A.T. Bowie, A.R., Vanden, R. Witte, H., and Brussa ard, C.P.D. 2011. Potential climate change impacts on microbial distribution and carbon cycling in the Australian Southern Ocean. Deep-Sea Res-II 58: 2150-2161.

Favre-Bonvin, J., Bernillon, J., Salin, N. and Arpin, N. 1987. Biosynthesis of mycosporines: mycosporine glutaminol in *Trichothecium roseum*. *Phytochemistry* 26: 2509–2514. doi: 10.1016/Soo31-9422(00) 83866-2.

Fay, P. and Lang, N.J. 1971. The heterocysts of the blue-green algae. I. Ultrastructural integrity after isolation. *Proc. R. Soc. Lond. (Biol.)*178: 185–192.

Fay, P., Stewart, W.D.P. Walsby, A.E. and Fogg. G.E. 1968. Is the heterocysts the site of nitrogen fixation in blue-green algae? *Nature* 220: 810–812.

Ferris, M.J., Sheehan, K.B., Kuhl, M., Cooksey, K., Wigglesworth-Cooksey, B., Harvey, R. and Henson, J.M. 2005. Algal species and light microenvironment in a low-pH, geothermal microbial mat community. *Appl. Environ. Microbiol.* 71: 7164–7171.

Flores, E., Herrero, A., Work, C.P. and Malolener, I. 2006. Is the periplasm continuous in filamentous multicellular cyanobacteria? *Trends in Microbiology* 14: 439–443. DOI: http://dx.doi.org/10.1016/j.tim.2006.08.007.

Fogg, G.E. & Wolfe, M. 1954. The nitrogen metabolism of blue-green algae (Myxophyceae) in 'Autotrophic Microorganisms'. The Fourth Symposium of the Society of General Microbiology. Pp. 99–125.

Fogg, G.E. 1949. Growth and heterocysts production in *Anabaena cylindrica* Lemm. II. in relation to carbon and nitrogen metabolism. *Ann. Bot.* 13: 241–259.

Fogg, G.E., Stewart, W.D.P., Fay, P. and Walsloy, A.E. 1973. The Blue-Green Algae, pp. 459. Academic Press, London & New York.

Forest, M. 1968. The approach of a modern algae taxonomist in D.F. JacKson, ed. Algae, man and the environment. Syracuse Univ. Press Publ. 185-200.

Fredrickson, C. and Bergman, B. 1997. Ultrastructural characterization of cells specialized for nitrogen function in a non-heterocystous cyanobacterium, *Trichodesmium* spp. *Protoplasma* 197: 76–85.

Fredrickson, C. and Bergman, B. 1997. Ultrastructural characterization of cells specialized for nitrogen function in a non-heterocystous cyanobacterium, *Trichodesmium* spp. *Protoplasma* 197: 76–85.

Fremy, P. 1927. Petite contribution a la flore des Myxophycees de I' Indes meridionale. *Arch. de Bot., Bull, mensuel*, n. 3: 46-47.

Fremy, P. 1942. Cyanophycees de K'Inde meridlonale, Blumea, Suppl. II (Dr. A.A. Weber van Bosse(Jubilee Volume). 21–40.

Fritsch, F.E and Dey, P.K. 1938. Nitrogen fixation by the Blue-green algae. *Nature, Lon.* 142: 878.

Fritsch, F.E. 1935. *The Structure and Reproduction of the Algae*, Vol. I. University Press, Cambridge.

Fritsch, F.E. 1945. *The Structure and Reproduction of Algae-II*. London: Cambridge University Press.

Fujisawa, T., Narikawa, R., Okamoto, S., Ehira, S., Yoshimura, H., Suzuki, I., Masuda, T., Mochimaru, M., Takaichi, S. and Awai, K. 2010. Genomic structure of an economically important cyanobacterium, *Arthrospira* (Spirulina) platensis NIES-39. *DNA Res.* 17: 85–103.

Gaidukov, N. 1902. Nber den Einfluss farbigen Lichtes auf die Färbung der Oscillarien. *Abh Preuss Akad Wiss* 5: 8–13.

Gallon, J.R., Kurz, W.G.W. and Larue, T.A. 1973. Isocitrate supported nitrogenase activity in *Gloeocapsa* sp. LB. 795. *Can. J. Microbiol.* 19: 461–465.

Ganapati, S.V. 1940. Studies on the Chemistry and Biology of the slow sand filters at the Madras water works. *Proc. Nat. Inst. Sci., India* 6: 237–300.

Ganapati, S.V. 1940. The ecology of a temple tank containing a permanent water bloom of *Microcystis aeruginosa* (Kutz.) Henfr. *J. Bombay Nat. Hist. Soc.* 42: 65–77.

Ganapati, S.V. 1943. An ecological study of a garden pond containing abundant zooplankton. *Proc. Indian Acad. Sci.* B. 17: 41–58.

Geethu, V. and Shamina, M. 2021. Filamentous cyanobacteria from Western Ghats of North Kerala, India. Bangladesh J. Pl. Taxon. 28(1): 83–95.

Geitler, L. 1932. Cynophyceae. In: Rabenhorst L. (ed.). Kryptogamenflora von Deutschland. Leipzig: Akad. Verlagsges: 673–1196.

Geitler, L. 1979. Einige kritische Bemerkungen zu neuen zussamen-fassenden Darstellung der Morphologie und Systematik der Cyanophyceen. Plant Syst. Evol. 132: 153–160.

Geraldes, V., de Medeiros, L.S., Lima, S.T., Alvarenga, D.O., Gacesa, R., Long, P.F., Fiore, M.F. and Pinto, E. 2020. Genetic and biochemical evidence for redundant pathways leading to mycosporine-like amino acid biosynthesis in the cyanobacterium *Sphaerospermopsis torques-reginae* ITEP-024. *Harmful Algae* 35: 177–187.

Geraldes, V., Jacinavicius, F.R., Genuário, D.B. and Pinto, E. 2020. Identification and distribution of mycosporine-like amino acids in Brazilian cyanobacteria using ultrahigh-performance liquid chromatography with diode array detection coupled to quadrupole time-of-flight mass spectrometry. *Rapid Commun. Mass Spectrom.* 34: e8634.

Ghadage, S.J. and Karande, V.C. 2020. Chroococcacean blue-green algae from the paddy fields of Satara District, Maharshtra, India. *J. Threatened Taxa.* 12(14).

Ghosh, S., Barinova, S. and Keshri, J.P. 2012. Diversity and seasonal variation of phytoplankton community in the Santragachi Lake, West Bengal, India. *Q-Science Connect* 3: 1–19.

Giani, D. and Krumbein, W.E. 1986. Growth characteristics of non-*heterocystous* cyanocactrium *Plectonema boryanum* with nitrogen as nitrogen sourse. *Arch. Microbiol.* 145: 259–265.

Gibbons, N.E., and Murray, R.G.E. 1978. Proposals concerning teh higher taxa of bacteria. Int. J. Syst. Bacterial 28:1-6.

Goiris, K., Van Colen, W., Wilches, I., León-Tamariz, F., De Cooman, L. and Muylaert, K. 2015. Impact of nutrient stress on antioxidant production in three species of microalgae. *Algal Res.* 7: 51–57.

Golden, J.W. and Yoon, H. 1998. Heterocyst formation in *Anabaena*. *Current Opinion in Microbiology*. 1(6): 623–629.

Goncalves, A.L. 2021. The use of Microalgae and Cynobacteria in the Improvement of Agriculture practices: A Review on their Biofertilising, Biostimulating and Biopesticide Roles. Appl. Sci. 11:871.

Gonzalves, E.A. 1947. The algal flora of the hot springs of Vajreswari near Bombay. *J. Univ. Bombay* 16: 22–27.

Gonzalves, E.A. and Gangla, K.S. 1949. Observations on the algae of paddy-field soils. J. Univ. Bombay 18(3): 51–59.

Gonzalves, E.A. and Joshi, D.B. 1943a. The algal flora of temporary water around Bombay – I. An ecological study of the algae in some rain water pools near Borivali. *J. Univ. Bombay* 11: 34–45.

Gonzalves, E.A. and Joshi, D.B. 1946. Freshwater algae near Bombay – I. The seasonal succession of algae in a tank at Bandra. *J. Bombay Nat. Hist. Soc.* 46(1): 155–176.

Goulart, H.R., Kimura, E.A., Peres, V.J., Couto, A.S., Durate, F.A.A. and Katzin, A.M. 2004. Terpenes arrest parasite development and inhibit biosynthesis of isoprenoids in *Plasmodium falciparum*. *Antimicrob. Agents Chemother.* 48: 2502–2509. doi: 10.1128/AAC.48.7.2502-2509.2004.

Goyal, S.K., Shahma, B.M. and Gupta, R.S. 1984. Algal flora of rice- field soils of Jammu and Kashmir State. *Phykos* 23(1&2): 59–64.

Goyel, S.K. 1982. Blue-green algae and rice cultivation. *Proc. Natl. Symp.*, BNF, 1ARI. New Delhi. P. 346–347.

Goyel, S.K., Sharma, B.M. and Gupta, R.S. 1984. Algal flora of rice field soils of Jammu and Kashmir states. *Phykos* 23: 59–64.

Grindberg, R.V., Ishoey, T., Brinza, D., Esquenazi, E., Coates, R.C., Liu, W.-T., Gerwick, L., Dorrestein, P.C., Pevzner, P. and Lasken, R. 2011. Single cell genome amplification accelerates identification of the apratoxin biosynthetic pathway from a complex microbial assemblage. *PLoS ONE* 6: e18565.

Grobbelaar, N., Huang, T.C., Lin, H.Y. and chow, T.J. 1986. Dinitrogen-fixing endogenous rhythm in synechoccus. RF-1. FEMS Microbial. Lett, 37:173-177

Gross, W. 2000. Ecophysiology of algae living in highly acidic environments. *Hydrobiologia* 433: 31–37.

Grover, I.S. and Pandhol, R.K. 1975. Algal flora of Paddy fields of Ludhiana and its adjacent areas. *Phykos* 14(1 & 2): 89–97.

Gugger, M.F. and Hoffmann, L. 2004. Polyphyly of true branching cyanobacteria (Stigonematales). *Int. J. Syst. Evol. Microbiol.* 54: 349–357.

Guiheneuf, F., Khan, A., Tran, L.S.P. 2016. Genetic engineering: a promising tool to engender physiological, biochemical and molecular stress resilience in green microalgae. Front. Plant Sci. 7:400.

Gupta, A.B. 1950. A contribution to the algal flora of Khajiar-Chamba State, Himachal Pradesh. *Proc. nat. Acad. Sci. India* 20(3): 109–115.

Gupta, A.B. 1956. A Contribution to the Algal flora of the Allahabad District. *J. Res. D.A.V. College, Kanpur* 3(1): 76–81.

Gupta, D. 1965. Some new records of blue-green algae from West Bengal. *Bull. Bot. Soc. Bengal* 19(1): 1–2.

Gupta, D. 1975. Some new records of blue-green algae from West Bengal-II. *Bull. Bot. Soc. Bengal.* 29: 29–31.

Gupta, D. and Sen, C. 1987. Observations on blue-green algae of paddy-field soils of Gangetic West Bengal I. Burdwan district. *Indian Agric.*. 31(3): 221–225.

Gupta, D. and Sen, C.R. 1987. Some Cyanophyceae from Gangetic delta of West Bengal-1. Hooghly district - A taxonomic enumeration. *J. Econ. Taxon. Bot.* 10(2): 381–385.

Gupta, M. & Pandey, D.C. 1974. On an interesting species of *Cylindrospermum* Kuetzing (Cyanophyceae). *Phykos* 13(1): 16–21.

Gupta, V. and Vyas, D. 2021. Antimicrobial effect of a cyclic peptide Nostophycin isolated from wastewater cyanobacteria, Nostoc calcicola. *Curr. Bot.* 12: 94–101.

Hagemann, M. 2002. Environmental stress, signalling and basic acclimation reactions: In: Solheim, R. *et al.* (eds.), cyanobacteria and Nitrogen fixation in Extreme Environments European Science Foundation CYANOFIX, p. 24. (Abstract.).

Harder, R. 1917a. Ernahrungsphysiologische Untersuchungen an Cyanophyceen hauptsachlish dem endophyti schen *Nostoc pruniforme*. *Z. Bot.* 9: 145–242.

Hayes, P.K., Batley, J. and Jekins, C. 2006. Gene exchange within populations of *Nodularia spumigena* in the Baltic Sea. In: Program and Abstract, 12th Internat. Symp. on Phototrophic Prokaryotes, Pau, p. 87. (Abstract.).

Herbert, R.A. and Codd, G.A. 1986. *Microbes in Extreme Environments*. Academic Press, London, UK.

Herdman, M. 1982. Evolution and generic properties of the genome. In: The Biology of Cyanobacteria. Carr. N.G. and Whitton, B.A. (eds.). Pp. 263–305. Blackwell, Oxford.

Hernández-Aquino, E. and Muriel, P. 2018. Beneficial effects of naringenin in liver diseases: Molecular mechanisms. *World J. Gastroenterol.* 24: 1679.

Hillwig, M.L., Fuhrman, H.A., Ittiamornkul, K., Sevco, T.J., Kwak, D.H. and Liu, X. 2014. Identification and characterization of a welwitindolinone alkaloid biosynthetic gene cluster in the stigonematalean cyanobacterium *Hapalosiphon welwitschii*. *Chem Bio Chem* 15: 665–669.

Hillwig, M.L., Zhu, Q. and Liu, X. 2014. Biosynthesis of ambiguine indole alkaloids in cyanobacterium *Fischerella ambigua*. *ACS Chem. Biol.* 9: 372–377.

Hirooka, S., Hirose, Y., Kanesaki, Y., Higuchi, S., Fujiwara, T., Onuma, R., Era, A., Ohbayashi, R., Uzuka, A. and Nozaki, H. 2017. Acidophilic green algal genome provides insights into adaptation to an acidic environment. *Proc. Natl. Acad. Sci. USA* 114: E8304–E8313.

Hoffmann, D., Hevel, J.M., Moore, R.E. and Moore, B.S. 2003. Sequence analysis and biochemical characterization of the nostopeptolide A biosynthetic gene cluster from Nostoc sp. GSV224. *Gene* 11: 171–180.

Hoffmann, L. 2005. Nomenclature of Cyanophyta/Cyanobacteria: round table on the unification of the nomenclature under the Botanical and Bacteriological Codes. *Algol. Stud.* 117: 13–29.

Huang, T.C. and Chow, J.J. 1988. Comparative studies of some nitrogen-fixing unicellular cyanobacteria isolated from rice fields. *J. Gen. Microbiol.* 134: 3089–3097.

Huang, T.C. and Chow, W.M. 1991. Setting the circadian nitrogen fixing rhythm of the prokaryotic *Synechoccus* sp. RF1, While its *nif* gene is repressed. *Plant Physiology* 96: 324–326.

Inger, L. 1970. Effect of two herbicides on nitrogen fixation by blue-green algae. *Sv. Bot. Tidskr.* 64: 460.

Ishida, K., Christiansen, G., Yoshida, W.Y., Kurmayer, R., Welker, M., Valls, N., Bonjoch, J., Hertweck, C., Börner, T. and Hemscheidt, T. 2007. Biosynthesis and structure of aeruginoside 126A and 126B, cyanobacterial peptide glycosides bearing a 2-carboxy-6-hydroxyoctahydroindole moiety. *Chem. Biol.* 14: 565–576.

Ishida, T., Watanabe., M.M., Sugiyama, J. and Yokota, A. 2001. Evidence for polyphyletic origin of the members of the orders of Oscillatoriales and Pleurocapsales as determined by 16S r DNA analysis. *FEMS Microbiol Lett.* 201: 79–82.

Ishihara, K., Watanabe, R., Uchida, H., Suzuki, T., Yamashita, M., Takenaka, H., Nazifi, E., Matsugo, S., Yamaba, M. and Sakamoto, T. 2017. Novel glycosylated mycosporine-like amino acid, 13-O-(β-galactosyl)-porphyra-334, from the edible cyanobacterium Nostoc sphaericum-protective activity on human keratinocytes from UV light. *J. Photochem. Photobiol. B Biol.* 172: 102–108.

Issa, A.A., Abd-Alla M.H. and Ohyama, T. 2014. Nitrogen fixing cyanobacteria: future prospect. Advances in Biology and Ecology of Nitrogen Fixation (ed. T. Ohyama), p. 23. In Tech, Rijeka, Croatia-European Union.

Issa, A.A., Abd-Alla, M.H. and Ohyama, T. 2014. Nitrogen fixing Cyanobacteria: future prospect. In: Advances in Biology and Ecology of Nitrogen Fixation, T. Ohyama (ed.) (Rijeka: In Tech).

Jahnke, L.L., Embaye, T., Hope, J., Turk, K.A., Van Zuilen, M., Des Marais, D.J., Farmer, J.D. and Summons, R.E. 2004. Lipid biomarker and carbon isotopic signatures for stromatolite-forming, microbial mat communities and *Phormidium* cultures from Yellowstone National Park. *Geobiology* 2: 31–47.

Jain, S., Prajapat, G., Abrar, M., Ledwani, L., Singh, A. and Agrawal, A. 2017. Cyanobacteria as efficient producers of mycosporine-like amino acids. *J. Basic Microbiol.* 57: 715–727.

Jancula, D. MiKula, P., MarsaleK, B., Rudoif, P., and Pochyly, F. 2014. Selective method for cyanobacterial bloom removal: hydraulic jet cavitation experience. Aquacult. Int. 22:509-521

Janssen, P.J., Morin, N., Mergeay, M., Leroy, B., Wattiez, R., Vallaeys, T., Waleron, K., Waleron, M., Wilmotte, A. and Quillardet, P. 2010. Genome sequence of the edible cyanobacterium *Arthrospira* sp. PCC 8005. *J. Bacteriol.* 192: 2465–2466.

Jensen, T.E. and Ayala, R.P. 1980. Microtubule-like inclusions in isolates of the blue-green bacteria *Anabaena* and *Nostoc*. *Cytolgia* 45: 315–326.

Jensen. T.E. and Ayala, R.P. 1976b. The fine structure of striated microtubules and sleeve bodies in several species of *Anabaena*. *J. Ultrastruct. Res.* 57: 185–193.

Jha, M.N., Jha, U.N., Ahmad, N. and Mallik, M.K. 1986. Cyanobacterial flora of Rice-field soils of Pusa and its adjoining areas. *Phykos* 25: 97–101.

Jimenez, J.I., Huber, U., Moore, R.E. and Patterson, G.M.L. 1999. Oxidized welwitindolinones from terrestrial *Fischerella* spp. *J. Nat. Prod.* 62: 569–572.

Kageyama, H. and Waditee-Sirisattha, R. 2018. Mycosporine-like amino acids as multifunctional secondary metabolites in cyanobacteria: From biochemical to application aspects. *Stud. Nat. Prod. Chem.* 59: 153–194.

Kallas, T., Coursin, T. and Rippka, R. 1985. Different organisation on *nif* genes in non-heterocystous and heterocystous cyanobacteria. *Plant. Mol. Biol.* 5: 321–329.

Kamat, N.D. 1974. Algae of Marathawada, Maharashtra. *Phykos* 13(1): 22–32.

Kamat, N.D. 1974. The Nostocales of Mysore State. *Phykos* 13(1): 33–37.

Kamat, N.D. and Harankhedkar, P.S. 1976. Bark algae of Nagpur, Maharashtra. Phykos 15(1 & 2): 53–57.

Kamat, N.D. and Patel, M.Z. 1973. Soil algae of rice-fields at different depths. *Botanique (India)* 4(2): 101–106.

Kandi, S., Godishala, V., Rao, P. and Ramana, K.V. 2015. Biomedical significance of terpenes: An insight. *Biomedicine* 3: 8–10.

Kantz, T. and Bold, H.C. 1969. Phycological studies. IX. Morphological and Taxonomic Investigations of *Nostoc* and *Anabaena* in culture. *Univ. Texas Publ.* 6924: 67.

Karatay, S.E., Dönmez, G. and Aksu, Z. 2017. Effective biosorption of phenol by the thermophilic cyanobacterium *Phormidium* sp. *Water Sci. Technol.* 76: 3190–3194.

Karmákek, J. and Kaŝtovský, J. 2003. Coincidences of structural and molecular characters in evolutionary lines of cyanobacteria. *Archi. Hydrobiol. Algol. Stud.* (Cyanobt. Res. 4) 109: 305–325.

Kaushik, B.D. 2014. Developments in Cyanobacterial Biofertiizers: Proc. Indian Natn. Sci. Acad., 80:379-388

Keeling, C.I. and Bohlmann, J. 2012. Plant terpenoids. In: N. Civjan (eds.), *Natural Products in Chemical Biology*. 1st Edition, New York, NY: Wiley), pp. 121–142.

Kehr, J.-C., Picchi, D.G. and Dittmann, E. 2011. Natural product biosyntheses in cyanobacteria: A treasure trove of unique enzymes. *Beilstein J. Org. Chem.* 7: 1622–1635.

Kenyon C.N., Rippka, R. and Stanier, R.Y. 1972. Fatty acid composition and physiological properties of some filamentous blue-green algae. *Archive of Microbiology* 83: 216–236. DOI: http://dx.doi.org/10.1007/bf00645123

Khamees, H.S., Gallon, J.R. and Chaplin, A.E. 1987. The pattern of acetylene reduction by cyanobacteria grown under alternating light and darkness. *British Phycol. J.* 22: 55–60.

Khan, M. 1985. Algal flora of Sultanpur (U.P.), India. *Phykos* 24: 52–57.

Khomutovska, N., de Los Ríos, A. and Jasser, I. 2021. Diversity and Colonization Strategies of Endolithic Cyanobacteria in the Cold Mountain Desert of Pamir. *Microorganisms* 9: 6.

Kim, H., Lantvit, D., Hwang, C.H., Kroll, D.J., Swanson, S.M., Franzblau, S.G. and Orjala, J. 2012. Indole alkaloids from two cultured cyanobacteria, *Westiellopsis* sp. and *Fischerella muscicola*. *Bioorganic Med. Chem.* 20: 5290–5295.

Kimura, M. 2000. Anaerobic microbiology in water logged rice fields. In: Bollag, J.M. & Stotzky, G. (eds.), *Soil Biochemistry*, Vol. 10. Marcel Dekker, New York, pp. 35–138.

Kirby, J., and Keasling, J.D. 2009. Biosynthesis of plant isoprenids: perspectives for microbial engineering. *Annu. Rev. Plant Biol.* 60: 335–355. doi: 10.1146/annurev.arplant. 043008.091955.

Kirk, G. 2004. *The Biogeochemistry of Submerged Soils*. Wiley, Chichester, 291 pp.

Kolte, S.O. and Goyal, S.K. 1985. Distributional pattern of Blue-green algae in rice-field soils of Vidarbha region of Maharashtra State, *Phykos* 24: 156–162.

Komarek,J.and Kastovsky,J.2003.Coincidences of structural and molecular characters in evolutionarylinesofcyanobacteria.Arch.Hydrobiol./Algol.Stud.(Cyanobact.Res.4)109:305-325

Komarek, J. 2006. Cyanobacterial taxonomy: current problems and prospects for the integration of traditional and molecular approaches. *Algae*. 21: 349–375.

Komárek, J., Kaŝtovský, J., Mareya, J. and Johansen, J.R. 2014. Taxonomic classification of cyanoprokaryotes (cyanobacterial genera) 2014, using a polyphasic approach. *Preslia*. 86: 295–335.

Kuiper. Goodman, T., Falconer, I and Fitzgerald, J. 1999. Human Health Aspects. In: Toxic Cyanobacteria in water. I. Chorus and J. Bartram (eds.) 1 Who and e and FN Spon, London:13-154

Kulasooriy, S.A., Lang, N.J. and Fay, P. 1972. The heterocysts of blue-green algae. III. Differentiation and nitrogenase activity. *Proc. R. Soc. Lond.* B181: 199–209.

Kulasooriya, S.A., Lang, N.J. and Fay P. 1972. The heterocysts of blue-green algae III differentiation and nitrogenase activity. *Proc. Royal Soc. London* B181: 199–209. DOI: http://dx.doi.org/10. 1098/rspb. 1972. 0046.

Kumar, H. 1970. Cyanophyceae of Sardhana, *Phykos* 9: 79–85.

Kumar, K., Mella-Hrrer, R.A. and Golden, J.W. 2009. Cyanobacterial heterocysts. *Cold Spring Harbor Perspectives of Biology* 2(4). A000315.

Kumazawa, S. and Mitsui, A. 1985. Comparative amperometric study of uptake-hydrogenese and hydrogenese photo production activities between heterocystous cyanobctrium *Anabaena cylindrica* B699 and non-heterocystous cyanobacterium *Oscillatoria* sp. *Strain Miami Bu7. Appl. Environmental Microbiol.* 50: 287–291.

Kusama, Y., Inoue, S., Jimbo, H., Takaichi, S., Sonoike, K., Hihara, Y. and Nishiyama, Y. 2015. Zeaxanthin and echinenone protect the repair of photosystem II from inhibition by singlet oxygen in *Synechocystis* sp. PCC 6803. *Plant Cell Physiol.* 56: 906–916.

Lane, D.J., Pace, B., Olsen, G.J., Stahl, D.A., Sogin, M.L. and Pace, N.R. 1985. Rapid determination of 16S ribosomal RNA sequences for phylogenetic analyses. Proceedings in National Academy of Science USA, 82: 6955–6959.

Lang, N.J. and Fay, P. 1971. The heterocysts of blue-green algae. II details of ultrastructure. *Proc. Royal Soc. London* B178: 193–203.

Lang, N.J. and Fay, P. 1971. The heterocysts of blue-green algae. II. Details of ultrastructure. *Proc. R. Soc. Lond.* B178: 193–203.

Lapage, S.P., Sneath, P.H.A., Lessel, E.F., Skermar, V.B.D., seeliger H.P.R. and Clark, W.A. 1975. Internation Code of nomenclature of bacteria. American Society for Microbiology, Washington, D.C.

Lefort, F., Calmin, G., Crovadore, J., Falquet, J., Hurni, J.-P., Osteras, M., Haldemann, F. and Farinelli, L. 2014. Whole-genome shotgun sequence of *Arthrospira platensis* strain Paraca, a cultivated and edible cyanobacterium. *Genome Announc.* 2: e00751-14.

Leikoski, N., Fewer, D.P., Jokela, J., Wahlsten, M., Rouhiainen, L. and Sivonen, K. 2010. Highly diverse cyanobactins in strains of the genus *Anabaena*. *Appl. Environ. Microbiol.* 76: 701–709.

Lewin, R.A. 1976. Prochlorophyta as a proposed new division of algae. *Nature* 261: 697–698.

Li, B., Sher, D., Kelly, L., Shi, Y., Huang, K., Knerr, P.J., Joewono, I., Rusch, D., Chisholm, S.W. and Van Der Donk, W.A. 2010. Catalytic promiscuity in the biosynthesis of cyclic peptide secondary metabolites in planktonic marine cyanobacteria. *Proc. Natl. Acad. Sci. USA* 107: 10430–10435.

Little, M.G. 1973. The zonation of marine supra-littoral blue-green alage. *Br. Phycol. J.* 8: 47–50.

Ljubic, A., Jacobsen, C., Holdt, S.L. and Jakobsen, J. 2020. Microalgae *Nannochloropsis oceanica* as a future new natural source of vitamin D3. *Food Chem.* 320, 126627.

Lopes, G., Clarinha, D. and Vasconcelos, V. 2020. Carotenoids from cyanobacteria: A biotechnological approach for the topical treatment of psoriasis. *Microorganisms* 8: 302.

Lundgren P., Soderback, E., Singer, A. Carpenter, E.J., and Berman, B: Katognymene: characterization of a novel marine diazotroph. J Phycol.2011,37:1052-16062

Maeda, K., Tamura, J., Okuda, Y., Narikawa, R., Midorikawa, T. and Ikeuchi, M. 2018. Genetic identification of factors for extracellular cellulose accumulation in the thermophilic cyanobacterium Thermo *Synechococcus vulcanus*: Proposal of a novel tripartite secretion system. *Mol. Microbiol.* 109: 121–134.

Magarvey, N.A., Beck, Z.Q., Golakoti, T., Ding, Y., Huber, U., Hemscheidt, T.K., Abelson, D., Moore, R.E. and Sherman, D.H. 2006. Biosynthetic characterization and chemoenzymatic assembly of the cryptophycins. Potent anticancer agents from *Nostoc* cyanobionts. *ACS Chem. Biol.* 1: 766–779.

Mahajan, S.K. 1987. Algal flora of Pawagiri Oon, Madhya Pradesh. *Phykos* 26: 61–62.

Maity, H. and Santra, S.C. 1985. Blue-green algal flora of 24-Parganas, West Bengal (India). *Phykos* 24: 46–51

Majid, M. and Khan T.A. 2014. Future of biofertilizers in Indian agriculture: an overview. *Int J. Agric. Food Res.* 3(3): 10–23. [Google Scholar]

Malathi, T. Babu, M., Mounika,t., Snehalatha, D. and Rao, B. 2014. Screening of cyanobacterial strains for antibacterial activity. PhyKos. 44(2):6-11.

Mandal, B., Vlek, P.L.G. and Mandal, L.N. 1999. Beneficial effect of blue-green algae and *Azolla* excluding supplying nitrogen, on wetland rice fields: a review. *Biol Fertil. Of Soils* 28: 329–342.

Marcozzi, C., Cumino, A.C. and Salerno, G.L. 2009. Role of NtcA, a cyanobacterial global nitrogen regulator in the regulation of sucrose metabolism gene expression in *Anabaena* sp. PCC7120. *Archives of Microbiology*. 191: 255–263. DOI: http://dx.doi.org/10.1007/s00203-008. 4050-y

Mareš, J., Hájek, J., Urajová, P., Kopecký, J. and Hrouzek, P. 2014. A hybrid non-ribosomal peptide/polyketide synthetase containing fatty-acyl ligase (FAAL) synthesizes the β-amino fatty acid lipopeptides puwainaphycins in the Cyanobacterium Cylindrospermum alatosporum. *PLoS ONE* 9: e111904.

Martens, G. Von. 1870a. A third list of Bengal Algae determined by G.V. Martens. *Proc. Asiatic Soc. Bengal* 39: 9–12.

Martens, G. Von. 1870b. A fourth list of Bengal Algae (determined by G.V. Martens, communicated by S. Kurz.). *Proc. Asiatic Soc. Bengal* 39: 257–260.

Martens, G. Von. 1871a. A fifth list of Bengal Algae (determined by G.V. Martens, communicated by S. Kurz). *Proc. Asiatic Soc. Bengal* 40: 170–173.

Martin-Figueroa, E., Navarro, F. and Florencio, F.J. 2000. The GS-GOGAT pathway is not operative in the heterocysts cloning and expression of gl SF gene from the cyanobacterium *Anabaena* sp. PCC 7172. *FEBS Letters* 476: 282-286.

Maryan, P.S., Eaday, R.R., Chaplin, A.E. and Gallon, J.R. 1986. Nitrogen fixation by the unicellular cyanobacterium *Gloeothece*. Nitrogenase synthesis is only transiently repressed by oxygen. *FEMS Microbiol. Lett.* 34: 251–255.

Majid, M. and Khan, T.A. 2014. Future of biofertilizers in Indian agriculture: an overview. *Int. J. Agric. and Food Res.* 3(3): 10–23.

Meffert, M.E., Oberhauser, R. and OverbecK, J. 1981. Morphology and taxonomy of Oscillatoria rederkei (Cyanophyta). Br. Phycol. J. 16:107-114

Meeks, J.C. and Elahi, J. 2002. Regulation of cellular differentiation in filamentous cyanobacteria in free-living and plant-associated symbiotic growth states. *Microbiology and Molecular Biology Reviews* 66: 94–121.

Micallef, M.L., Sharma, D., Bunn, B.M., Gerwick, L., Viswanathan, R. and Moffitt, M.C. 2014. Comparative analysis of hapalindole, ambiguine and welwitindolinone gene clusters and reconstitution of indole-isonitrile biosynthesis from cyanobacteria. *BMC Microbiol.* 14: 1–18.

Mihali, T.K., Kellmann, R., Muenchhoff, J., Barrow, K.D. and Neilan, B.A. 2008. Characterization of the gene cluster responsible for cylindrospermopsin biosynthesis. *Appl. Environ. Microbiol.* 74: 716–722.

Mishra, G.P. and Purohit, J. 1979. A preliminary list of soil algae from Sagar. *Phykos* 18(1 & 2): 63–67.

Mishra, S. and Kaushik, B.D. 1989. Growth promoting substances of Cyanobacteria. I: Vitamins and their influence on rice plant. *Proc. Ind. Sci. Acad. B* 55: 295–300.

Mishra, U. and Pabbi, S. 2004. Cyanobacteria: A Potential Biofertilizer for Rice. *Resonance*. DOI: 10.1007/BF02839213

Misra, U. and Pabbi, S. 2004. Cyanobacteria: a potential biofertilizer for rice. *Resonance*. 9(6): 6–10. DOI: 10.1007/BFO 2839213. [Google Scholar]

Mitra, A.K. 1951. The algal flora of certain Indian soils. *Indian J. agric. Sci.* 21: 357 –373.

Mitsui, A.,Kumazawa, Takahashi, A., Ikemota, H, Cao, S. and Arai, T. 1986. Strategy by which nitrogen-fixing unicellular cyanobacteria grow photo autotrophically. Nature.323:720-722.

Mo, S., Krunic, A., Pegan, S.D., Franzblau, S.G. and Orjala, J. 2009. An antimicrobial guanidine-bearing sesterterpene from the cultured cyanobacterium *Scytonema* sp. *J. Nat. Prod. 72:* 2043–2045.

Mo, S., Krunic, A., Santarsiero, B.D., Franzblau, S.G. and Orjala, J. 2010. Hapalindole-related alkaloids from the cultured cyanobacterium *Fischerella ambigua*. *Phytochemistry* 71: 2116–2123.

Moffitt, M.C. and Neilan, B.A. 2004. Characterization of the nodularin synthetase gene cluster and proposed theory of the evolution of cyanobacterial hepatotoxins. *Appl. Environ. Microbiol.* 70: 6353–6362.

Mohanty, R.C. 1982. Blue-green algae of Bhubaneswar and its adjoining regions-II. *Phykos* 21: 96–98.

Mohanty, R.C. 1984. Blue-green algae of Bhubaneswar and its adjoining regions - III. *Phykos* 23(1 & 2): 90–95.

Monteiro e Silva, S.A., Calixto, G.M.F., Cajado, J., De Carvalho, P.C.A., Rodero, C.F., Chorilli, M. and Leonardi, G.R. 2017. Gallic acid-loaded gel formulation combats skin oxidative stress: Development, characterization and ex vivo biological assays. *Polymers* 9: 391.

Moore, R.E., Cheuk, C. and Patterson, G.M.L. 1984. Hapalindoles: New alkaloids from the blue-green alga *Hapalosiphon fontinalis. J. Am. Chem. Soc.* 106: 6456–6457.

Moore, R.E., Cheuk, C., Yang, X.Q.G., Patterson, G.M.L., Bonjouklian, R., Smitka, T.A., Mynderse, J.S., Foster, R.S., Jones, N.D. and Swartzendruber, J.K. 1987. Hapalindoles, antibacterial and antimycotic alkaloids from the cyanophyte *Hapalosiphon fontinalis. J. Org. Chem.* 52: 1036–1043.

Mueller, D.R., Vincent, W.F., Bonilla, S. and Laurion, I. 2005. Extremotrophs, extremophiles and broadband pigmentation strategies in a high arctic ice shelf ecosystem. *FEMS Microbiol. Ecol.* 53: 73–87.

Mukhopadhyay, A. and Chatterjee, P. 1981 A check list of blue-green algae from the paddy fields of 24-Parganas and Howrah districts of West Bengal - I. *Phykos* 20(1 & 2): 81–84.

Mukhopadhyay, S.T. and Naskar, N. 2015. Diversity of cyanobacteria from Barasat, Kolkata, West Bengal, India. *IJIRR:* 107–113.

Mukund, S. and Sivasubramanian, V. 2014. Anticane activity of Oscillatoria terebriformis cyanobacteria in human lung cancer cell lime A: 549.Int. J. Apple. Biol. Pharma. 5:2

Mulvihill, E.E., Assini, J.M., Sutherland, B.G., DiMattia, A.S., Khami, M., Koppes, J.B., Sawyez, C.G., Whitman, S.C. and Huff, M.W. 2010. Naringenin decreases progression of atherosclerosis by improving dyslipidemia in high-fat–fed low-density lipoprotein receptor–null mice. *Atertio. Thromb. Vasc. Biol.* 30: 742–748.

Murray, R.G.E. 1986. Microbial structure as an aid to microbial classification and taxonomy. Spisy Pri doved.Falk.Univ. J.E. PurKyne Brne.43:245-252.

Murry, M.A. and Wolk, C.P. 1989. Evidence that the barrier to the penetration of oxygen into the heterocysts depends upon two layers of the envelope. *Archives of Microbiology* 151: 469–474. DOI: http://dx.doi.org/10.1038/scientific american1203-52

Murry, M.A., Horne, A.J. and Bennemann, J.R. 1984. Physiological studies of oxygen protective mechanisms in the heterocysts of Anabaena cylindrical. *Applied Environmental Microbiology* 47: 449–454.

Nadeau, T.L. and Castenholz, R.W. 2000. Characterization of psychrophilic oscillatorians (cyanobacteria) from Antarctic meltwater ponds. *J. Phycol.* 36: 914–923.

Nägeli, C. 1849. *Gattungen Einzelliger Algen: Physiologisch und Systematisch Bearbeitet*, Friedrich Schulthess: Zürich, Switzerland.

Nandagopal, P., Steven, A.N., Chan, L.-W., Rahmat, Z., Jamuluddin, H. and Noa, N.I.M. 2021. Bioactive metabolites produced by cyanobacteria for growth adaptation and their pharmacological properties. Biology 10, 1061.https://doi.org/10.3390/biology10101061.

Nandkar, P.B., Marathe, K.V. and Motikhaye, B.G. 1983. Algae in sewage containing pools and oxidation ponds. *Phykos* 22: 37–45.

Nasar, S.A.K. and Datta Munshi, J. 1976. On the algal flora of some ponds of Bhagalpur, India. *Phykos* 15(1 & 2): 49–52.

Naskar, N., Naskar, K.R. and Sen, C.R. 2008. Brackish water Oscillatoriaceae from North 24-Pargans, West Bengal, India. *Bangladesh J. Plant Taxon.* 15(1): 31–38.

Naskar, N.M., Naskar, K.R. and Sen, C.R. 2006. A systematic account of Chroococcales (Myxophyceae) from brackish water wetlands of North 24 Parganas District, West Bengal. *Environ. Ecol.* 24S(3): 655–657.

Nayak, S. and Prasanna, R. 2007. Soil pH and its role in Cyanobacterial Abundance and Diversity in Rice Field Soils. *App. Ecol. Environ. Res.* 5: 103–113.

Nowruzi, B., Khavari-Nejad, R.A. Nejad sattari, T. Sivonen, K., Fewer, D2017 A Proposal for the unification of tow cyanobacterial strains of Nostoc as the same species. Rostaniha.17(2):161-172

Nubel, U., Gracia-Pichel, F. and Muyzer, G. 1997. PCR primers to amplify 16S r RNA genes from cyanobacteria. *Applied and Environmental Microbiology* 63(8): 3327–3332.

Nikitina, K.A., Yudina, T.G. and Gusev, M.V. 1979. Morphological heterogeneity of the cyanobacterium anabaena variabilis under different conditions of its growth and destruction. Microbial. 48(5):783-879.

Nishizawa, T., Asayama, M., Fujii, K., Harada, K.-I. and Shirai, M. 1999. Genetic analysis of the peptide synthetase genes for a cyclic heptapeptide microcystin in Microcystis spp. *J. Biochem.* 126: 520–529.

Nishizawa, T., Ueda, A., Asayama, M., Fujii, K., Harada, K.-I., Ochi, K. and Shirai, M. 2000. Polyketide synthase gene coupled to the peptide synthetase module involved in the biosynthesis of the cyclic heptapeptide microcystin. *J. Biochem.* 127: 779–789.

Ogino, J., Moore, R.E., Patterson, G.M.L. and Smith, C.D. 1996. Dendroamides, new cyclic hexapeptides from a blue-green alga. Multidrug-resistance reversing activity of dendroamide A. *J. Nat. Prod.* 59: 581–586.

Okino, T., Matsuda, H., Murakami, M. and Yamaguchi, K. 1995. New microviridins, elastase inhibitors from the blue-green alga *Microcystis aeruginosa*. *Tetrahedron* 51: 10679–10686.

Okuda, A. and Yamaguchi, M. 1952. Algae and atmospheric nitrogen fixation in paddy fields. II. Relation between the growth of blue-green algae and physical or chemical properties of soil and effect of soil treatments and inoculation on the nitrogen fixation. *Mem. Res. Inst. Food Sci.* 4: 1–11.

Oren, A and Ventura, S. 2017. The current status of cyanobacterial nomenclature under the "prokaryotic" and the "botanical" code. Antoni-C van Leeuwenhock. 110: 1257-1269. Springer International Publishing, Switzerland.

Oren, A. and Tindall, B.J. 2005. Nomenclature of the Cyanophyta/Cyanobacteria/ Cyanoprokaryotes under the International code of Nomenclature of Prokaryotes. *Algological Studies* /Archiv fiir Hydrobiologie,117: 39-52.

Oren, A. and Tindall, B.J. 2005. Nomenclature of the cyanophyta/cyanobacteria/ cyanoprokaryaotes under the International Code of Nomenclature of Prokaryotes. Algological studies/Archive for Hydrobiologie, Supplement Volumes No. 117. p. 39–52.

Otsuka, S., Suda, S., Shibata, S., Oyaizu, H., Matsumoto, S. and Watanabe, M.M. 2001. A proposal for the unification of five species of the cyanobacterial genus *Microcystis* Kützing ex Lemmerman 1907 under the rules of the Bacteriological Code. *Int. J. Syst. Evol. Microbiol.* 51: 873–879.

Pabbi, S. 2008. *Cyanobacterial biofertilizers. J. Ecofriendly Agric.* 3(2): 95–111.

Pabbi, S. 2015. Blue green algae: a potential biofertilizer for rice. In: Sahoo D.B., Seekbach, J. editors. *The Algae World*. New Delhi, Springer India Pvt. Ltd., 2015. Pp. 449–466 [Google Scholar]

Pabbi, S. 2015. Blue-green algae: a potential biofertilizer for rice. In: Saho, D.B., Seckbach, J. (eds.), *The algae World*. Springer, New Delhi, pp. 449–466.

Pace, N.L. 1997. A molecular view of microbial diversity and the biosphere. *Science*. 276: 734–740.

Padan, E. and Cohen, Y. 1982. Anoxygenic photosynthesis. In: The biology of Cyanobacteria. N.G. Carr and B.A. Whitton (eds.). Blackwell Sceientific Publications, Ltd., Oxford. pp. 215–238.

Pal, S. and Yadav, A.K. 1974. Some Cyanophyceae from Saharanpur district - A taxonomic enumeration. *Phykos* 13(1): 38–47.

Pal, T.K. and Santra, S.C. 1984. New additions to the algal flora of Murshidabad, West Bengal. *Phykos* 23(1 & 2): 139–141.

Pal, U.C. and Santra, S.C. 1985. Algal flora of Midnapore, West Bengal (India): 1. Cyanophyceae. *Phykos* 24: 12–17.

Pandey, D.C. 1965. A study of the algae from paddy soils of Ballia and Ghazipur districts of Uttar Pradesh, India. Cultural and ecological considerations. *Nova Hedwigia* 9: 299–334.

Pandey, D.C. 1965. A study on the algae from paddy field soils of Ballia and Ghazipur District of U.P., India. *Nova Hedwigia* 9: 299–334.

Pandey, U.C. 1982. Additions to the algal flora of Rohilkhand Division VI. Cyanophyceae. *Phykos* 21: 137–140.

Pandey, U.C. and Pandey, D.C. 1982. Additions to the algal flora of Allahabad VIII Cyanophyceae. *Phykos* 21: 76–79.

Pandhol, R.K. and Grover, I.S. 1976. Algal flora of Ludhiana and its adjacent areas. *Phykos* 15(1 & 2): 81–87.

Pankratz, H.S. and Bowen, C.C. 1963. Cytology of blue-green algae. I. The cells of *Symploca muscorum*, *Am. J. Bot.* 50: 387–399.

Papendorf, O., Konig, G.M., Wright, A.D. 1998. Hierridin b and 2,4-dirnethoxy-6-hepatadecylphecyphenol, secondary metabolites from the cyanobacterium Phormidium ecto earpi with antiplasmodial activity. Phytochemistry 49:2383-2386.

Parameswaran. 1993. *Light responses and physiological characterization of cyanobacteria (Blue green algae) from rice fields.* Ph.D. thesis, University of Madras.

Park, A., Moore, R.E. and Patterson, G.M.L. 1992. Fischerindole L, a new isonitrile from the terrestrial blue-green alga *Fischerella muscicola*. *Tetrahedron Lett.* 33: 3257–3260.

Park, H.-J., Cho, J.-H., Hong, S.-H., Kim, D.-H., Jung, H.-Y., Kang, I.-K. and Cho, Y.-J. 2018. Whitening and anti-wrinkle activities of ferulic acid isolated from *Tetragonia tetragonioides* in B16F10 melanoma and CCD-986sk fibroblast cells. *J. Nat. Med.* 72: 127–135.

Parukutty, P.R. 1939. Collection of Algae from Assam. *Proc. Indian Acad. Sci.* B. 9(5): 229–235.

Parukutty, P.R. 1940. The Myxophyceae of the Travancore State, India. *Proc. Indian Acad. Sci.* B. 11(3): 117–124.

Pathak, J., Ahmed, H., Singh, S.P., Häder, D.-P. and Sinha, R.P. 2019. Genetic regulation of scytonemin and mycosporine-like amino acids (MAAs) biosynthesis in cyanobacteria. *Plant Gene* 17: 100172.

Patil, P.L. & Satav, S.D. 1986. A study of nitrogen fixing blue- green algae from rice fields of western Maharashtra. *Phykos* 25: 113–116.

Patipong, T., Hibino, T., Waditee-Sirisattha, R. and Kageyama, H. 2019. Induction of antioxidative activity and antioxidant molecules in the halotolerant cyanobacterium *Halothece* sp. PCC 7418 by temperature shift. *Nat. Prod. Commun. 14:* 1934578X19865680.

Pattanaik, B. and Lindberg, P. 2015. Terpenoids and their biosynthesis in cyanobacteria. *Life* 5: 269–293. doi: 10.3390/life5010269

Pearl, H.W., Prufert, L.E. and Ambrose, W.W. 1991. Contemporaneous nitrogen fixation and oxygenic photosynthesis in the non-heterocystous rna forming cyanobacterium *Lynbgya aestuarii*. *App. Eviron. Microbiol.* 57: 3086–3092.

Pearson. H.W., Malin, G. and Howsley, R. 1981. Physiological studies on *in vitro* nitrogenase activity by axenic cultures of the blue-green alga *Microcoleus chthonoplastes.* Br. Phycol. J. 16: 139.

Pedersen, D. and Miller, S.R. 2017. Photosynthetic temperature adaptation during niche diversification of the thermophilic cyanobacterium *Synechococcus* A/B clade. *ISME J.* 11: 1053–1057.

Pereira, I., Ortega, R., Barrientos, L., Moya, M., Reyes, G. and Kramm, V. 2009. Development of a biofertilizer based on filamentous nitrogen-fixing cyanobacteria for rice crops in Chile. *J. Appl. Phycol.* 21(1): 135–144.

Perkerson, R.B., Johansen, J.R., KováčiK, L., Brand, J, Kaŝtovský, J. and Casamatta, D.A. 2011. A unique pseudanabaenalean. (Cyanobacteria) genus *Nodosilinea* gen. nov. based on morphological and molecular data. based on morphological and molecular data. *J. Phycol.* 47: 1397–1412.

Peschek, G.A., Villgrater, K. and Watsy, N.M. 1991. Respiratory protection of the nitrogenase in dinitrogen fixing cyanobacteria. *Plant and Soil.* 137: 17–24.

Pillay, A.R. and Tchan, Y.T. 1970. The use of soil alga for the bioassay of study of triallate (Herbicide). *Soil Biol.* 12: 20-22.

Portmann, C., Blom, J.F., Gademann, K. and Jüttner, F. 2008. Aerucyclamides A and B: Isolation and synthesis of toxic ribosomal heterocyclic peptides from the cyanobacterium *Microcystis aeruginosa* PCC 7806. *J. Nat. Prod.* 71: 1193–1196.

Portmann, C., Blom, J.F., Kaiser, M., Brun, R., Jüttner, F. and Gademann, K. 2008. Isolation of aerucyclamides C and D and structure revision of microcyclamide 7806A: Heterocyclic ribosomal peptides from *Microcystis aeruginosa* PCC 7806 and their antiparasite evaluation. *J. Nat. Prod.* 71: 1891–1896.

Prain, D. 1905. The Vegetation of the Districts Hooghly, Howrah and 24 Parganas. *Rec. bot. Surv. India* 3: 143–339.

Prasad, B.N. 1952. Some Nostocaceae from Uttar Pradesh. *J. Indian bot. Soc.* 31(4): 358–361.

Prasad, B.N. and Mehrotra, R.K. 1980. Blue-green algae of paddy fields of U.P. *Phykos* 19: 121–128.

Prasad, B.N. and Mehrotra, R.K. 1980. Blue-green algae of Paddy fields of Uttar Pradesh. *Phykos* 19(1): 121–128.

Prasad, B.N. and Srivastava, M.N. 1986. Some Cyanophyceae from Andaman and Nicobar Islands. *Phykos* 25: 88–96.

Prasad, B.N. and Srivastava, P.N. 1968. Systematic and Ecological studies on algae of alkaline (Usar) soils. *Phykos* 7: 102–111.

Prasanna, R., Nain, L., Tripathi, R., Gupta, V., Chaudhary, V., Middha, S. and Joshi, M. 2008. Evaluation of fungicidal activity of extracellular filtrates of cyanobacteria-possible role of hydrolytic enzymes. *J. basic Microbiology* 48(3): 186–194.

Prescott, G.W. 1969. *The algae: a review*. Thomas Nelson and Sons., London.

Pringsheim, E.G. 1967. Phycology in the field and in the laboratory. *J. Phycol.* 3: 93–95.

Ramaswamy, A.V., Sorrels, C.M. and Gerwick, W.H. 2007. Cloning and biochemical characterization of the hectochlorin biosynthetic gene cluster from the marine cyanobacterium *Lyngbya majuscula*. *J. Nat. Prod.* 70: 1977–1986.

Ramos, V., Castelo-Branco, R., Leao, P.N., Martins, J., Carvalhal-Gomes, S., Sobrinho da Silva, F., Mendonca Filho, J.G. and Vasconcelos, V.M. 2017. Cyanobacterial diversity in microbial mats from the hypersaline lagoon system of Araruama, Brazil: An in-depth polyphasic study. *Front. Microbiol.* 8: 1233.

Ramos, V., Morais. and Vasconcelos, V.M. 2017. A curated database of cyanobacterial strains relevant for modern taxonomy and phylogenetic studies. *Sci. Data* 4: 170054. 10.1038/sdata.2017.54

Randhawa, M.S. 1936. Occurrence and distribution of the freshwater algae of North India. *Proc. Indian Acad. Sci.* B. 4(1): 36–44.

Rao, C.B. 1936. The Myxophyceae of the United Provinces, India II. *Proc. Indian Acad. Sci.* B. 3(2): 165–174,

Rao, C.B. 1937a. A new species of *Anabaena* (*Anabaena ambigua* sp. nov.). *Proc. Indian Acad. Sci.* B. 5(3): 101–108.

Rao, C.B. 1937b. The Myxophyceae of the United Provinces, India III. Proc. Indian Acad. Sci. B. 6(6): 339–375.

Rao, C.B. 1938a. The Myxophyceae of the Orissa Province, India-I. *Proc. Indian Acad. Sci.* B. 8: 157–170.

Rao, C.B. 1938b. The Myxophyceae of the Madras Presidency, India-I. *J. Indian bot. Soc.* 17: 81–96.

Rao, C.S. 1939. The Myxophyceae of the Bihar Province, India-I. *Proc. Indian Acad. Sci.* B. 9(3): 142–150.

Rao, C.S. 1940. The Myxophyceae of the Delhi Province, India-I. *Proc. Indian Acad. Sci.* B. 11(3): 125–131.

Rath, B. and Priyadarshani, I. 2013. Antibacterial and antifungal activity of marine cyanobacteria from Odisha coast. Int. J. Curr. Trends. Res., 2:248-251.

Rao, D.L.N. and Burns, R.G. 1990. The effect of surface growth of blue-green algae and bryophytes on some microbiological biochemical and physical soil properties. *Biol Fertil. Soils* 9: 239–244.

Reddy, P.M., Yumnam, D.D. and Imchen, T.Y. 1986. Investigations on the blue-green algae on North-East India's Distribution and habitat preferences. *Phykos* 25: 148–158.

Řeháková, K., Johansen, J.R., Casamatta, D.A., Xueson, L. and Vincent, J. 2007. Morphological and molecular characterization of selected desert oil cyanobacteria: three species new to science including *Mojavia pulchra* gen. et sp. nov. *Phycologia* 46: 481–502.

Reinl, K.L., Sterner, R.W., Lafrancois, B.M. and Brovold, S. 2020. Fluvial seeding of cyanobacterial blooms in oligotrophic Lake Superior. *Harmful Algae* 100: 101941.

ResenchuK, N.V. and Mikhailo, A.A. 1980. Morphological variability of trichomes of the clone culture Spirulina platensis strain 603K under various conditions fo cultivation. Bot. 2h. (Leningrad) 65 (6): 805-815. (in Russian)

Reynaud, P.A. and Franche, C. 1986. Isolation and characterization of non-heterocystous tropical cyanobacteria growing on nitrogen free medium. MIRCE N. J., 2:427-443.

Richter, J.M., Ishihara, Y., Masuda, T., Whitefield, B.W., Llamas, T., Pohjakallio, A. and Baran, P.S. 2008. Enantiospecific total synthesis of the hapalindoles, fischerindoles, and welwitindolinones via a redox economic approach. *J. Am. Chem. Soc.* 130: 17938–17954.

Rippka R., Nelson, A., Kunisawa, R & Cohen-Bazire, G. 1971. Nitrogen fixation by unicellular blue-green algae. *Archives of Microbiology* 76: 341–348. DOI: http://dx.doi.org/10.1007/fb00408530

Rippka, R. 1988. Recognition and identification of Cyanobacteria, p. 28–67. In: L. Packer and A.N. Glazer (eds.), *Methods in Enzvmology*. Vol. 167. Academic Press Inc., New York.

Rippka, R., Deruelles, J., Waterbury, J.B., Herdman M. and Stanier, R. 1979. Generic assignments, strain histories and properties of pure cultures of cyanobacteria. *J. Gen. Microbiol.* 111: 1–61.

Rippka, R., Waterbury, J.B. and Chohen-Bazire, G. 1974. A cyanobacterium which lacks thylakoids. *Arch. Microbiol.* 100: 419–436.

Roger, P.A. 1985. Unicellular mucilaginous blue-green algae (BGA) impressive blooms but deceptive biofertilizers. *International Rice Research News.* 10: 27–28.

Romans, K.M., Carpenter, E.J. and Bergman, B. 1994. Buoyancy regulation in the colonial diazotrophic cyanobaterium *Trichodesmium tenue*: ultrastructure and storage of carbohydrates, polyphosphate and nitrogen. *J. Phycol.* 30: 935–942.

Rott, E. 2016.(ed) 20th Cyanophyta/Cyanobacteria Research Symposium. Program & Abstracts Volume. Institute of Botany, University of Innsbruck, 78 pp. ISBN 978-3-9500090-6-4.

Rouhiainen, L., Jokela, J., Fewer, D.P., Urmann, M. and Sivonen, K. 2010. Two alternative starter modules for the non-ribosomal biosynthesis of specific anabaenopeptin variants in Anabaena (Cyanobacteria). *Chem. Biol.* 17: 265–273.

Rudi, K., Skulberg, O.M. and Jakobsen, K.S. 1998. Evolution of Cyanobacteria by Exchange of Genetic Material among Phyletically Related Strains. *J. Bacteriol.* 180(13): 3453–3461 doi: 10.1128/jb.180.13.3453-3461.1998.

Saha, L.C. and Pandit, B. 1987. Pond and riverine algae of Bhagalpur. *Phykos* 26: 152–158.

Saito, T. and Hattori, A. 1982. Aerobic nitrogen fixation by the marine non-heterocystous cyanobacterium *Trichodesmium* (*Oscillatoria*) sp. its protective mechanism against oxygen. *Mar. Biol.* 70: 251–254.

Santra, S.C. 1987. Airborne algae of Kalcutta metropolis. *Phykos* 26: 71–74.

Santra, S.C. 1993. Biology of rice fields blue-green algae. Daya Publishing House, Delhi.

Santra, S.C., Pal, U.C., Maity, H. and Bandyopadhyaya, G. 1988. Blue-green algae in saline habitats of West Bengal: A systematic account. *Biol. Mem.* 14(1): 81–108.

Sardeshpande, J.S. and Goyal, S.K. 1981. Distributional pattern of BGA in rice field soils of Konkan region of Maharashtra. *Phykos* 20: 102–106.

Sardeshpande, J.S. and Goyal, S.K. 1981. Effect of pH on growth and nitrogen fixation by Blue-green algae. *Phykos* 20(1 & 2): 107–113.

Sarma, T.A. and Kanta, S. 1978. Algal flora of Patiala and its environs I. *Phykos* 17(1 & 2): 105–111.

Sarma, T.A., Kanta, S. and Sunita. 1979. Algal flora of Patiala and its environs - Cyanophyceae II. *Phykos* 18(1 & 2): 13–19.

Saxena, P.N., Tewari, A. and Khan, M.A. 1970. Studies on the seasonal variation of Algal flora of sewage. *Phykos* 9: 57–62.

Schirrmeister, B.E., Anisimova, M., Antonelli, A. and Bagheri, C.H. 2011. Evolution of cyanobacterial morphotypes: taxa required for improved phylogenomic approaches. *Communicative and Integrative Biology* 4: 424–427.

Schmidle, W. 1900. Ueber einige von Professor Hansgirg in Ostin-diensesairmette Susswasserai p. en. *Hedwigia* 22: 160–190. (not seen in original)

Schmidt, E.W., Nelson, J.T., Rasko, D.A., Sudek, S., Eisen, J.A., Haygood, M.G. and Ravel, J. 2005. Patellamide A and C biosynthesis by a microcin-like pathway in Prochloron didemni, the cyanobacterial symbiont of *Lissoclinum patella*. *Proc. Natl. Acad. Sci. USA* 102: 7315–7320.

Schopf, J.W. 1974a. The development and diversification of the Precambrian life. *Origin of life* 5: 119–135

Schopf, J.W. 1974b. Paleobiology of the Precambrian, the age of the bluer-green algae. *Evol. Biol.* 7: 1–43.

Schopf, J.W. 1993. Microfossils of the early Archean Apex period: new evidence of the antiquity of life. *Science* 260: 640–646.

Schopf, J.W. 1996. Cyanobacteria: pioneers of the early Earth. *Nova Hedwigia* 112: 13–32.

Schopf, J.W. 2006. Fossil evidence of Archaean life. *Phil. Trans. R. Soc.* B 361: 869–885.

Schwabe, G.H. 1962 blaual genprobleme. Schweitz. Z. Hydrol. 24:207-222

Seki, H., Ozawa, H. and Ichimura, S.1981. Temperature dependence of filament length of Anabaena spiroides var. crassa. Hydrabiol. 83 (13):419-424.

Semary, N.A.E. and Fouda, M. 2015. Anticancer activity of Cyan three sp. strain extracts from Egypt: First record. Asian Pacific Journal of Tropical Biomedicine. Vol 5 Issue 12: 992-995.

Sen, C.R. 1987. Observations on Blue-green algae of paddy field soil of Gangetic West Bengal I, Burdwan district. *Indian Agric.* 31(3): 221–225.

Sen, C.R. 2001. Association and succession of blue-green algae at different growth stages of rice crop. *Phykos* 40(1&2): 121–128.

Sen, C.R. 2005. The genus *Phormidium* Kutz. from lower Gangetic Plains of West Bengal. In: Keshri, J.P. and Kargupta, A.N. (eds.), *Glimpses of Indian Phycology*, pp. 205–209. Bishen Singh Mahendra Pal Singh, Dehra Dun.

Sen, C.R. 2006. Some Cyanophyceae from Greater Kalcutta I: A taxonomic enumeration. J. Econ. Taxon. Bot. 30(1): 885–890.

Sen, C.R. and Gupta, D. 1998. The genus *Oscillatoria* Vaucher from Lower Gangetic Plains of West Bengal. *Phykos* 37(1 & 2): 89–93.

Sen, C.R. and Gupta, D. 1998. The genus Oscillatoria Voucher from Lower Gangetic Plains of West Bengal. *Phykos* 37(1&2): 89–93.

Sen, Chittaranjan & Gupta, D. 1987. Chroococcales Wettstein (Myxophyceae) from Gangetic delta of West Bengal: A taxonomic enumeration, *J. Bengal Natural History Society*. New Series 6(1): 66–69.

Sen, Chittaranjan & Gupta, D. 1987. The genus *Oscillatoria* Voucher from greater Kalcutta. *Bull. Bot. Soc. Bengal* 41: 41–45.

Shalaby, E.A., Shanab., S.M.M., Singh, V. 2010. Salt stress enhancement of antioxidant and antiviral efficiency of Spirulina platensis. J. Med Plants Rs. 4:2622-2632.

Sharma, A. 2010. Cyanobacterial flora from freshwater habitat of Bokaro Thermal Power Station (Dist. Bokaro), Jharkhand, India. *Int. J. Pl. Sci.* 5(2): 679–681.

Shih, P.M., Wu, D., Latifi, A., Axen, S.D., Fewer, D.P., Talla, E., Calteau, A., Cai, F., De Marsac, N.T. and Rippka, R. 2013. Improving the coverage of the cyanobacterial phylum using diversity-driven genome sequencing. *Proc. Natl. Acad. Sci. USA* 110: 1053–1058.

Shirani, M.K. and Gamage, W. 2020. The 16S r RNA gene and cyanobacterial taxonomy; current problems and future prospects. *Journal of the University of Ruhuna* 8(2): 60. D01:10.4038/jur.V8i2.7945.

Shishido, T.K. Humisto, A. JoKela, J., Liu, L., Wahlsten, M., TamraKar, A., Fewer, D.P. Permi, P., Andreote, A.P.D., Fiore, M.F. and Sivonen, K. 2015. Antifungal compounds from cyanobacteria. Mar. Drugs. 13: 2124-2140.

Siegesmund, M., Johansen, J.R., Karsten, U. and Friedl, T. 2008. *Coleofasciculus* gen. nov. (Cyanobocteria): morphologicl ad molecular criteria for revision of the genus *Microcoleus* Gomont. *J. Phycol.* 44: 1572–1585.

Singh, S. 2014. A review on possible elicitor molecules of cyanobacteria: their role in improving plant growth and providing tolerance against biotic or abiotic stress. *J. Appl. Microbiol.* 117(5): 1221–1244.

Singh, B., Kumar, A. and Malik, A.K. 2017. Flavonoids biosynthesis in plants and its further analysis by capillary electrophoresis. *Electrophoresis 38:* 820–832.

Singh, D.P., Khattar, J.I.S., Kaur, G., Gupta, M., Singh, Y. and Gulati, A. 2015. Effect of pretilachior on nitrogen uptake and assimilation by the cyanobacterium *Desmonostoc muscorum* PUPCCC 405.10. *Acta Physiolgiae Plantarum* 37(9): 1–14.

Singh, D.P., Prabha, R., Meena, K.K., Sharma, L. and Sharma, A.K. 2014. Induced accumulation of polyphenolics and flavonoids in cyanobacteria under salt stress protects organisms through enhanced antioxidant activity. *Am. J. Plant Sci.* 2014: 43916.

Singh, J.S. 2015b. Plant-microbe interactions: a viable tool for agricultural sustainability. *Appl. Soil Ecol.* 92: 45–46.

Singh, K.P. and Chaturvedi, U.K. 1970. Myxophyceae of the Rohilkhand division, U.P., India-II. *Phykos* 9: 36–40.

Singh, P.K. 1973. Effect of pesticides on blue-green algae. *Arch. Microbiol.* 89: 317 –320.

Singh, P.K. 1975. Fertilizers tolerance of blue-green algae and their effect on heterocyst differentiation. *Phykos* 14: 81–88.

Singh, P.K. 1981a. Potentiality of blue-green algae and *Azolla* biofertilizers in rice cultivation in India. In: *Current Perspective in Nitrogen Fixation*. A.H. Gibson and W. E. Newton eds., Canberra, Australian Academy of Sciences. p. 499.

Singh, P.K. 1981b. Use of *Azolla* and blue-green algae in rice cultivation in India. In: Associative $N_2$-fixation, Vol. II (ed. P.B. Vose and A.P. Ruschel), Florida, CRC Press. pp. 183–195.

Singh, P.K. 1982. *Azolla* and blue-green algae biofertilizer technology of rice. *Indian Farming* 32: 3–8.

Singh, P.K. 1985. Nitrogen fixation by the blue-green algae in paddy fields. In: *Rice Research in India.* Pp. 344-362, ICAR, New Delhi.

Singh, R., Singh, R.P and Singh, D.V. 2015. Distribution of Cyanobacteria (Blue-green algae) in Rice fields of Varanase. *Int. J. of Adv. Res.* 3: 1055–1060.

Singh, R., Singh, R.P. and Singh, D.V. 2015. Distribution of Cyanobacteria (Blue-green algae) in Rice fields of Varanasi. *Int. J. Adv. Res.* 3(8): 1055–1060.

Singh, R.K., Stevens, S.E. and Bryant, D.A. 1987. Molecular cloning and physical mapping of the nitrogenase structural genes from the filamentous non-heterocystous cyanobacterium *Pseudoanabaena* sp. *PCC. 7409. FEMS Microbiol. Lett.* 48: 53–58.

Singh, R.N. 1939. An investigation into the algal flora of paddy field soils of the United Provinces-I. *Indian J. Agri. Sci.* 9(10): 55–77.

Singh, R.N. 1942. The fixation of elementary nitrogen by some commonest blue-green algae from paddy field soils of United Provinces and Bihar. *Indian J. Agri. Sci.* 12: 743–756.

Singh, R.N. 1961. Role of blue-green algae in nitrogen economy of Indian agriculture, ICAR, New Delhi.

Singh, R.N. 1961. Role of blue-green algae in Nitrogen Economy of Indian Agriculture. ICAR, New Delhi.

Singh, S.M. and Elster, J. 2007. Cyanobacteria in Antarctic lake environments. In: Seckbach, J. (ed.), *Algae and Cyanobacteria in Extreme Environments. Cellular Origin, Life in Extreme Habitats and Astrobiology*, Springer, Dordrecht, The Netherlands, Volume 11, pp. 303–320.

Singh, V.K., Singh, S.K., Singh, K.P., Verma, H., Pandey K.D., Singh, P.K. and Kumar, A. 2020. Impact of pesticides applications on the growth and function of cyanobacterial Biology. 1st edition, Elsevier.

Singh, V.P. 1941. On a collection of algae from Chamba State (Punjab) I. *Proc. Indian Acad. Sci.* B. 14: 250–255.

Singh,V.P,Saxena,P.N.,Tewari,A.andKhan,M.A.1970:Studies on the seasonal variation of Algal flora of sewage.*Phykos 9:57-62.*

Sinha, B.D. and Srivastava, N.K. 1980. Algal flora of Balmikinagar, Bihar. *Phykos* 19(2): 171–174.

Sinha, J.P. & Mukherjee, D. 1975. On blue-green algae from the paddy fields of Bankura district of West Bengal-II. *Phykos* 14(1 & 2): 119–120.

Sinha, J.P. & Mukherjee, D. 1984. Blue-green algae from the paddy fields of Bankura district of West Bengal-III. *Phykos* 23(1 & 2): 142–143.

Sinha, J.P. and Mukherjee, D. 1975. On Blue-green algae from the paddy fields of Bankura district of West Bengal-I. *Phykos* 14(1 & 2): 117–118.

Sivanen, K. and Jones, G. 1999. Cyanobacterial toxins In Toxic Cyanobacterial in water:(1. Chrus and J. Bartram, eds.) Who and E and F N Spon, London:41-111

Sivonen, K., Leikoski, N., Fewer, D.P. and Jokela, J. 2010. Cyanobactins–ribosomal cyclic peptides produced by cyanobacteria. *Appl. Microbiol. Biotechnol.* 86: 1213–1225.

Skleryk, R.S., Tyrrell., P.N. and Espie, G.S. 1997. Photosynthesis and inorganic Carbon accumulation in the cyanobacterium *Chlorogleopsis* sp. Strain A TCC27193. *Physiol. Plant* 99: 81–88.

Smarda, J., Caslavaska J. and Komarek, J. 1979. Cell wall structure of *Synechocystis aquatilis* (Cyanophyceae). *Arch. Hydrobiol.* [*Suppl.*] 56: 154–165.

Smitka, T.A., Bonjouklian, R., Doolin, L., Jones, N.D., Deeter, J.B., Yoshida, W.Y., Prinsep, M.R., Moore, R.E. and Patterson, G.M.L. 1992. Ambiguine isonitriles, fungicidal hapalindole-type alkaloids from three genera of blue-green algae belonging to the Stigonemataceae. *J. Org. Chem.* 57: 857–861.

Smoker, J.A. and Barnum, S.R. 1988. Analysis of *nif* K,D,H genes and their expression in the cyanobacterium *Plectonema boryanum.* In: Bothe, H., de Bruiji, F.J. and Newton, W.E. Fisher (eds.), *Nitrogen fixation. Hundred years after*. Stuttgart.

Somashekar, R.K. 1983. Algal flora of river Cauvery, Karnataka I. Cyanophyceae and Chlorophyceae. *Phykos* 22: 73–80.

Song, W., Zhao, C., Zhang, D., Mu, S. and Pan, X. 2016. Different resistance to UV-B radiation of extracellular polymeric substances of two cyanobacteria from contrasting habitats. *Front. Microbiol.* 7: 1208.

Sood, A., Singh, P.K., Kumar, A., Singh, R., and Prasanna, R. 2011. Growth and biochemical characterization of associations between Cyanobionts and wheat seedlings in co-culturing experiments. Biologia. 66 (1):104-110

Spiller. H. and Shanmugam, K.T. 1987. Physiological conditions for nitrogen fixation in a unicellular marine cyanobacterium *Synechcoccus* sp. strain SFI. *J. Bacteriol.* 169: 5379–5384.

Srivastava, P.N. 1967. On some Blue-green algae from Central Himalayas. *Phykos* 6: 110–114.

Srivastava, V. and Sen, N. 1987. Fresh water algae of the river Narmada near Jabbalpur. *Phykos* 26: 78–81.

Stal, L.J. and Krumbein, W.E. 1985. Oxygen protection of nitrogenase in the aerobically nitrogen-fixing non-heterocystous cyanobacterium *Oscillatoria* sp. *Arch. Microbiol.* 143: 72–76.

Stam, W.T. and Holleman, H.C. 1979. Cultures of Phormidium, Plectonema, Lyngbya and Synechococcus (Cyanophyceae) under different conditions: their growth and morphological variability Acta. Bot. Neerl. 28(1), 45–66.

O Stam, W.T. 1980. Relationship between a number of filamentous blue-green algal strains (Cyanophyceae) revealed by DNA-DNA hybridization. Arch. Hydrobiol. Supply.56(4):351-374

Stanier, R.Y. and Cohen-Bazire, G. 1977. Phototrophic prokaryotes: The Cyanobacteria. *Annu. Rev. Microbiol.* 31: 225–274.

Stanier, R.Y., Ingraham, J.L. , Wheelis, M.L. and Painter, P.R. 1990. The Microbial World (Fifth edition), Pp. 360–369. Prentice- Hall of India Private Ltd., New Delhi-110 001.

Stanier, R.Y., Kunisawa, R., Mandel, M. and Cohen-Bazire, G. 1971. Purification and properties of unicellular Blue-green algae (Order - Chroococcales). *Bacteriol. Rev.* 35: 171–205.

Stanier, R.Y., Pfennig, N. and Truper, H.G. 1981. Introduction to the phototrophic prokaryotes. In: M.P. Starr, H. Stolp, H.G. Truper, A. Balows and H.G. Schlegel (eds.), *The Prokaryotes.* Vol. I. Springer-Verlag, Berlin, Chap. 7 (pp. 197–211).

Stanier, R.Y., Sistrom, W.R., Hansen, T.A., Whitton, B.A. Castenholz, R.W., Pfenning, N., et al., 1978. Proposal to place the nomenclature of the cyanobacteria (blue-green algae) under the rules of the international code of nomenclature of bacteria. *Int. J. Syst. Bacteriol.* 28: 335–336.

Sterner, R.W., Reinl, K.L., Lafrancois, B.M., Brovold, S. and Miller, T.R. 2020. A first assessment of cyanobacterial blooms in oligotrophic Lake Superior. *Limnol. Oceanogr.* 65: 2984–2998.

Stewart, W.D.P. 1970. Algal fixation of atmospheric nitrogen *Pl. and Soil* 32: 555–558.

Stewart, W.D.P. 1973. Nitrogen fixation by photosynthetic microorganisms. *Ann. Rev. Microbiol.* 27: 283–316.

Stewart, W.D.P. 1980. Some aspects of structure and function in nitrogen fixing cyanobacteria. *Ann. Rev. Microbiol.* 134: 497–536.

Stewart, W.D.P. and Lex, M. 1970. Nitrogenase activity in the blue-green alga *Plectonema boryanum* strain 549. *Arch. Microbiol.* 74: 250–260.

Stratmann, K., Moore, R.E., Bonjouklian, R., Deeter, J.B., Patterson, G.M.L., Shaffer, S., Smith, C.D. and Smitka, T.A. 1994. Welwitindolinones, unusual alkaloids from the blue-green algae *Hapalosiphon welwitschii* and *Westiella intricata*. Relationship to fischerindoles and hapalinodoles. *J. Am. Chem. Soc.* 116: 9935–9942.

Suda, S., Watanabe, M.M., Otsuka, S., MahaKahant, A., Yongmanitchai, W., Nopartnaraporn, N. Liu, Y and Day J.G. 2002. Taxonimic revision of water-bloom-forming species of owscillatorioid cyanobacteria. Int. J. Syst. Evol. Microbial. 52:1577-1595.

Strunecky O., KomareK, J., Johansen J.R., LuKesova, A. and Elster, J. 2013. Molecular and morphological criteria for revision of the genus Micro coleus (Oseillatoriales, cyanobacteria). J. Phycol. 49:1167-1180

Sudek, S., Haygood, M.G., Youssef, D.T.A. and Schmidt, E.W. 2006. Structure of trichamide, a cyclic peptide from the bloom-forming cyanobacterium Trichodesmium erythraeum, predicted from the genome sequence. *Appl. Environ. Microbiol.* 72: 4382–4387.

Sugumar, M., Sevanan, M. and, Sekar, S. 2019. Neuroprotective effect of naringenin against MPTP-induced oxidative stress. *Int. J. Neurosci.* 129: 534–539.

Summons, R.E., Jahnke, L.L., Hope, J.M. and Logan, G.A. 1999. 2-Methylhopanoids as biomarkers for cyanobacterial oxygenic photosynthesis. *Nature* 400: 554–557.

Taha, H.S., El Bahr, M.K. and Seif, E.L.N.M.M. 2009. In Vitro Studies on Egyption Catharanthus roseus (L.). Ii. Effect of Biotic and Abiotic Stress on Indole Alkaloids Production. *J. Appl. Sci. Res.* 3: 3137–3144.

Talai-Mukhopadhyay, S. and Naskar, N. 2015. Diversity of Cyanobacteria from Barasat, Kolkata, West Bengal, India. *Int. J. Innov. Res. Rev.* ISSN: 2347-4424 (on line) January-March, pp. 107–113.

Talpasayi, E.R.S. 1962. The Myxophyceae of the Kumaon hills, U.P., India-II. *Proc. Indian Acad. Sci.* B. 55: 251–255.

Tandeau de Marsac N. 1977 Occurrence and nature of Chromatic adaptation in Cyanobacteria. *J Bacteriol* 130(1): 82–91.

Tang, W. and Van Der Donk, W.A. 2012. Structural characterization of four prochlorosins: A novel class of lantipeptides produced by planktonic marine cyanobacteria. Biochemistry 51: 4271–4279. [Google Scholar] [CrossRef]

Tarar, J.L. and Salpekar, C.R. 1978. Effect of Zinc phosphide (Rodenticide) on soil algae. *Curr. Sci.* 47: 789.

Tarar, J.L. and Salpekar, C.R. 1979. Studies on the effect of six insecticides on the soil algae of cotton field. *Proc. 66th Ind. Sci. Cong. Abst.* 18: 8.

Tarar, J.L. and Shewale, T.H. 1984. Studies on the effects of some fungicides on soil algae of paddy field. *Phykos.* 23(1 & 2): 191–201.

Tarento, T.D.C., McClure, D.D., Vasiljevski, E., Schindeler, A., Dehghani, F. and Kavanagh, J.M. 2018. Microalgae as a source of vitamin K1. *Algal Res. 36:* 77–87.

Thangaraj, B., Rajasekar, D.P., Vijayaraghavan, R., Garlapati, D., Devanesan, A.A., Lakshmanan, U. and Dharmar, P. 2017. Cytomorphological and nitrogen metabolic enzyme analysis of psychrophilic and mesophilic *Nostoc* sp.: A comparative outlook. *Biotech* 7: 107.

Thomas, J. 1970. Absence of pigments of photosystem II of photosynthesis in heterocysts of a blue-green alga. *Nature* 228: 181–183. DOI: http://dx.doi.org/10.doi.org/10.1038/22818160

Thomas, N.V. and Kim, S.K. 2011. Potential pharmacological applications of polyphenolic derivatives from marine brown algae. *Environ. Toxicol. Pharmacol.* 32: 325–335. doi: 10.1016/j.etap.2011.09. 004

Tidgewell, K., Engene, N., Byrum, T., Media, J., Doi, T., Valeriote, F.A. and Gerwick, W.H. 2010. Evolved diversification of a modular natural product pathway: Apratoxins F and G, two cytotoxic cyclic depsipeptides from a Palmyra collection of *Lyngbya bouillonii. ChemBioChem* 11: 1458–1466.

Tillett, D., Dittmann, E., Erhard, M., Von Döhren, H., Börner, T., and Neilan, B.A. 2000. Structural organization of microcystin biosynthesis in *Microcystis aeruginosa* PCC 7806: An integrated peptide–polyketide synthetase system. *Chem. Biol.* 7: 753–764.

Torres, A.M.R. and Flaherty, Lawrance M.O. 1976. Influence of pesticides on *Chlorella, Chloroococcum, Stigeoclonium, Tribonema* and *Oscillatoria*. *Phycologia* 15: 25–36.

Trivedy, R.K. 1982. Some observations on algal flora of Jaipur, Rajasthan. *Phykos* 21: 160–163.

Tuffery, A.A. 1969. Light and Electron Microscopy of the sheath of a Blue-green Alga. Microbiology 57: Issue 1.

Turner, W.B. 1892. The Freshwater Algae of East India. *K, Sy, Yetensk Akad. Handb.* 25(5): 1–187.

Turner, P.C., Gammie, A.J., HollinraKe, k. and Codd, G.A. 1990. Preumonia associated with cyanobacteria-A review. J. Acute Medicine 5:15-23

Tyagi, S., Singh, R.K. and Tiwari, S.P. 2021. Anti-enterococcal and anti-oxidative potential of a thermophilic cyanobacterium, *Leptolyngbya* sp. HNBGU 003. *Saudi J. Biol. Sci.* 28: 4022–4028.

Varma, A.K. 1965. Cultural studies of some members of Chroococcales. *Phykos* 4: 3–9.

Vasishta, P.C. 1968. Thermal Cyanophyceae of India-7. *Phykos* 7: 198–241.

Venkataraman, G.S. 1957. A list of marine Myxophyceae from Cape Comorin (Kanya Kumari). *J. Indian Bot. Soc.* 36(4): 472–474.

Venkataraman, G.S. 1966. Algalization. *Phykos* 5: 164–174.

Venkataraman, G.S. 1972. *Algal biofertilizers and rice cultivation*. Today and Tomorrows Publishers, Delhi.

Venkataraman, G.S. 1981. Economics and energetics of blue-green algal contribution to rice. *Curr. Sci.* 50: 94–96.

Venkataraman, G.S. and Goyal, S.K. 1969a. Influence of BGA on the high yielding paddy variety IR-80. *Sci. Cult.* 35: 58.

Venkataraman, G.S. and Goyal, S.K. 1969b. Some recent observations on the effect of nitrogen fixing BGA on crop plants. *Mikrobiologiya* 38: 709–712.

Venkataraman, G.S. and Rajyalakshmi, B. 1971a. Tolerance of blue-green algae to pesticides. *Curr. Sci.* 40: 143

Venkataraman, G.S. and Rajyalakshmi, B. 1971b. Relative tolerance of blue-green algae to pesticides. *Indian J. Agric. Sci.* 42: 119.

Vidal, L., Ballot, A., Sandra, M., Azevedo, F.O., Padisák, J. and Welker, M. 2021. Introduction to cyanobacteria. *Toxic Cyanobacteria in Water*. Ed. 2nd. Page 49. World Health Organization.

Vijaya Kumar, S. and Menakha, M. 2015. Pharmaceutical applications of cyanobacteria-A review J. Acute Medicine 5: 15-23

Vítek, P., Ascaso, C., Artieda, O., Casero, M.C. and Wierzchos, J. 2017. Discovery of carotenoid red-shift in endolithic cyanobacteria from the Atacama Desert. *Sci. Rep.* 7: 11116.

Volk, R.-B. 2005. Screening of microalgal culture media for the presence of algicidal compounds and isolation and identification of two bioactive metabolites, excreted by the cyanobacteria *Nostoc insulare* and *Nodularia harveyana*. *J. Appl. Phycol.* 17: 339–347.

Wacklin, P., Hoffmann, L. and KomareK, J. 2009: Nomenelatural Validation of the genetically revised cyanobacterial genus *Dolickospermum* (Relfs ex Bornet et Flahault) comb.nove. Fottea 9:59-64

Walsby, A.E. 1978. The properties and buoyancy-providing role of gas vacuoles in *Trichodesmium* Ehrenberg. *Br. Phycol. J.* 13: 103–116.

Walsby, A.E. 1985. The permeability of Heterocysts to the Gases Nitrogen and Oxygen. *Proc. Royal Soc. London*. Series B, Biological Sciences. 226.(1244): 345–366.

Walton, K. and Berry, J.P. 2016. Indole alkaloids of the *Stigonematales* (Cyanophyta): Chemical diversity, biosynthesis and biological activity. *Mar. Drugs* 14: 73.

Wang, C.Y. and Wang, F.Z. 1988. The relationship between nitrogen fixation and several kinds of energy metabolism by *Gloeocapsa* sp. *Acta Botanica Sinica*. 30: 269–277.

Wang, H., Fewer, D.P., Holm, L., Rouhiainen, L. and Sivonen, K. 2014. Atlas of nonribosomal peptide and polyketide biosynthetic pathways reveals common occurrence of nonmodular enzymes. *Proc. Natl. Acad. Sci. USA* 111: 9259–9264.

Watanabe, A. 1956. On the effect of atmospheric nitrogen fixing BGA on the yield of rice. *Bot. Mag. Tokyo* 69: 530.

Watanabe, A: 1962. Effect of nitrogen fixing blue-green alga tolypothrix tenuis on the nitrogen tenuis on the nitrogen fertility of paddy soils and on the crop yield of rice plant. J. Genl. Microbiology 8:85

Watanabe, A. and Yamamoto 1971. Algal nitrogen fixation in the tropics. *Plant soil*. special vol. 403–413.

Watanabe, A. and Yamamoto, Y. 1970. Mass culturing, preservation and transportation of the nitrogen-fixing BGA. Proceedings of the Second Symposium on Nitrogen fixation & Nitrogen Cycle. Sendai, Japan, pp. 22–28,

Waterbury, J.B. and Stanier, R.Y. 1978. Patterns of growth and development in pleurocapsalean cyanobacteria. *Microbiol. Rev.* 42: 2–44.

Waterbury, J.B. and Stanier, R.Y. 1989. Order Pleurocapsales Geitler 1925, emend. Waterbury and Stanier 1978, p. 1746–1770. In: J.T. Staley, M.P. Bryant, N. Pfennig and J.G. Holt (eds.), *Bergey's manual of systematic bacteriology*, Vol. 3. Williams and Wilkins, Baltimore.

Waterbury, J.B., Stanier, R.Y. and Rippka, R. 1989. The Order Chroococcales Wettstein 1924, emend. Rippaka *et al.* 1979, p. 1728–1746. In: J.T. Staley, M.P, Bryant, N. Pfennig and J.G. Holts (eds.), *Bergey's manual of systematic bacteriology*, Vol. 3. Williams and Wilkins, Baltimore.

Weissharr, H. and Boger, P. 1983. Nitrogenase activity of the non-heterocystous cyanobacterium *Phormidium foveolarum*. *Arch. Microbiol.* 136: 270–274.

Welker, M. and Von Döhren, H. 2006. Cyanobacterial peptides–nature's own combinatorial biosynthesis. *FEMS Microbiol. Rev.* 30: 530–563.

West. G.S. 1916. Algae (Vol. I) Botany handbooks, Cambridge, a treatise on the British freshwater algae. Cambridge University Press, London.

Whitton, B.A. and Potts, M. 2000. The ecology of cyanobacteria: their diversity in time and space. Heidelberg, Springer.

Wilcox, M., Mitchinson, G.J. and Smith, R.J. 1973b. Pattern formation in blue-green alga, *Anabaena* II controlled proheterocyst regression. *J. Cell. Sci.* 13: 637–649.

William, S., Bachofen, R. and Zeyer, J. 2003. Molecular characterization of endolithic cyanobacteria inhabiting exposed dolomite in central Switzerland. *Environ. Microb.* 5(7): 618–627.

Williams, S.B., Vakonakis, I., Golden, S.S and Liwang, A.C. 2002. Structure and function from the Circadian Clock protein Kai A of *Synechococcus elongatus*: a potential clock input mechanism. *Proc. Natl. Acad. Sci. USA.* 99(24): 15357–62.

Wilmotte, A. 1994. Molecular evolution and taxonomy of the cyanobacteria. In: Bryant D.A.(eds.). The molecular biology of cyanobacteria. Dordrecht: Kluwer Academic Publishers: 1–25.

Wolk, C.P. 1996. Heterocyst formation. *Annu. Rev. Genet.* 30: 59–78 [Pub Med]

Wose, C. 1987. Bacterial Evolution. *Microbiol Rev.* 51: 221–271.

Wyatt and Silvey, J.K.G. 1969. Nitrogen fixation by *Gloeocapsa*. *Science (Washington)*. 165: 906–908.

Zapomelova, E., Pumann, P., Koop, R. and Janecek, E. 2012. Bibliographically interesting planKtonic Nostocales (Cyanobacteria) in the Czech Republic and their polyphasic evaluation resulting in taxonomic revisions of *Anabaena bergii* Ostenfeld 1908 *Chrysosporum* gen. nov.); and A. Tenericaulis Nygaard 1949 (*Dolichospermum tenericaule* comb. nova). –Hydrobiologia. 698: 353-365.

Zhang, J., Xie, Z., Jiang, X., and Wang, Z. 2014. Control of Cyanobacterial blooms via synergistic effects of pulmonates and submerged plants. Clean Soil Air Water. 42: 1-6.

Zhang, J., Wang, Z., Song, Z.Y., Xie, Z.C., Li,L., and Song, L.R. 2012. Bioaccumulation of microcystins in two freshwater gastropods from a cyanobacteria bloom Plateau Lake, Lake Dianchi. Environ. Pollut. 164: 227-234.

Zehnder, A. 1973. 6 Symopsium ueber Frangen der cyanophyten taxonomi in Kastanienbaum, 3-6 August 1972. Schw. Zeit. Hydrol. 35(1): 121-135.

Zhang, L.C., Chen, Y.F., Che, W.L. and Zhang, C.C. 2008. Existence of periplasmic barriers preventing green fluorescent protein diffusion from cell to cell in the cyanobacterium *Anabaena* sp. Strain PCC 7120. *Molecular Microbiology* 70: 814–823.

Zhu, Z., Fu, F., Qu, P., Mak, E.W.K., Jiang, H., Zhang, R., Zhu, Z., Gao, K. and Hutchins, D.A. 2020. Interactions between ultraviolet radiation exposure and phosphorus limitation in the marine nitrogen-fixing cyanobacteria Trichodesmium and Crocosphaera. *Limnol. Oceanogr.* 65: 363–376.

Ziemert, N., Ishida, K., Quillardet, P., Bouchier, C., Hertweck, C., de Marsac, N.T. and Dittmann, E. 2008. Microcyclamide biosynthesis in two strains of Microcystis aeruginosa: From structure to genes and vice versa. Appl. Environ. Microbiol. 74: 1791–1797. [Google Scholar] [CrossRef][Green Version]

# INDEX